Fundamentals of Electrical and Computer Engineering

Volume I

Fundamentals of Electrical and Computer Engineering

Volume I

Dr. Samir I. Abood
Dr. Muna H. Fayyadh

CWP

This edition has been published by Central West Publishing PTY LTD, (ABN 13 683 898 722) Australia

For more information about the books published by Central West Publishing PTY LTD, please visit https://centralwestpublishing.com

Disclaimer

Every effort has been made by the publisher, editors and authors while preparing this book, however, no warranties are made regarding the accuracy and completeness of the content. The publisher, editors and authors disclaim without any limitation all warranties as well as any implied warranties about sales, along with fitness of the content for a particular purpose. Citation of any website and other information sources does not mean any endorsement from the publisher, editors and authors. For ascertaining the suitability of the contents contained herein for a particular lab or commercial use, consultation with the subject expert is needed. In addition, while using the information and methods contained herein, the practitioners and researchers need to be mindful for their own safety, along with the safety of others, including the professional parties and premises for whom they have professional responsibility. To the fullest extent of law, the publisher, editors and authors are not liable in all circumstances (special, incidental, and consequential) for any injury and/or damage to persons and property, along with any potential loss of profit and other commercial damages due to the use of any methods, products, guidelines, procedures contained in the material herein.

NATIONAL LIBRARY OF AUSTRALIA

A catalogue record for this book is available from the National Library of Australia

ISBN (print): 978-1-922617-71-2

About Authors

Samir I. Abood received his BS and MS from the University of Technology, Baghdad, Iraq, in 1996 and 2001; respectively. He got his Ph.D. in the Electrical and Computer Engineering Department at Prairie View A & M University. From 1997 to 2001, he worked as an engineer at the University of Technology. From 2001 to 2003, he was a professor at the University of Baghdad and Al-Nahrain University. From 2003 to 2016, Mr. Abood was a Middle Technical University / Baghdad-Iraq professor. He is an Electrical and Computer Engineering Professor at Prairie View A&M University in Prairie View. He is the author of 30 papers and 12 books. His main research interests are sustainable power and energy systems, microgrids, power electronics and motor drives, digital PID Controllers, digital methods for electrical measurements, digital signal processing, and control systems.

Muna H. Fayyadh has received her BS and MS from Al-Nahrain University, Baghdad, Iraq, in 1998 and 2002, respectively. She got her Doctoral degree in computer science - Cybersecurity concentration in 2019 from Colorado Tech University. She has worked as an instructor at Al-Nahrain university from 2001-2005. She worked as an IT specialist at the Ministry of water resources in Iraq from 2003-2013. From 2014-2018 she worked as an IT instructor for a private school in Houston, Tx. From 2018 -2020 she worked at the City of Houston, an adjunct professor (Grad level) at American Intercontinental University and Houston community college. She is the author of 15 papers and three books. Her main research interests are cybersecurity and intrusion detection techniques, network security, machine learning, and AI.

Preface

Fundamentals of Electrical and Computer Engineering is a comprehensive textbook designed to provide students with a solid foundation in the core concepts and principles of electrical and computer engineering. This book is intended for students who are new to the field and those who are interested in refreshing their knowledge.

The book is organized into 22 chapters in two volumes, covering topics such as electrical units and constants, mathematical engineering, concepts of probability and statistics, engineering economics, electrical materials, electrical circuits, circuit analysis and network, communication engineering, digital signal processing, digital communications, power electronics, power engineering, power system analysis, electrical machines, magnetic field, electronics, digital systems, control engineering systems, computer systems, computer networks, software engineering, and ethics of practicing engineering. Volume I contains Chapters 1-11, whereas Volume II comprises of Chapters 12-22.

Each chapter provides clear explanations of the fundamental concepts, followed by numerous examples and exercises to reinforce the understanding of the material. The book also includes illustrations and diagrams to help students visualize abstract concepts and ideas. In addition, the book features several special sections that highlight the practical applications of the concepts being discussed, their historical development, and the ethical considerations associated with their use.

Chapter 1: Electrical Units and Constants: This chapter introduces the fundamental units of electrical measurements, and the physical constants used in electrical engineering. Students will learn about the various systems of units, including SI units, and how to convert between different units. The chapter also covers voltage, current, and resistance principles.

Chapter 2: Mathematical Engineering: This chapter provides an overview of the mathematical concepts used in electrical and computer engineering. Students will learn about calculus, differential equations, complex numbers, and linear algebra and how these concepts are applied to engineering problems. The chapter also covers using software tools such as MATLAB to solve mathematical problems.

Chapter 3: Concepts of Probability and Statistics: This chapter introduces the basic principles of probability theory and statistics. Students will learn about probability distributions, random variables, statistical inference, and hypothesis testing. The chapter also covers probability and statistics in engineering applications, such as reliability analysis and quality control.

Chapter 4: Engineering Economics: This chapter overviews economic principles applied to engineering projects. Students will learn about the time value of money, cash flow analysis, and cost estimation. The chapter also covers economic analysis in decision-making and project evaluation.

Chapter 5: Electrical Materials: This chapter covers the properties and characteristics of materials used in electrical engineering, such as conductors, insulators, and semiconductors. Students will learn about the behavior of materials under different electrical conditions and how these properties are used to design electrical devices.

Chapter 6: Electrical Circuits: This chapter introduces electrical circuits, including circuit elements such as resistors, capacitors, and inductors. Students will learn about circuit analysis techniques, including Kirchhoff's and Ohm's laws, and how to solve simple circuits.

Chapter 7: Circuit Analysis and Network: The chapter covers more advanced circuit analysis techniques, including mesh analysis, nodal analysis, and Thevenin's and Norton's theorems. Students will also learn about network analysis techniques, including impedance and admittance.

Chapter 8: Communication Engineering: This chapter introduces communication systems and the principles of analog and digital modulation and demodulation. Students will learn about signal transmission, noise, and error control coding.

Chapter 9: Digital Signal Processing: This chapter covers the theory and practice of digital signal processing. Students will learn about discrete-time signals and systems, Fourier analysis, digital filtering, and the design of digital signal processing systems.

Chapter 10: Digital Communications: This chapter covers the principles of digital communication systems, including the design of modulation and demodulation schemes, error detection and correction techniques, and channel coding.

Chapter 11: Power Electronics: This chapter introduces power electronics, including the design and operation of power electronic devices such as diodes, transistors, and thyristors. Students will also learn about power electronic circuits and their applications in power conversion and control.

In recognition of requirements by the Accreditation Board for Engineering and Technology (ABET) on integrating computer tools, the use of MATLAB® is encouraged in a student-friendly manner.

Overall, Fundamentals of Electrical and Computer Engineering is an essential resource for any student seeking to develop a strong foundation in this exciting and rapidly evolving field. Whether used as a primary textbook for a course or as a reference for self-study, this book will provide students with the knowledge and skills they need to succeed as electrical and computer engineers.

Table of Contents

Chapter 1

Electrical Units and Constants

The International System of Units (SI) and the US Customary System (USCS) are two commonly used measurement systems, including in electrical engineering.

The SI is a metric system of measurement used worldwide and based on a set of base units, which are used to define all other units. The SI base units include the meter (m) for length, the kilogram (kg) for mass, the second (s) for time, the ampere (A) for electric current, the kelvin (K) for temperature, the mole (mol) for the amount of substance, and the candela (cd) for luminous intensity.

In electrical engineering, some commonly used SI units include the volt (V) for electric potential or electromotive force, the ampere (A) for electric current, the ohm (Ω) for electrical resistance, the farad (F) for electrical capacitance, the henry (H) for electrical inductance, and the watt (W) for electrical power.

The US Customary System (USCS), also known as the imperial system, is a system of measurement used primarily in the United States. The USCS includes units such as the inch (in) for length, the pound (lb) for mass, the second (s) for time, the foot (ft) for length, the yard (yd) for length, and the mile (mi) for length.

In electrical engineering, some commonly used USCS units include the volt (V) for electric potential or electromotive force, the ampere (A) for electric current, the ohm (Ω) for electrical resistance, the farad (F) for electrical capacitance, the henry (H) for electrical inductance, and the watt (W) for electrical power. This chapter explains the International System of Units (SI) and the US Customary System (USCS).

1.1 Units and Symbols

Table 1.1 Commonly used equivalents

METRIC PREFIXES		
Multiple	Prefix	Symbol
10^{-18}	atto	a
10^{-15}	femto	f
10^{-12}	pico	p
10^{-9}	nano	n
10^{-6}	micro	μ
10^{-3}	milli	m
10^{-2}	centi	c
10^{-1}	deci	d
10^{1}	Deka	da
10^{2}	hecto	h
10^{3}	kilo	k
10^{6}	mega	M
10^{9}	Giga	G
10^{12}	tera	T
10^{15}	Peta	P
10^{18}	exa	E

1.2 Temperature Conversions

Conversion temperature from Celsius to Fahrenheit

$$ºF = 1.8\,(ºC) + 32$$

Conversion temperature from Fahrenheit to Celsius

$$K = ºC + 273.15$$

Conversion temperature from Celsius to Kelvin

$$ºC = (ºF - 32)/1.8$$

1.3 Fundamental Electrical Constants

Table 1.2 Fundamental electrical constants

Quantity	Value	Unit
Electron charge(e)	1.6022×10^{-19}	$C\ (coulombs)$
Faraday constant *(F)*	96,485	$coulombs\ /(mol)$
Speed of light (*c*)	299,792,458	m/s
Electron-volt (eV)	1.6022×10^{-19}	$joule\ (J)$
electron mass	$9.1093837015 \times 10^{-31}$	kg
The Newtonian constant of gravitation	6.67430×10^{-11}	$m^3 \cdot kg^{-1} \cdot s^{-2}$
Planck constant (h)	$6.62607015 \times 10^{-34}$	J·s
vacuum electric permittivity(ϵ_0)	$8.8541878128 \times 10^{-12}$	$F \cdot m^{-1}$
vacuum magnetic permeability(μ_0)	$1.25663706212 \times 10^{-6}$	$N \cdot A^{-2}$
The Newtonian constant of gravitation	$1.602176634 \times 10^{-19}$	$\boldsymbol{C}$
fine-structure constant	$7.2973525693(11) \times 10^{-3}$	
alpha particle mass	$6.6446573357e-27$	kg
alpha particle mass-energy equivalent	$5.9719201914e-10$	J
alpha particle mass energy equivalent in MeV	3727.3794066	MeV
Angstrom star	1.00001495e-10	m
The atomic mass unit-kilogram relationship	1.6605390666e-27	kg
The atomic unit of current	0.00662361823751	A
The atomic unit of electric dipole moment	8.4783536255e-30	C m
The atomic unit of electric potential	27.211386245988	V
The atomic unit of force	8.2387234983e-08	N

classical electron radius	2.8179403262e-15	m
kilogram-electron volt relationship	5.609588603e+35	eV
neutron molar mass	0.0010086649156	kg mol^-1
proton-muon mass ratio	8.88024337	
proton mass	1.67262192369e-27	kg

1.4 Plank Constants

Table 1.3 Plank Constants

Planck constant	6.62607015e-34 J Hz^-1
Planck constant in eV/Hz	4.135667696e-15 eV Hz^-1
Planck length	1.616255e-35 m
Planck mass	2.176434e-08 kg
Planck mass energy equivalent in GeV	1.22089e+19 GeV
Planck temperature	1.416784e+32 K
Planck time	5.391247e-44 s

1.5 Newtonian Constants

Table 1.4 Newtonian constants

A Newtonian constant of gravitation	6.6743e-11 m^3 kg^-1 s^-2
A Newtonian constant of gravitation over h-bar c	6.70883e-39 (GeV/c^2)^-2

1.6 Molar Constant

Table 1.5 Molar constant

gas constant	8.314462618 J mol^-1 K^-1

molar mass constant	0.00099999999965 kg mol^-1
the molar mass of carbon-12	0.0119999999958 kg mol^-1
molar Planck constant	3.990312712e-10 J Hz^-1 mol^-1
molar volume of an ideal gas (273.15 K, 100 kPa)	0.02271095464 m^3 mol^-1
molar volume of an ideal gas (273.15 K, 101.325 kPa)	0.02241396954 m^3 mol^-1
molar volume of silicon	1.205883199e-05 m^3 mol^-1

1.7 Conversion Between Units

Table 1.6 Conversion between units

From	Multiply By	To Obtain
joule (J)	9.478 × 10–4	Btu
J	0.7376	ft-lbf
J	1	Newton•m (N•m)
J/s	1	watt (W)
kilogram (kg)	2.205	pound-mass (lbm)
kW	3,413	Btu/hr
kilowatt (kW)	1.341	horsepower (hp)
kWh	1.341	hp-hr
tesla	1,000	weber/m2
Electron-volt (eV)	1.6022×10^{-19}	joule (J)

1.8 Fundamentals of Electrical Symbols

Electrical symbols are the standard technique to represent an electrical circuit. It makes the graphical representation easier to work on and implement. The electrical symbols represent various circuit components, devices, and functionalities. It helps to show the details of an electrical diagram so that the engineer can adequately plan a circuit before working on it.

There are a variety of electrical symbols, including Common electronic symbols and historical electronic symbols. The users can also follow the standards, including the IEEE, IEC (International Electrotechnical Commission), Std., ANSI, JIC, Australian Standard, and more.

Table 1.7 Fundamentals of electrical symbols

Symbols	Description
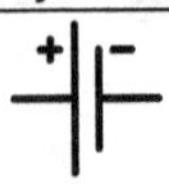 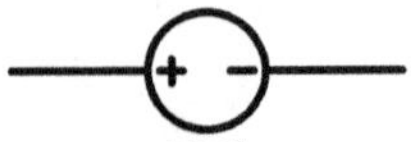 DC	Battery (DC Source) A battery symbol has two disjointed and disproportionate parallel lines. The lines are present to signify the series of cells in the battery.
	AC Source
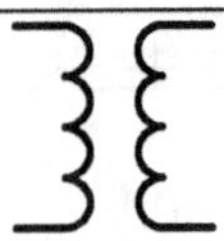 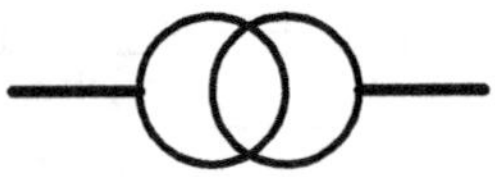	Two windings transformer
 R Fig. a  R Fig. b	Resistor Resistors in an electrical diagram look like wavy lines with pointed ends. The resistors control the current flow by dividing voltage, terminating transmission lines, and more. **Resistor (IEEE) Fig. a** It symbolizes a fixed resistor, looks like

	waves with pointed heads, and connects to two points at the end. **Resistor (IEC) Fig. b** The resistor is a two-terminal device, and the IEC standard resistor symbol looks like a bar connected to two points.
 C	Capacitor A capacitor symbol has two terminals with two plates. A curved surface has a lower voltage, identifying a capacitor as polarized.
L 	Inductor The Inductor is an electronic device that stores electronic energy as magnetic energy.
	Electric Ground: A ground symbol or terminal works as electrical shock protection. It is a zero potential reference point where an electrician measures the current.

	Lamp Load
 t	Switch: Switches are diverse, for example, single-pole single-throw, pushbutton, dip, relay, and more. When it is closed, a switch connects a circuit and disconnects the circuit when open.
	Static load
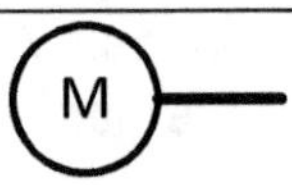	Dynamic load (Motor): A motor is an electronic device that transforms electric energy into a mechanical one.
	Diode: A diode is a device that allows the current to flow in a single direction after being polarized with an anode and a cathode side.
 Fuse	**Fuse:** The Fuse protects a circuit from getting burnt by disconnecting it when the current flow exceeds the set limit. The Fuse has a wire which gets melt while the connection gets disconnected.

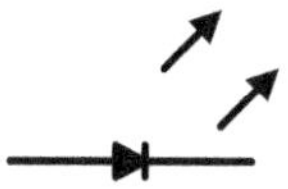	**Diode LED:** The LED looks similar to a usual diode symbol with small arrows that indicate light radiation.

Multiple Choice Questions

1) The SI unit for electric current is
A. joule
B. watt
C. volt
D. ampere

Ans: D

2) The instrument which is used to measure electric current is called
A. ammeter
B. cell
C. battery
D. galvanometer

Ans: A

3) SI unit of resistance is
A. volts
B. amps
C. ohms
D. watts

Ans: C

4) SI unit of power is
A. volts
B. amps
C. ohms

D. watts

Ans: D

5) SI unit of voltage is
A. volts
B. amps
C. ohms
D. watts

Ans: A

6) SI unit of inductance is:
A. farad
B. henry
C. weber
D. tesla

Ans: B
7) SI unit of capacitance is
A. farad
B. henry
C. weber
D. tesla

Ans: A

8) SI unit of magnetic flux is:
A. farad
B. henry
C. weber
D. tesla

Ans: C

9) SI unit of magnetic flux density is:
A. farad
B. henry
C. weber

D. tesla

Ans: D

10) SI unit of charge is:
A. farad
B. henry
C. coulomb
D. tesla

Ans: C

11) SI unit of energy is
A. farad
B. joule
C. coulomb
D. tesla

Ans: B

12) SI unit of frequency is
A. farad
B. hertz
C. weber
D. tesla

Ans: B

13) SI unit of electric conductance is
A. siemens
B. becquerel
C. tesla
D. farad

Ans: A

14) The relationship between current, voltage, and resistance is expressed by _____ Law.
A. Newton's

B. Ohm's
C. Coulomb's
D. Stefan's

Ans: B

Chapter 2

Mathematical Engineering

Mathematical engineering, also known as mathematical modeling, is a branch of engineering that involves using mathematical tools and techniques to solve complex engineering problems. Mathematical engineering is used in various engineering fields, including electrical, mechanical, chemical, and civil engineering.

Mathematical engineering involves creating mathematical models of real-world systems and processes to understand and predict their behavior. These models may be based on differential equations, linear algebra, probability theory, optimization theory, or other mathematical disciplines.

Once a mathematical model has been created, mathematical engineers use numerical methods and computer simulations to solve the model and make predictions about the behavior of the real-world system. Mathematical engineering can design and optimize engineering systems, simulate and analyze the effects of various scenarios on a system, and identify potential problems and risks.

Examples of applications of mathematical engineering include designing and optimizing power systems, predicting the behavior of chemical processes, optimizing manufacturing processes, and simulating the flow of fluids in pipes and channels.

Mathematical engineering is a powerful tool for solving complex engineering problems and plays a critical role in many areas of modern engineering.

2.1 Analytic Geometry

Analytic Geometry and **Coordinate Geometry** study geometric figures using the coordinate plane or coordinates in space.

Using the Cartesian coordinate system, geometric shapes (such as curves) can be described by algebraic equations, namely equations satisfied by the coordinates of the points on the shape. For example, the circle of radius two may be described by the equation $x^2 + y^2 = 4$.

Points in the Cartesian coordinate system

In mathematics, the **Cartesian coordinate system** uniquely determines each **point in the plane** P(x,y) through an ordered pair of numbers, usually called the **x-coordinate** and the **y-coordinate** of the point. To define the coordinates, two perpendicular directed lines (**the x-axis or abscissa** and the **y-axis or ordinate**) are specified, as well as **the unit length**, marked off on the two axes. The point at which the axes cross is called **the origin**. The standard way of orienting the axes: the positive x-axis pointing right and the positive y-axis pointing up (the x-axis being the "first" and the y-axis the "second" axis). Examples:

A(1,4)   B(5,2)   C(–3,5) D(0,5) E(1,–3)

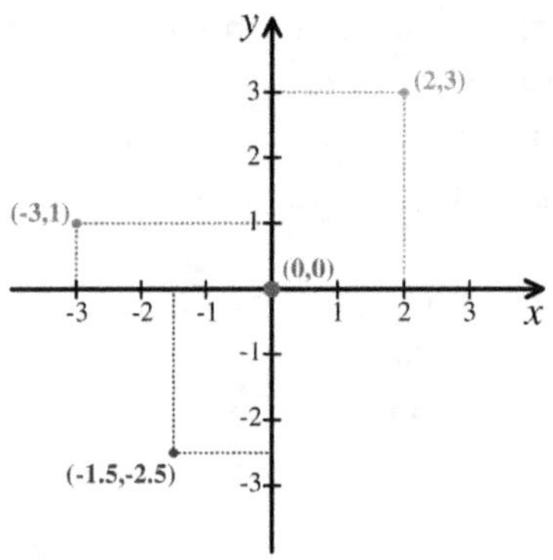

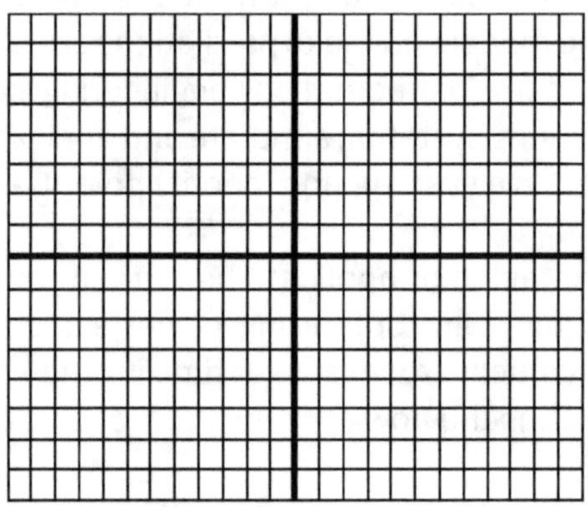

Directed line segments

Fixed vectors are usually shown as arrows, called **directed line segments.**

If the direction is from point $A(a_x,a_y)$ to $B(b_x, b_y)$, the directed line segment is denoted by $\overrightarrow{AB}$.

Point *A* is called **the initial point, tail, base, start, or origin.**
Point *B* is called **the terminal point, head, tip, endpoint, or destination.**

The length of the arrow represents the vector's **magnitude**:

norm or magnitude $\left|\overrightarrow{AB}\right| = \sqrt{(b_x - a_x)^2 + (b_y - a_y)^2}$ (Pythagoras)

The direction in which the arrow points represents the vector's **direction**:

Slope or gradient $m_{AB} = \frac{b_y - a_y}{b_x - a_x}$ $\left(\frac{rise}{run}\right)$

Examples: $\overrightarrow{AB}$ $\overrightarrow{CD}$ $\overrightarrow{GC}$ $\overrightarrow{EF}$

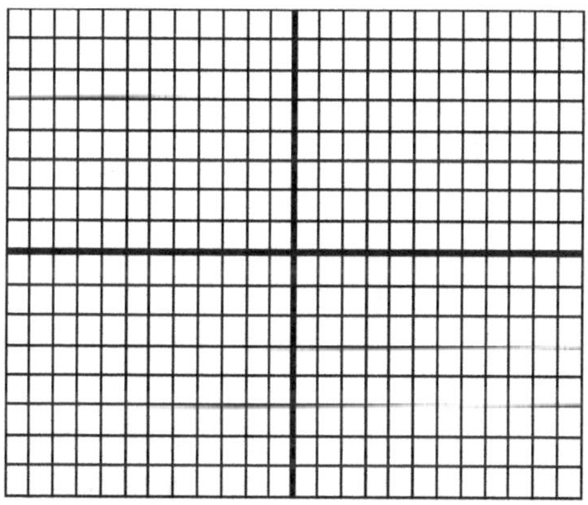

$$\left|\overrightarrow{AB}\right| = \sqrt{(7-5)^2 + (2-7)^2} = \sqrt{29}$$

$$m_{AB} = \frac{2-7}{7-5} = \frac{-5}{2}$$

$$\left|\overrightarrow{CD}\right| = \sqrt{(-6+4)^2 + (5-0)^2} = \sqrt{29}$$

$$m_{CD} = \frac{5-0}{-6+4} = \frac{-5}{2}$$

$$\left|\overrightarrow{GC}\right| = \sqrt{(-4+7)^2 + (0+2)^2} = \sqrt{13}$$

$$m_{GC} = \frac{0+2}{-4+7} = \frac{2}{3}$$

$$\left|\overrightarrow{\boldsymbol{EF}}\right| = \sqrt{(\mathbf{4}-\mathbf{1})^2 + (-\mathbf{1}+\mathbf{3})^2} = \sqrt{\mathbf{13}}$$

$$m_{EF} = \frac{-1+3}{4-1} = \frac{2}{3}$$

$\overrightarrow{AB}$ and $\overrightarrow{CD}$ are parallel with the same direction and opposite sense. $\overrightarrow{GC}$ and $\overrightarrow{EF}$ are parallel and have equal direction and sense. They are said to be equivalent.

Let A(a_x,a_y) and B(b_x,b_y) be points in the plane; we call the **coordinates of the vector $\overrightarrow{\boldsymbol{AB}}$** to $B - A = (b_x - a_x, b_y - a_y)$

Two directed line segments (fixed vectors) are **equivalent** if they have the same magnitude, direction, and sense. That happens when they have the same coordinates.

Let A(a_x,a_y) , B(b_x,b_y), C(c_x,c_y) and D(d_x,d_y) be points in the plane and consider $\overrightarrow{AB}$ and $\overrightarrow{CD}$

$\overrightarrow{AB} \cong \overrightarrow{CD} \Leftrightarrow \begin{cases} b_x - a_x = d_x - c_x \\ \quad and \\ b_y - a_y = d_y - c_y \end{cases} \Leftrightarrow \overrightarrow{AB}$ and $\overrightarrow{CD}$ have the same coordinates.

Free vectors

Each class of equivalent fixed vectors determines a **free vector**. The **coordinates of the free vector** are those of all the fixed vectors of the class. Each fixed vector of the class is said to be an instance of the free vector. $\overrightarrow{\boldsymbol{PQ}}\,(\boldsymbol{v}_x, \boldsymbol{v}_y)$ is an instance of a free vector $\vec{\boldsymbol{v}}(\boldsymbol{v}_x, \boldsymbol{v}_y)$

Examples: $\vec{\boldsymbol{u}}(\mathbf{2},-\mathbf{5})$,$\vec{\boldsymbol{v}}(-\mathbf{2},\mathbf{5})$,$\vec{\boldsymbol{w}}(\mathbf{3},\mathbf{2})$

$\overrightarrow{\boldsymbol{AB}}$ is an instance of $\vec{\boldsymbol{u}}$

$\overrightarrow{\boldsymbol{CD}}$ is an instance of $\vec{\boldsymbol{v}}$

$\overrightarrow{\boldsymbol{GC}}$ and $\overrightarrow{\boldsymbol{EF}}$ are instances of $\vec{\boldsymbol{w}}$

A **free vector** consists of a magnitude and a direction (direction and sense).

Vectors are usually denoted in boldface, as **v**; other conventions include $\vec{v}$ or v̲, especially in handwriting; alternately, some use a tilde (~) or a wavy underline drawn beneath the symbol, which is a convention for indicating boldface type.

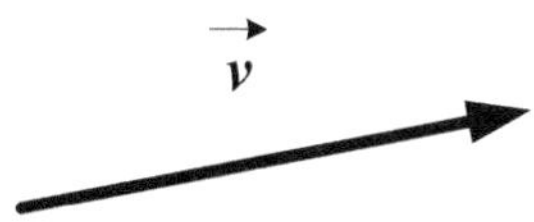

Vectors are usually shown in graphs or other diagrams as arrows. The length of the arrow represents the vector's magnitude, while the direction in which the arrow points represents the vector's direction.

The length of the arrow represents the vector's **magnitude**:

norm or **magnitude** of $\vec{v}(v_x, v_y)$ $\qquad |\vec{\boldsymbol{v}}| = \sqrt{\boldsymbol{v}_x^2 + \boldsymbol{v}_y^2}$ (Pythagoras)

The direction in which the arrow points represents the vector's **direction**:

Slope or **gradient** of $\vec{v}(v_x, v_y)$ $\qquad m_{\vec{v}} = \frac{v_y}{v_x} \qquad \left(\frac{rise}{run}\right)$

Examples: $\quad |\vec{\boldsymbol{u}}| = \sqrt{2^2 + (-5)^2} = \sqrt{29} \qquad m_{\vec{u}} = \frac{-5}{2}$

$|\vec{\boldsymbol{v}}| = \sqrt{(-2)^2 + 5^2} = \sqrt{29} \qquad m_{\vec{v}} = \frac{-5}{2}$

$|\vec{\boldsymbol{w}}| = \sqrt{3^2 + 2^2} = \sqrt{13} \qquad m_{\vec{w}} = \frac{2}{3}$

Note: Any vector **v** can be written as $\boldsymbol{v} = vx\boldsymbol{i} + vy\boldsymbol{j}$ with real numbers v_x and v_y, which are uniquely determined by $\boldsymbol{v}$ ($\boldsymbol{i}$ and $\boldsymbol{j}$ are the unit vectors parallel to the *x*-and *y*-axes respectively).

Note: Whenever we say 'vector', without specifying free or fixed, we mean 'free vector'

Note: If we add separately x-coordinates and y-coordinates, we can see that $\boldsymbol{P} + \overrightarrow{\boldsymbol{PQ}} = \boldsymbol{Q}$, so when placed at *P*, $\overrightarrow{\boldsymbol{PQ}}$ describes how to get from *P* to *Q*.

Slope intercept form

$$Slope(m) = \frac{y_2 - y_1}{x_2 - x_1}$$

Example: Find the slope of the line through each pair of points.

a. **(8, -7) and (5, -3).**
b. **(-5, 9) and (5, 11).**

Solution

$$Slope(m) = \frac{y_2 - y_1}{x_2 - x_1}$$

a. $\mathbf{(8, -7)}$ ***and*** $\mathbf{(5, -3)} \rightarrow (x_1, y_1)$ ***and*** (x_2, y_2) **so** $\boldsymbol{m} = \frac{\mathbf{-3-(-7)}}{\mathbf{5-8}} = -\frac{\mathbf{4}}{\mathbf{3}}.$

b. **(-5, 9) and (5, 11)** $\rightarrow \boldsymbol{m} = \mathbf{1/5}.$

Example: For each graph: Write the equation of the line in SLOPE-INTERCEPT FORM

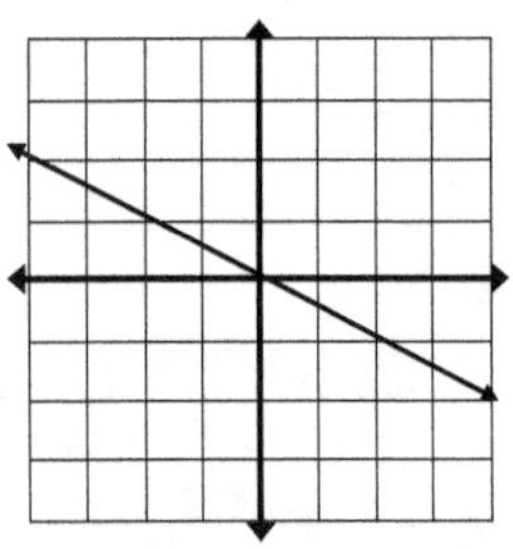
(a)

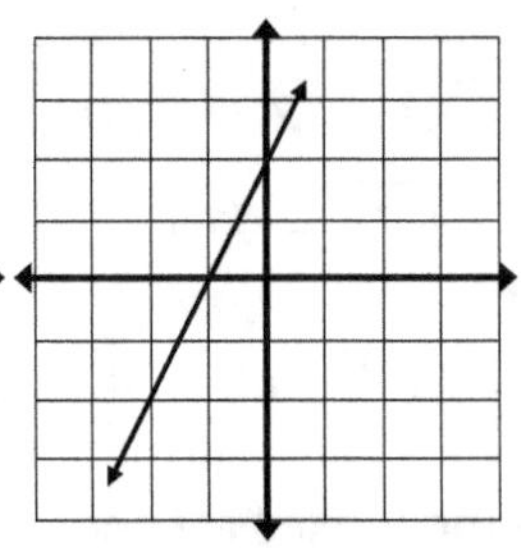
(b)

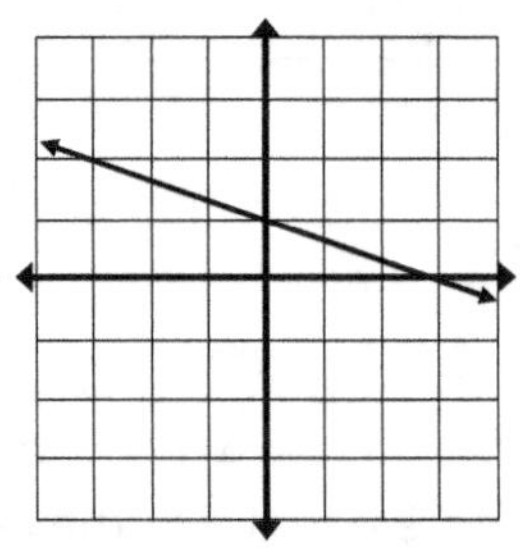
(c)

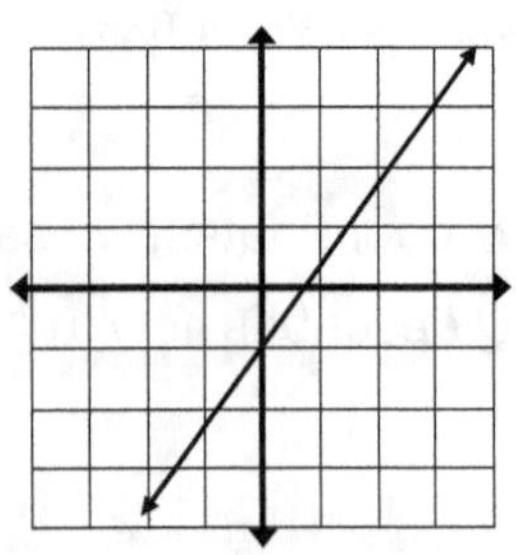
(d)

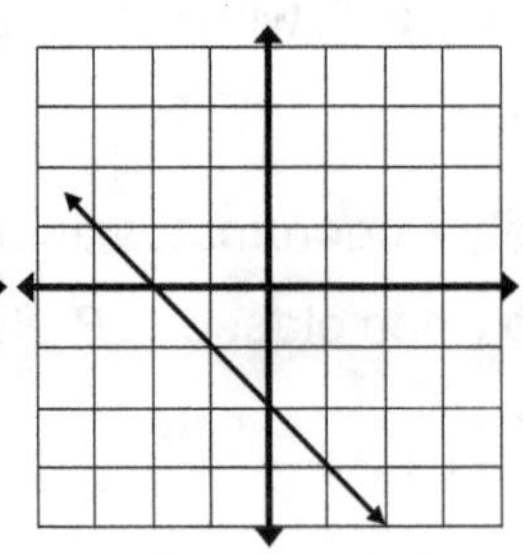
(e)

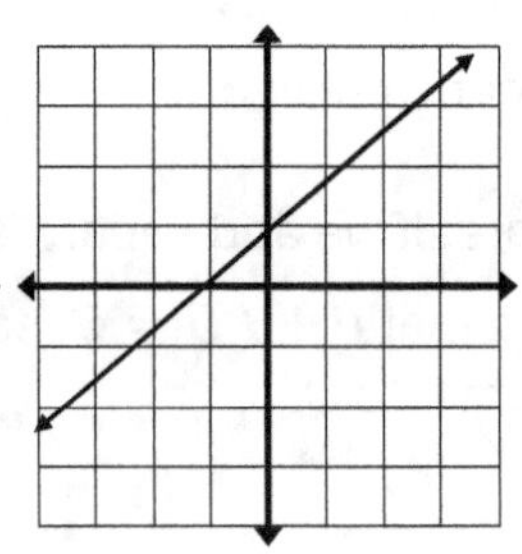
(f)

Solution:

According to the slope equation $y = mx + b$, the solution as indicated in the table

	Slope(m)	b	Equation	Slope(m)	b	Equation
a	$-\frac{1}{2}$	0	$y = -\frac{1}{2}x$	$\frac{4}{3}$	-1	$y = \frac{4}{3}x - 1$
b	$\frac{2}{1}$	2	$y = 2x + 2$	$-\frac{2}{2}$	-2	$y = -1x - 2$
c	$-\frac{1}{3}$	1	$y = -\frac{1}{3}x + 1$	$\frac{1}{1}$	1	$y = x + 1$

Example: In each linear equation, identify the slope (m) and the y-intercept (b)

a. $-\frac{5}{9} - \frac{8}{7}x = y$ b. $y = 4 + \frac{2}{5}x$ c. y = 5x – 6
d. $6 - \frac{7}{2}x = y$ e. $y = \frac{3}{2}x - \frac{21}{8}$ $f.\ y = \frac{5}{3} - x$

Solution: According to the slope equation $y = mx + b$, the solution as indicated in the table

Equation	Slope(m)	b	Equation	Slope(m)	b
$-\frac{5}{9} - \frac{8}{7}x = y$	$-\frac{8}{7}$	$-\frac{5}{9}$	$6 - \frac{7}{2}x = y$	$-\frac{7}{2}$	6
$y = 4 + \frac{2}{5}x$	$\frac{2}{5}$	4	$y = \frac{3}{2}x - \frac{21}{8}$	$\frac{3}{2}$	$-\frac{21}{8}$
y = 5x – 6	5	– 6	$y = \frac{5}{3} - x$	-1	$\frac{5}{3}$

Example: Find the equation of the line in slope-intercept form

$$(y = mx + b)$$

a. The slope of 4 and y-intercept of -5
b. (-4, 5) with slope = 6
c. $m = -7/5$ through (4, 6)
d. $Slope = -3/5$ and (0, -4).
e. b = 6 and m = -7
f. m = -2 and (4, 4)
g. Slope= $^4/_3$ through (4, 3)
h. (4, -5) and no slope
i. Slope = -6 and (-3, 8)

Solution:

a. $y = mx + b \rightarrow y = 4x - 5$.
b. $y = mx + b \rightarrow y = 6x + b$ substitute the point $(-4, 5)$ in y equation $5 = 6 \times (-4) + b$ get $b = 29$ so, $y = 6x + 29$.
c. $y = mx + b \rightarrow y = -7/5\,x + b$ substitute the point $(4, 6)$ in y equation $6 = -7/5 \times (4) + b$ get $b = 11.6$ so, $y = -\frac{7}{5}x + 11.6$.
d. $y = mx + b \rightarrow y = -\frac{3}{5}x + b$ substitute the point $(0, -4)$, $-4 = -\frac{3}{5}(0) + b$
   get $b = -4$,so $y = -\frac{3}{5}\,x - 4$.
e. $y = mx + b \rightarrow y = -7x + 6$
f. $y = mx + b \rightarrow y = -2x + b \rightarrow y = -2x + 12$.
g. $y = mx + b \rightarrow y = \frac{4}{3}x + b \rightarrow y = \frac{4}{3}x - \frac{7}{3}$.
h. A vertical **line** will have **no slope,** and "the **slope** is **undefined**" or "the **line** has **no slope**" **means** that the **line** is vertical. (By the way, all vertical **lines** are of the form "x = some **number**", and "x = some **number**" **means** the **line** is vertical.
i. $y = mx + b \rightarrow y = -6x + b \rightarrow y = -6x - 10$.

Example: Graph the line for the equation:

$$y = 6x - 3$$

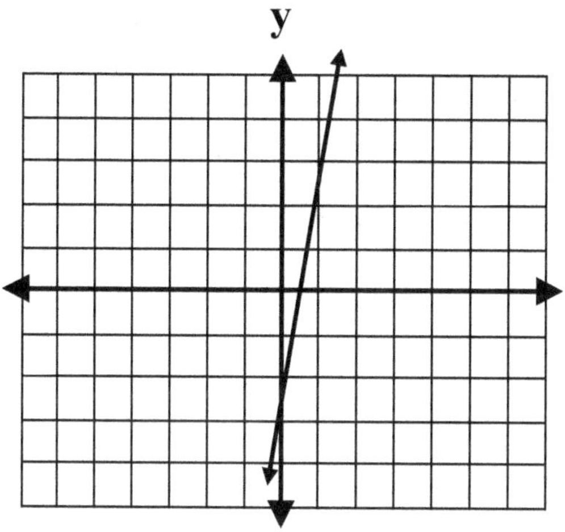

Solution:

Slope = 6
Y-Intercept = -3

Example: At the car rental company, you must pay a flat rate of $150 and then a daily fee of $20. Write a linear equation to describe the total cost, y, of renting the car for *x* days. What is the cost of renting a car for 13 days?

Solution:

$y = mx + b \rightarrow y = 20x + 150$

For 13 days, the cost is $y = 20 \times 13 + 150 = \$410$

Example: A membership to the gym cost $30 per person in 2018. The membership cost has increased by an average of $7.5 per person yearly since 2018. Write a linear equation for the cost of a gym membership for one person since 2018. What is the cost of gym membership in 2028?

Solution:

$y = mx + b \rightarrow y = 7.5x + 30$

For ten years, the cost is $y = 7.5 \times 10 + 30 = \$105$

2.2 Linear Algebra (Matrix Operations)

A matrix is a rectangular array or square of numbers of rows and columns. The *size* of a matrix is it is $(no.of\ \ rows) \times (no.of\ columns)$.

Example: Here are some various matrices with different sizes:

$A = \begin{bmatrix} 6 & 10 & 3 & 3 \\ 2 & 1 & 4 & -1 \end{bmatrix}$ (Size: 2×4) $\qquad B = \begin{bmatrix} 3 & 20i \\ 1/5 & -1 \\ 4i & -3/5 \end{bmatrix}$ (Size: 3×2)

$C = [4 \quad -3 \quad 7i \quad 2]$ (Size: 1×4) $\qquad D = \begin{bmatrix} -0.2 \\ 0 \\ 3 \end{bmatrix}$ (Size: 3×1)

a_{ij} (represents the element in the ith row and j^{th}column of the matrix *A*)

Thus, we have for

$A = \begin{bmatrix} -5 & 6 & 0 & 1 \\ 3 & 2 & -8 & 1 \end{bmatrix}$ $(a_{12} = 6, a_{23} = -8)$ and for $C = [2 \quad -2 \quad 9 \quad 3]$ $(c_{12} = -2, c_{14} = 3)$

A general matrix A of size $m \times n$ has the form.

$$A = \begin{bmatrix} a_{11} & a_{12} & \cdots & a_{1n} \\ a_{21} & a_{22} & \cdots & a_{2n} \\ \vdots & \vdots & \cdots & \vdots \\ a_{m1} & a_{m2} & \cdots & a_{mn} \end{bmatrix}$$

A *square matrix* is a matrix where (# rows) = (# columns) or *m* = *n*. In this case, the matrix is a square matrix of order *n*. Here are some examples of square matrices:

$\begin{bmatrix} 1 & -2 \\ -7 & 2 \end{bmatrix}$ (2× 2 square matrix of order 2) $\begin{bmatrix} -7 & 1 & 20 \\ 56 & 56 & 91 \\ 1/12 & 0 & 90 \end{bmatrix}$ (3×3 square matrix of order 3).

2.2.1 Equality of Matrices

Two matrices *A* and *B* are equal (written *A* = *B*) if they have the same size and their corresponding entries are equal, that is $a_{ij} = b_{ij}$ for all *i* and *j*.

Example: Given the matrices $A = \begin{bmatrix} 6 & 6 & 7 \\ -1/2 & -8 & 5 \end{bmatrix}$, $B = \begin{bmatrix} -3 & 3 & 1 \\ -2 & -2 & 6 \end{bmatrix}$, $C = \begin{bmatrix} 6 & 6 & 7 \\ -1/2 & -8 & 5 \end{bmatrix}$, and $D = \begin{bmatrix} -5 & -4 \\ 6 & 7 \\ 4 & 5 \end{bmatrix}$, then ***A*=*C***.
However, $A \neq B$ since these matrices do not have equal corresponding entries and $A \neq D$ since the matrix sizes are not equal (A is a 2×3 matrix and D is a 3×2 matrix).

2.2.2 Special Types of Matrices – Vectors

A *row vector* is a matrix with one row. A column vector is a matrix with one column.

Example: The matrix $C = [3 \quad -4 \quad 8 \quad 1]$ is a 1×4 row vector and the matrix $D = \begin{bmatrix} 0 \\ 1 \\ 4 \end{bmatrix}$ is a 3×1 column vector.

2.2.3 Addition, Subtraction, and Scalar Multiplication of Matrices

Matrices can only be added and subtracted if they have the same size. To add or subtract matrices, we add or subtract each corresponding component.

Example: Given the 2×3 matrices $A = \begin{bmatrix} 2 & 1 & 1 \\ -1 & -1 & 4 \end{bmatrix}$ and $B = \begin{bmatrix} 2 & -3 & 4 \\ -3 & 1 & -2 \end{bmatrix}$, we see that

$$\begin{aligned} A + B = \begin{bmatrix} 2 & 1 & 1 \\ -1 & -1 & 4 \end{bmatrix} + \begin{bmatrix} 2 & -3 & 4 \\ -3 & 1 & -2 \end{bmatrix} &= \begin{bmatrix} 2+2 & 1+(-3) & 1+4 \\ -1+(-3) & -1+1 & 4+(-2) \end{bmatrix} \\ &= \begin{bmatrix} 4 & -2 & 5 \\ -4 & 0 & 2 \end{bmatrix} \end{aligned}$$

and

$A - B = \begin{bmatrix} 2 & 1 & 1 \\ -1 & -1 & 4 \end{bmatrix} - \begin{bmatrix} 2 & -3 & 4 \\ -3 & 1 & -2 \end{bmatrix} = \begin{bmatrix} 2-2 & 1-(-3) & 1-4 \\ -1-(-3) & -1-1 & 4-(-2) \end{bmatrix} = \begin{bmatrix} 0 & 4 & -3 \\ 2 & -2 & 6 \end{bmatrix}$.

Example: Given the 2×2 matrix $A = \begin{bmatrix} 2 & 3 \\ 1 & 0 \end{bmatrix}$ and the 3×2 matrix $B = \begin{bmatrix} 1 & 0 \\ 2 & 1 \\ 1 & 1 \end{bmatrix}$, the matrix $A + B$ is not defined since A and $\boldsymbol{B}$ are not the same size.

When working with matrices, numbers are referred to as *scalars*. To multiply a matrix by a scalar, we multiply each matrix entry by the given scalar.

Example: Let $A = \begin{bmatrix} -2 & 3 \\ 1 & 2 \end{bmatrix}$; then $3A = 3\begin{bmatrix} -2 & 3 \\ 1 & 2 \end{bmatrix} = \begin{bmatrix} (3)(-2) & (3)(3) \\ (3)(1) & (3)(2) \end{bmatrix} = \begin{bmatrix} -6 & 9 \\ 3 & 6 \end{bmatrix}$.

Example: Let $A = \begin{bmatrix} -3 & 2 \\ 2 & 1 \\ 5 & 0 \end{bmatrix}$ and $B = \begin{bmatrix} 0 & -5/2 \\ 1 & 2 \\ 3 & -1 \end{bmatrix}$, then

$5A - 2B = 5\begin{bmatrix} -3 & 2 \\ 2 & 1 \\ 5 & 0 \end{bmatrix} - 2\begin{bmatrix} 0 & -5/2 \\ 1 & 2 \\ 3 & -1 \end{bmatrix} = \begin{bmatrix} -15 & 10 \\ 10 & 5 \\ 25 & 0 \end{bmatrix} - \begin{bmatrix} 0 & -5 \\ 2 & 4 \\ 6 & -2 \end{bmatrix} = \begin{bmatrix} -15 & 15 \\ 8 & 1 \\ 19 & 2 \end{bmatrix}$.

Matrix addition and scalar multiplication have certain basic properties that we now summarize.

2.2.4 Addition and Scalar Multiplication Properties of Matrices

Let ***A***, ***B***, and ***C*** be matrices of the same size and *c* and *d* be scalars.
1. ***A*** + ***B*** = ***B*** + ***A*** (matrix addition is commutative)
2. $(A + B) + C = A + (B + C)$ (matrix addition is associative)
3. *cd*(**A**) = *c*(*d***A**).
4. $c(A + B) = cA + c\mathbf{B}$

2.2.5 Matrix Multiplication

One must first understand how to multiply a row vector times a column vector to understand matrix multiplication. If $A = [a_{11}, a_{12}, \dots, a_{1n}]$ is a $1 \times n$ row vector and $B = \begin{bmatrix} b_{11} \\ b_{21} \\ \vdots \\ b_{n1} \end{bmatrix}$ is a $n \times 1$column vector, then the product of ***A*** and ***B*** is the scalar produced by multiplying each corresponding entry of ***A*** and ***B*** and adding. That is

$$AB = [a_{11,}a_{12}, \dots, a_{1n}] \begin{bmatrix} b_{11} \\ b_{21} \\ \vdots \\ b_{n1} \end{bmatrix} = a_{11}b_{11} + a_{12}b_{21} + \cdots + a_{1n}b_{n1}.$$

Example: If $A = [1 \quad -1 \quad 0 \quad 4]$ and $B = \begin{bmatrix} 3 \\ -5 \\ 1 \\ -1 \end{bmatrix}$, then

$$AB = [1 \quad -1 \quad 0 \quad 4] \begin{bmatrix} 3 \\ -5 \\ 1 \\ -1 \end{bmatrix} = (1)(3) + (-1)(-5) + (0)(1) + 4(-1) = 3 + 5 + 0 - 4 = 4.$$

Multiplication of matrices, in general, involves multiple multiplications of rows and columns. If **A** is a $m \times n$matrix and **B** is a $n \times p$matrix, the product ***C=AB*** is the matrix where each element c_{ij} is made up of the product of the ith the row of the left matrix ***A*** multiplied to the jth column of the right matrix ***B***. That is,

$c_{ij} = (row\ i\ of\ A) \cdot (column\ j\ of\ B).$

Note! For the matrix product ***AB*** to exist, the number of columns of the left matrix ***A*** must equal the number of rows in the right matrix ***B***. The size of the product will be
(the number of rows in ***A***) × (the number of columns in ***B***).
It can be easier to see this by examining the following:

$$C = \underset{m\times n}{A} \underset{\substack{must\\equal}}{\overset{\cdot}{<--->}} \underset{n\times p}{B}$$

Example: If $A = \begin{bmatrix} 1 & -3 \\ 0 & 2 \end{bmatrix}$ and $B = \begin{bmatrix} 1 & 0 & -2 \\ -1 & 2 & 3 \end{bmatrix}$ Find the product ***AB***.

The 2 ×3 matrix ***AB*** is computing by performing the following row column multiplications:

$$AB = \begin{bmatrix} (row\ 1A) \cdot (column\ 1B) & (row\ 1A) \cdot (column\ 2B) & (row\ 1A) \cdot (column\ 3B) \\ (row\ 2A) \cdot (column\ 1B) & (row\ 2A) \cdot (column\ 2B) & (row\ 2A) \cdot (column\ 3B) \end{bmatrix}$$

Performing these multiplications, we obtain the following:

$$AB = \begin{bmatrix} 1 & -3 \\ 0 & 2 \end{bmatrix} \begin{bmatrix} 1 & 0 & -2 \\ -1 & 2 & 3 \end{bmatrix} =$$
$$\begin{bmatrix} (1)(1) + (-3)(-1) & (1)(0) + (-3)(2) & (1)(-2) + (-3)(3) \\ (0)(1) + (2)(-1) & (0)(0) + (2)(2) & (0)(-2) + (2)(3) \end{bmatrix} =$$
$$\begin{bmatrix} 4 & -6 & -11 \\ -2 & 4 & 6 \end{bmatrix}.$$

Example: For the matrices, $A = \begin{bmatrix} 3 & 0 \\ 1 & 1 \\ -1 & 0 \end{bmatrix}$, $B = \begin{bmatrix} 1 & -1 & 0 \\ 2 & -5 & 1 \end{bmatrix}$, obtain the product AB and BA.

$$AB = \begin{bmatrix} 3 & 0 \\ 1 & 1 \\ -1 & 0 \end{bmatrix} \begin{bmatrix} 1 & -1 & 0 \\ 2 & -5 & 1 \end{bmatrix} =$$

$$\begin{bmatrix} (3)(1)+(0)(2) & (3)(-1)+(0)(-5) & (3)(0)+(0)(1) \\ (1)(1)+(1)(2) & (1)(-1)+(1)(-5) & (1)(0)+(1)(1) \\ (-1)(1)+(0)(2) & (-1)(-1)+(0)(-5) & (-1)(0)+(0)(1) \end{bmatrix} =$$

$$\begin{bmatrix} 3 & -3 & 0 \\ 3 & -6 & 1 \\ -1 & 1 & 0 \end{bmatrix}.$$

and

$$BA = \begin{bmatrix} 1 & -1 & 0 \\ 2 & -5 & 1 \end{bmatrix} \begin{bmatrix} 3 & 0 \\ 1 & 1 \\ -1 & 0 \end{bmatrix} =$$

$$\begin{bmatrix} (1)(3)+(-1)(1)+(0)(-1) & (1)(0)+(-1)(1)+(0)(0) \\ (2)(3)+(-5)(1)+(1)(-1) & (2)(0)+(-5)(1)+(1)(0) \end{bmatrix} =$$

$$\begin{bmatrix} 2 & -1 \\ 0 & -5 \end{bmatrix}.$$

$$AB \neq BA$$

Example: Consider the matrices $A = \begin{bmatrix} 1 & -1 & 0 & 2 \\ 2 & -5 & 1 & 5 \end{bmatrix}$ and $B = \begin{bmatrix} 3 & 0 \\ 1 & 1 \\ -1 & 0 \end{bmatrix}$,

The product BA is the 3×4 matrix given by

$$BA = \begin{bmatrix} 3 & 0 \\ 1 & 1 \\ -1 & 0 \end{bmatrix} \begin{bmatrix} 1 & -1 & 0 & 2 \\ 2 & -5 & 1 & 5 \end{bmatrix} = \begin{bmatrix} 3 & -3 & 0 & 6 \\ 3 & -6 & 1 & 7 \\ -1 & 1 & 0 & -2 \end{bmatrix}.$$

2.2.6 Multiplicative Properties of Matrices

Let ***A***, ***B***, and ***C*** be matrices whose sizes are multiplicatively compatible, c a scalar.

1. $(AB)C = A(BC)$ (matrix multiplication is associative)
2. $A(B + C) = AB + AC$.
3. $(A + B)C = AC + BC$.
4. $c(AB)$ = (cA) ***B*** = ***A***(c***B***)

2.2.7 Addition and Multiplicative Identity Matrices

The *zero* matrix is a $m \times n$ matrix made up of all zeros. The following examples represent $2 \times 2, 3 \times 2, 2 \times 4$ zero matrices

$$\underset{2\times 2}{\begin{bmatrix} 0 & 0 \\ 0 & 0 \end{bmatrix}}, \underset{3\times 2}{\begin{bmatrix} 0 & 0 \\ 0 & 0 \\ 0 & 0 \end{bmatrix}}, \underset{2\times 4}{\begin{bmatrix} 0 & 0 & 0 & 0 \\ 0 & 0 & 0 & 0 \end{bmatrix}}$$

The zero matrix represents the additive identity matrix of all matrices under the addition operation. If ***A*** is a $m \times n$ matrix and 0 is a $m \times n$ zero matrix, then

A **+ 0 = 0 + A = A.**

The multiplicative identity matrix, usually referred to just the *identity* matrix ***I***, is the $n \times n$ matrix defined by

$$I = \begin{bmatrix} 1 & 0 & 0 & \cdots & 0 \\ 0 & 1 & 0 & \cdots & 0 \\ 0 & 0 & 1 & \cdots & 0 \\ \vdots & \vdots & \vdots & \ddots & 0 \\ 0 & 0 & 0 & \cdots & 1 \end{bmatrix}.$$

I has 1's on the main diagonal, and 0's elsewhere. The following represent a 2×2, 3×3, and 4×4 identity matrices:

$$\underset{2\times 2}{\begin{bmatrix} 1 & 0 \\ 0 & 1 \end{bmatrix}}, \underset{3\times 3}{\begin{bmatrix} 1 & 0 & 0 \\ 0 & 1 & 0 \\ 0 & 0 & 1 \end{bmatrix}}, \underset{4\times 4}{\begin{bmatrix} 1 & 0 & 0 & 0 \\ 0 & 1 & 0 & 0 \\ 0 & 0 & 1 & 0 \\ 0 & 0 & 0 & 1 \end{bmatrix}}.$$

Since ***I*** serve as the multiplicative identity, if ***A*** is a $m \times n$ matrix, then ***AI*** **=** ***IA*** **= A.**

Note that ***I*** is always a square matrix. That is, the number of rows equals the number of columns. Of course, the size of ***I*** depends on the size of ***A*** when multiplying on the left and right, as the next example demonstrates.

Example: Given the 2×3matrix $A = \begin{bmatrix} 1 & 2 & 0 \\ -1 & 1 & 2 \end{bmatrix}$, to be multiplicatively compatible, the size of the identity I on the left must be 2×2 and on the right 3×3. This gives

$$IA = \begin{bmatrix} 1 & 0 \\ 0 & 1 \end{bmatrix}\begin{bmatrix} 1 & 2 & 0 \\ -1 & 1 & 2 \end{bmatrix} = \begin{bmatrix} 1 & 2 & 0 \\ -1 & 1 & 2 \end{bmatrix} = A \quad \text{and} \quad AI = \begin{bmatrix} 1 & 2 & 0 \\ -1 & 1 & 2 \end{bmatrix}\begin{bmatrix} 1 & 0 & 0 \\ 0 & 1 & 0 \\ 0 & 0 & 1 \end{bmatrix} = \begin{bmatrix} 1 & 2 & 0 \\ -1 & 1 & 2 \end{bmatrix} = A.$$

2.2.8 Upper Triangular and Diagonal Matrices

A square matrix is *upper triangular* if all entries below the main diagonal are zero. A *diagonal* matrix is a special case of an upper triangular matrix where entries above and below the main diagonal are zero. The general form of each matrix looks like:

$$\underset{upper\ triangular}{\begin{bmatrix} a_{11} & a_{12} & a_{13} & \cdots & a_{1n} \\ 0 & a_{22} & a_{23} & \cdots & a_{2n} \\ 0 & 0 & a_{33} & \cdots & a_{3n} \\ \vdots & \vdots & \begin{matrix} 0 \\ \vdots \end{matrix} & \ddots & \vdots \\ 0 & 0 & 0 & \cdots 0 & a_{nn} \end{bmatrix}} \quad \underset{diagonal}{\begin{bmatrix} b_{11} & 0 & 0 & \cdots & 0 \\ 0 & b_{22} & 0 & \cdots & 0 \\ 0 & 0 & b_{33} & 0 \cdots & 0 \\ 0 & 0 & 0 & \ddots & \begin{matrix} \vdots \\ 0 \end{matrix} \\ 0 & 0 & 0 & \cdots 0 & b_{nn} \end{bmatrix}}$$

The matrix $A = \begin{bmatrix} -2 & 1 & 3 \\ 0 & 1 & 4 \\ 0 & 0 & 3 \end{bmatrix}$ is an example of a upper triangular matrix while $B = \begin{bmatrix} 5 & 0 & 0 & 0 \\ 0 & 2 & 0 & 0 \\ 0 & 0 & -3 & 0 \\ 0 & 0 & 0 & -1 \end{bmatrix}$ is an example of a diagonal matrix.

Two important facts are evident when concerned with upper triangular and diagonal matrices:

1. The sum or difference of two upper triangular matrices is an upper triangular matrix and the sum or
   difference of two diagonal matrices is a diagonal matrix.

2. The product of two upper triangular matrices is an upper triangular matrix and the product of two

diagonal matrices is a diagonal matrix.

Example: Given the two upper triangular matrices $A_1 = \begin{bmatrix} -2 & 1 & 3 \\ 0 & 1 & 4 \\ 0 & 0 & 3 \end{bmatrix}$ and $A_2 = \begin{bmatrix} 3 & -1 & 0 \\ 0 & 1 & 1 \\ 0 & 0 & 2 \end{bmatrix}$ and two diagonal matrices $B_1 = \begin{bmatrix} 5 & 0 & 0 & 0 \\ 0 & 2 & 0 & 0 \\ 0 & 0 & -3 & 0 \\ 0 & 0 & 0 & -1 \end{bmatrix}$ and $B_2 = \begin{bmatrix} 5 & 0 & 0 & 0 \\ 0 & 2 & 0 & 0 \\ 0 & 0 & -3 & 0 \\ 0 & 0 & 0 & -1 \end{bmatrix}$, one can readily verify the following sums and products:

$A_1 + A_2 = \begin{bmatrix} 1 & 0 & 3 \\ 0 & 2 & 5 \\ 0 & 0 & 5 \end{bmatrix}$, $A_1 \cdot A_2 = \begin{bmatrix} -6 & 3 & 7 \\ 0 & 1 & 9 \\ 0 & 0 & 6 \end{bmatrix}$, $B_1 + B_2 = \begin{bmatrix} 10 & 0 & 0 & 0 \\ 0 & 4 & 0 & 0 \\ 0 & 0 & -6 & 0 \\ 0 & 0 & 0 & -2 \end{bmatrix}$, and $B_1 \cdot B_2 = \begin{bmatrix} 25 & 0 & 0 & 0 \\ 0 & 4 & 0 & 0 \\ 0 & 0 & 9 & 0 \\ 9 & 9 & 0 & 1 \end{bmatrix}$

2.2.9 Determinants

The determinant of a 2×2 matrix $A = \begin{bmatrix} a_{11} & a_{12} \\ a_{21} & a_{22} \end{bmatrix}$ is given by $\det(\boldsymbol{A}) = a_{11}a_{22} - a_{21}a_{12}$.

Example: The determinant of the matrix $A = \begin{bmatrix} -3 & 1 \\ 5 & 2 \end{bmatrix}$ is:

$$det(A) = (-3)(2) - (5)(1) = -6 - 5 = -11.$$

Example: The determinant of the matrix $A = \begin{bmatrix} 2 & -2 \\ 4 & 3 \end{bmatrix}$ is:

$$det(A) = (2)(3) - (4)(-2) = 6 - -8 = 6 + 8 = 14.$$

A fact to note is that the determinant of a 1×1 matrix is defined as the entry of the matrix. For example, if $A = [3]$, then $det(A) = 3$. Another important note to make is that is possible to take determinants of larger size square matrices (larger than 2×2). However, this will not be necessary for most purposes except when dealing with diagonal and upper triangular matrices defined above. If we have an upper triangular and diagonal matrix of the form:

$$A = \begin{bmatrix} a_{11} & a_{12} & \cdots & a_{1n} \\ 0 & a_{22} & \cdots & a_{2n} \\ \vdots & \vdots & \ddots & \vdots \\ 0 & 0 & 0 \cdots 0 & a_{nn} \end{bmatrix} \text{ and } B = \begin{bmatrix} b_{11} & 0 & \cdots & 0 \\ 0 & b_{22} & \cdots & 0 \\ \vdots & \vdots & \ddots & \vdots \\ & & & 0 \\ 0 & 0 & \cdots 0 & b_{nn} \end{bmatrix},$$

then the determinant is defined to be the product of the main diagonal elements, that is

$det(A) = a_{11} \cdot a_{22} \cdots a_{nn}$ and $det(B) = b_{11} \cdot b_{22} \cdots b_{nn}$.

Example: Given the upper triangular matrix $A = \begin{bmatrix} -2 & 1 & 3 \\ 0 & 1 & 4 \\ 0 & 0 & 3 \end{bmatrix}$ and diagonal matrix $B = \begin{bmatrix} 5 & 0 & 0 & 0 \\ 0 & 2 & 0 & 0 \\ 0 & 0 & -3 & 0 \\ 0 & 0 & 0 & -1 \end{bmatrix}$, we see that $det(A) = (-2)(1)(3) = -6$ and $det(B) = (5)(2)(-3)(-1) = 30$.

An important property concerning the determinant of the product of two matrices is the following:

$$det(AB) = det(A)\, det(B)$$

Example: For the matrices $A = \begin{bmatrix} 1 & 2 \\ 2 & 4 \end{bmatrix}$ and $B = \begin{bmatrix} -1 & 2 \\ 3 & 0 \end{bmatrix}$, one can verify that $AB = \begin{bmatrix} 5 & 2 \\ 10 & 4 \end{bmatrix}$ and compute $det(A) = 0$, $det(B) = -6$, and $det(AB) = 0$. Thus one can see that $0 = det(AB) = det(A)\, det(B) = (-6)(0) = 0$.

However, this distributive property does not, in general, hold for the sum of two matrices, that is, in general:

$$det(A + B) \neq det(A) + det(B)$$

Example: For the matrices $A = \begin{bmatrix} 1 & 2 \\ 2 & 4 \end{bmatrix}$ and $B = \begin{bmatrix} -1 & 2 \\ 3 & 0 \end{bmatrix}$, one can verify that $A + B = \begin{bmatrix} 0 & 4 \\ 5 & 4 \end{bmatrix}$ and compute $det(A) = 0$, $det(B) = -6$, and $det(A + B) = -20$. Thus one can see that $-20 = det(A + B) \neq det(A) + det(B) = -6 + 0 = -6$.

2.2.10 Matrix Inverses

Matrices defined with addition have an obvious inverse. Given a matrix ***A***, its additive inverse is defined to be the matrix **–*A***, where all the elements of A are negated. Note that ***A* + (-*A*) = 0**, where **0** is the zero matrix. We are more interested, however, in the multiplicative inverse of a matrix, or *inverse* for short.

The inverse of a $n \times n$ matrix A, if it exists, is denoted by A^{-1}, and is defined to be the $n \times n$ matrix where
$AA^{-1} = A^{-1}A = I$,
Where I is the $n \times n$ identity matrix, it can be shown that. A^{-1} is unique. Note the inverse only exists for square matrices where the row and column numbers are the same.

Example: To show that $A^{-1} = \begin{bmatrix} 3/5 & 1/5 \\ -2/5 & 1/5 \end{bmatrix}$ is the inverse of the matrix $A = \begin{bmatrix} 1 & -1 \\ 2 & 3 \end{bmatrix}$, one computes
$A^{-1}A = \begin{bmatrix} 3/5 & 1/5 \\ -2/5 & 1/5 \end{bmatrix} \begin{bmatrix} 1 & -1 \\ 2 & 3 \end{bmatrix} = \begin{bmatrix} 3/5 + 2/5 & -3/5 + 3/5 \\ -2/5 + 2/5 & 2/5 + 3/5 \end{bmatrix} =$
$\begin{bmatrix} 1 & 0 \\ 0 & 1 \end{bmatrix}$. A similar calculation can be shown for $AA^{-1} = I$.

Given a matrix A, how do we know if A^{-1} exists, and if it does, how can we calculate A^{-1}? For matrices larger than 2×2, the recommended method involves row reduction. However, for 2×2 matrices, there is a method based on the following:

FACT: For a given 2×2 matrix$A = \begin{bmatrix} a_{11} & a_{12} \\ a_{21} & a_{22} \end{bmatrix}$, the matrix A^{-1} exists only if $det(A) \neq 0$. If this is so, the inverse of the matrix is defined to be:

$A^{-1} = \frac{1}{det(A)}\begin{bmatrix} a_{22} & -a_{12} \\ -a_{21} & a_{11} \end{bmatrix}.$

Note that the matrix part of the formula for A^{-1} switching the main diagonal elements and negating the back diagonal elements is obtained.

Example: Given the matrix $A = \begin{bmatrix} 2 & 1 \\ 3 & 3 \end{bmatrix}$, we see that $det(A) = (2)(3) - (3)(1) = 3 \neq 0$. Hence the inverse exists and $A^{-1} = \frac{1}{3}\begin{bmatrix} 3 & -1 \\ -3 & 2 \end{bmatrix} = \begin{bmatrix} 1 & -1/3 \\ -1 & 2/3 \end{bmatrix}$.

Example: Given the matrix$A = \begin{bmatrix} 1 & 2 \\ 2 & 4 \end{bmatrix}$, one can quickly see that $det(A) = 4 - 4 = 0$. Therefore, A^{-1} does not exist.

Example: Given the diagonal matrix $B = \begin{bmatrix} 5 & 0 & 0 & 0 \\ 0 & 2 & 0 & 0 \\ 0 & 0 & 0 & 0 \\ 0 & 0 & 0 & -1 \end{bmatrix}$, one can see that $det(B) = (5)(2)(0)(-1) = 0$. Thus B^{-1} does not exist. For the upper triangular matrix $A = \begin{bmatrix} -2 & 1 & 3 \\ 0 & 1 & 4 \\ 0 & 0 & 3 \end{bmatrix}$, one can see that $det(A) = (-2)(1)(3) = -6 \neq 0$. Thus A^{-1} exists, although row-reduction methods are needed to verify that $A^{-1} = \begin{bmatrix} -1/2 & 1/2 & -1/6 \\ 0 & 1 & -4/3 \\ 0 & 0 & 1/3 \end{bmatrix}$.

2.3 Trigonometry and Vector Analysis

Quadratic formulas

The roots of the quadratic equation ax2 + bx + c = 0

$$x_1, x_2 = \frac{-b \pm \sqrt{b^2 - 4ac}}{2a}$$

Trigonometric identities

$$sin(-x) = -\sin x$$
$$cos(-x) = \cos x$$

$$\sec x = \frac{1}{\cos x}, \quad \text{cec}\, x = \frac{1}{\sin x}$$

$$\tan x = \frac{\sin x}{\cos x}, \quad \cot x = \frac{1}{\tan x}$$

$$sin(x \pm 90°) = \pm \cos x$$
$$cos(x \pm 90°) = \mp \sin x$$
$$sin(x \pm 180°) = -\sin x$$
$$cos(x \pm 180°) = -\cos x$$

$$\cos^2 x + \sin^2 x = 1$$
$$\frac{a}{\sin A} = \frac{b}{\sin B} = \frac{c}{\sin C} \ (law\ of\ sines)$$

$$a^2 = b^2 + c^2 - 2bc \cos A \ (law\ of\ cosines)$$

$$\frac{tan\frac{1}{2}(A-B)}{tan\frac{1}{2}(A+B)} = \frac{a-b}{a+b} \text{ (law of tangents)}$$

$$\sin(x \pm y) = \sin x \cos y \pm \cos x \sin y$$
$$\cos(x \pm y) = \cos x \cos y \mp \sin x \sin y$$
$$\tan(x \pm y) = \frac{\tan x \pm \tan y}{1 \mp \tan x \tan y}$$
$$2 \sin x \sin y = \cos(x - y) - \cos(x + y)$$
$$2 \sin x \cos y = \sin(x + y) - \sin(x - y)$$

$$2 \cos x \cos y = \cos(x + y) - \cos(x - y)$$
$$\sin 2x = 2 \sin x \cos x$$
$$\cos 2x = \cos^2 x - \sin^2 x = 2\cos^2 x - 1 = 1 - 2\sin^2 x$$
$$\tan 2x = \frac{2 \tan x}{1 - \tan^2 x}$$
$$\cos^2 x = \frac{1+\cos(2x)}{2}$$
$$\sin^2 x = \frac{1-\cos(2x)}{2}$$

$a \cos x + b \sin x = K \cos(x + \theta)$, where $K = \sqrt{a^2 + b^2}$ and $\theta = tan^{-1}\left(\frac{-b}{a}\right)$

$e^{\pm jx} = \cos x \pm j \sin x \ (Euler's\ formula)$

$$\cos x = \frac{e^{jx} + e^{-jx}}{2}$$
$$\sin x = \frac{e^{jx} - e^{-jx}}{2j}$$

Trigonometric substitution

Form	Trig Sub	Identity
$\sqrt{a^2 + x^2}$	$x = a \tan \theta$	$1 + \tan^2 \theta = \sec^2 \theta$
$\sqrt{a^2 - x^2}$	$x = a \sin \theta$	$1 - \sin^2 \theta = \cos^2 \theta$
$\sqrt{x^2 - a^2}$	$x = a \sec \theta$	$\sec^2 \theta - 1 = \tan^2 \theta$

Hyperbolic functions

$$\sinh x = \frac{1}{2}(e^x - e^{-x})$$
$$\cosh x = \frac{1}{2}(e^x + e^{-x})$$
$$\tanh x = \frac{\sinh x}{\cosh x}$$
$$\coth x = \frac{1}{\tanh x}$$
$$\mathrm{csc\,h}\, x = \frac{1}{\sinh x}$$
$$\mathrm{sec\,h}\, x = \frac{1}{\cosh x}$$

$$\sinh(x \pm y) = \sinh x \cosh y \pm \cosh x \sinh y$$
$$\cosh(x \pm y) = \cosh x \cosh y \pm \sinh x \sinh y$$
$$\tan(x \pm y) = \frac{\tan x \pm \tan y}{1 \mp \tan x \tan y}$$

2.4 Calculus

2.4.1 Derivatives

where the prime indicates differentiation.

If U = U(x), V = V(x), and a = constant,

$$\frac{d}{dx}(aU) = a\frac{dU}{dx}$$

$$\frac{d}{dx}(UV) = U\frac{dV}{dx} + V\frac{dU}{dx}$$

$$\frac{d}{dx}\left(\frac{U}{V}\right) = \frac{V\frac{dU}{dx} - U\frac{dV}{dx}}{V^2}$$

$$\frac{d}{dx}(aU^n) = naU^{n-1}$$

$$\frac{d}{dx}(a^U) = a^U \text{1na}\frac{dU}{dx}$$

$$\frac{d}{dx}(e^U) = e^U\frac{dU}{dx}$$

$$\frac{d}{dx}(\sin U) = \cos U\frac{dU}{dx}$$

$$\frac{d}{dx}(\cos U) = -\sin U\frac{dU}{dx}$$

$$\frac{d}{dx}\tan U = \frac{1}{\cos^2 U}\frac{dU}{dx}$$

2.4.2 Indefinite Integrals

If U = U(x), V = V(x), and a = constant,

$$\int adx = ax + C$$

$$\int UdV = UV - \int VdU$$

$$\int U^n\, dU = \frac{U^{n+1}}{n+1} + C, n \neq 1$$

$$\int \frac{dU}{U} = \text{1n}U + C$$

$$\int x^2 e^{ax} dx = \frac{e^{ax}}{a^3}(a^2x^2 - 2ax + 2) + C$$

$$\int 1\text{n}x dx = x\,1\text{n}x - x + C$$

$$\int sin\, a\, x dx = -\frac{1}{a} cos\, a\, x + C$$

$$\int cos\, a\, x dx = \frac{1}{a} sin\, a\, x + C$$

$$\int sin^2\, a\, x dx = \frac{x}{2} - \frac{sin\, 2\, ax}{4a} + C$$

$$\int cos^2\, a\, x dx = \frac{x}{2} + \frac{sin\, 2\, ax}{4a} + C$$

$$\int x\, sin\, a\, x dx = \frac{1}{a^2}(sin\, a\, x - ax\, cos\, a\, x) + C$$

$$\int x\, cos\, a\, x dx = \frac{1}{a^2}(cos\, a\, x + ax\, sin\, a\, x) + C$$

$$\int x^2\, sin\, a\, x dx = \frac{1}{a^3}(2ax\, sin\, a\, x + 2\, cos\, a\, x - a^2x^2\, cos\, a\, x) + C$$

$$\int x^2\, cos\, a\, x dx = \frac{1}{a^3}(2ax\, cos\, a\, x - 2\, sin\, a\, x + a^2x^2\, sin\, a\, x) + C$$

$$\int e^{ax}\, sin\, b\, x dx = \frac{e^{ax}}{a^2 + b^2}(a\, sin\, b\, x - b\, cos\, b\, x) + C$$

$$\int e^{ax}\, cos\, b\, x dx = \frac{e^{ax}}{a^2 + b^2}(a\, cos\, b\, x + b\, sin\, b\, x) + C$$

$$\int sin\, a\, x\, sin\, b\, x dx = \frac{sin(a - b)\, x}{2(a - b)} - \frac{sin(a + b)\, x}{2(a + b)} + C, a^2 \neq b^2$$

$$\int sin\, a\, x\, cos\, b\, x dx = -\frac{cos(a - b)\, x}{2(a - b)} - \frac{cos(a + b)\, x}{2(a + b)} + C, a^2 \neq b^2$$

$$\int cos\, a\, x\, cos\, b\, x dx = \frac{sin(a - b)\, x}{2(a - b)} + \frac{sin(a + b)\, x}{2(a + b)} + C, a^2 \neq b^2$$

$$\int \frac{dx}{a^2 + x^2} = \frac{1}{a} tan^{-1} \frac{x}{a} + C$$

$$\int \frac{x^2 dx}{a^2 + x^2} = x - a\, tan^{-1} \frac{x}{a} + C$$

$$\int \frac{dx}{(a^2 + x^2)^2} = \frac{1}{2a^2}\left(\frac{x}{x^2 + a^2} + \frac{1}{a} tan^{-1} \frac{x}{a}\right) + C$$

2.4.3 Definite Integrals

If m and n are integers,

$$\int_0^{2\pi} \sin a\,x\,dx = 0$$

$$\int_0^{2\pi} \cos a\,x\,dx = 0$$

$$\int_0^{\pi} \sin^2 a\,x\,dx = \int_0^{\pi} \cos^2 a\,xdx = \frac{\pi}{2}$$

$$\int_0^{\pi} \sin m\,x \sin n\,xdx = \int_0^{\pi} \cos m\,x \cos n\,xdx = 0, m \neq n$$

$$\int_0^{\pi} \sin m\,x \cos n\,xdx = \begin{cases} 0, & m+n = \text{even} \\ \dfrac{2m}{m^2-n^2}, & m+n = \text{odd} \end{cases}$$

$$\int_0^{2\pi} \sin m\,x \sin n\,xdx = \int_{-\pi}^{\pi} \sin m\,x \sin n\,xdx = \begin{cases} 0, & m \neq n \\ \pi, & m \neq n \end{cases}$$

$$\int_0^{\infty} \frac{\sin a\,x}{x} dx = \begin{cases} \dfrac{\pi}{2}, & a > 0 \\ 0, & a = 0 \\ -\dfrac{\pi}{2}, & a < 0 \end{cases}$$

$$\int_0^{\infty} \frac{\sin^2 x}{x} dx = \frac{\pi}{2}$$

$$\int_0^{\infty} \frac{\cos b\,x}{x^2+a^2} dx = \frac{\pi}{2a} e^{-ab}, a > 0, b > 0$$

$$\int_0^{\infty} \frac{x \sin b\,x}{x^2+a^2} dx = \frac{\pi}{2} e^{-ab}, a > 0, b > 0$$

$$\int_0^{\infty} \sin c\,xdx = \int_0^{\infty} \sin c^2\,xdx = \frac{1}{2}$$

$$\int_0^{\pi} \sin^2 n\,xdx = \int_0^{\pi} \sin^2 x\,dx = \int_0^{\pi} \cos^2 n\,xdx = \int_0^{\pi} \cos^2 x\,dx = \frac{\pi}{2}, n$$
$$= \text{an integer}$$

$$\int_0^{\pi} \sin m\,x \sin n\,xdx = \int_0^{\pi} \cos m\,x \cos n\,xdx = 0, m$$
$$\neq n, m, n \text{ integers}$$

$$\int_0^{\pi} \sin m\,x \cos n\,xdx = \begin{cases} \dfrac{2m}{m^2-n^2}, & m+n = \text{odd} \\ 0, & m+n = \text{even} \end{cases}$$

$$\int_{-\infty}^{\infty} e^{\pm j2\pi tx} dx = \delta(t)$$

$$\int_{0}^{\infty} x^n e^{-ax} dx = \frac{n!}{a^{n+1}}$$

$$\int_{0}^{\infty} e^{-a^2x^2} dx = \frac{\sqrt{\pi}}{2a}, a > 0$$

$$\int_{0}^{\infty} x^{2n} e^{-ax^2} dx = \frac{1 \bullet 3 \bullet 5 \bullet\bullet\bullet (2n-1)}{2^{n+1}a^n}\sqrt{\frac{\pi}{a}}$$

$$\int_{0}^{\infty} x^{2n+1} e^{-ax^2} dx = \frac{n!}{2a^{n+1}}, a > 0$$

2.5 Complex Numbers

The ability to handle complex numbers is important in signals and systems. Although calculators and computer software packages such as MATLAB are now available to manipulate complex numbers, students should know how to handle them by hand.

2.5.1 Representation of Complex Numbers

A complex number z may be written in *rectangular form* as

$$z = x + jy$$

where $j = \sqrt{-1}$; x is the real part of z while y is the imaginary part; that is

$$x = Re(z), \qquad y = Im(z)$$

The complex number z is

$$j = \sqrt{-1}$$
$$\frac{1}{j} = -j$$
$$j^2 = -1$$
$$j^3 = j \cdot j^2 = -j$$

$$j^4 = j^2 \cdot j^2 = 1$$
$$j^5 = j \cdot j^4 = j$$

A second way of representing the complex number z is by specifying its magnitude r and angle θ it makes with the real axis. This is known as the *polar form*. It is given by

$$z = |z|\angle\theta = r\angle\theta$$

where

$$r = \sqrt{x^2 + y^2}, \theta = tan^{-1}\frac{y}{x}$$

or

$$x = r\cos\theta\,, y = r\sin\theta$$

that is,

$$z = x + jy = r\angle\theta = r\cos\theta + jr\sin\theta$$

In converting from rectangular to polar , we must exercise care in determining the correct value of θ. These are the four possibilities:

$$z = x + jy, \quad \theta = tan^{-1}\frac{y}{x}\ (1st\ quadrant)$$
$$z = -x + jy, \theta = 180^o - tan^{-1}\frac{y}{x}\ (2nd\ quadrant)$$
$$z = -x - jy, \quad \theta = 180^o + tan^{-1}\frac{y}{x}\ (3rd\ quadrant)$$
$$z = x - jy, \theta = 180^o + tan^{-1}\frac{y}{x}\ (4th\ quadrant)$$

if x and y are positive.

The third way of representing the complex number x is the *exponential form*:

$$z = re^{j\theta}$$

This is almost the same as the polar form because we use the same magnitude r and the angle θ.
The three forms of representing a complex number are summarized as follows:

$z = x + jy, (x = r\,cos\,\theta\,, y = r\,sin\,\theta)$ Rectsngular form
$z = r\angle\theta, (r = \sqrt{x^2 + y^2}, \theta = tan^{-1}\frac{y}{x})$ Polar form
$z = re^{j\theta}, (r = \sqrt{x^2 + y^2}, \theta = tan^{-1}\frac{y}{x})$ Exponential form

2.5.2 Mathematical Operations

Two complex numbers $z_1 = x_1 + jy_1$ and $z_2 = x_2 + jy_2$ are equal if and only their real parts are equal, and their imaginary parts are equal, that is

$$x_1 = x_2, y_1 = y_2$$

The complex conjugate of the complex number z = x + jy is

$$z^* = x - jy = r\angle - \theta = re^{-j\theta}$$

Thus, the complex conjugate of a complex number is found by replacing every j with –j.

Given two complex number $z_1 = x_1 + jy_1 = r_1\angle\theta_1$ and $z_2 = x_2 + jy_2 = r_2\angle\theta_2$, their sum is

$$z_1 + z_2 = (x_1 + x_2) + j(y_1 + y_2)$$

and their difference is

$$z_1 - z_2 = (x_1 - x_2) + j(y_1 - y_2)$$

While adding and subtracting complex numbers in rectangular form is more convenient, the product and quotient of two complex numbers are best done in polar or exponential form. For their product,

$$z_1 z_2 = r_1 r_2 \angle\theta_1 + \theta_2$$

Alternatively, using the rectangular form

$$z_1 z_2 = (x_1 + jy_1)(x_2 + jy_2) = (x_1x_2 - y_1y_2) + j(x_1y_2 + x_2y_1)$$

For their quotient,

$$\frac{z_1}{z_2} = \frac{r_1}{r_2}\angle(\theta_1 - \theta_2)$$

Alternatively, using the rectangular form,

$$\frac{z_1}{z_2} = \frac{x_1 + jy_1}{x_2 + jy_2}$$

We rationalize the denominator by multiplying both the numerator and denominator. z_2^*.

$$\frac{z_1}{z_2} = \frac{(x_1 + jy_1)(x_2 - jy_2)}{(x_2 + jy_2)(x_2 - jy_2)} = \frac{x_1x_2 + y_1y_2}{x_2^2 + y_2^3} + j\frac{x_2y_1 - x_1y_2}{x_2^2 + y_2^3}$$

2.5.3 Euler's Formula

Euler's formula is an important result in complex variables. We derive it from the series expansion of $e^x, \cos\theta$, and $\sin\theta$. We know that

$$e^x = 1 + x + \frac{x^2}{2!} + \frac{x^3}{3!} + \frac{x^4}{4!} + \cdots$$

Replacing x by jθ gives

$$e^{j\theta} = 1 + j\theta - \frac{\theta^2}{2!} - j\frac{\theta^3}{3!} + \frac{\theta^4}{4!} + \cdots$$

Also,

$$\cos\theta = 1 - \frac{\theta^2}{2!} + \frac{\theta^4}{4!} - \frac{\theta^6}{6!} + \cdots$$

$$\sin\theta = \theta - \frac{\theta^3}{3!} + \frac{\theta^5}{5!} - \frac{\theta^7}{7!} + \cdots$$

so that

$$\cos\theta + j\sin\theta = 1 + j\theta - \frac{\theta^2}{2!} - j\frac{\theta^3}{3!} + \frac{\theta^4}{4!} + j\frac{\theta^5}{5!} - \cdots$$

$$e^{j\theta} = \cos\theta + j\sin\theta$$

This is known as Euler's formula.

$$\boldsymbol{\cos\theta = Re(\,e^{j\theta}), \qquad \sin\theta = Im(\,e^{j\theta})}$$

and that

$$|e^{j\theta}| = \sqrt{\cos^2\theta + \sin^2\theta} = 1$$

Replacing θ by -θ gives

$$e^{-j\theta} = \cos\theta - j\sin\theta$$

$$\cos\theta = \frac{1}{2}\left(e^{j\theta} + e^{-j\theta}\right)$$

$$\sin\theta = \frac{1}{2j}\left(e^{j\theta} - e^{-j\theta}\right)$$

The following identities are useful in dealing with complex numbers.

If $z = x + jy = r\angle\theta$, then

$$zz^* = |z|^2 = x^2 + y^2 = r^2$$

$$\boldsymbol{\sqrt{z} = \sqrt{x + jy} = \sqrt{r}e^{j\theta/2} = \sqrt{r}\angle\theta/2}$$

$$\boldsymbol{z^n = (x + jy)^n = r^n\angle n\theta = r^n(\cos n\,\theta + j\sin n\,\theta)}$$

$$\boldsymbol{z^{1/n} = (x + jy)^{1/n} = r^{1/n}\angle\theta/n + 2\pi k/n, k = 0, 1, 2, \ldots, n - 1}$$

$\boldsymbol{ln(re^{j\theta}) = ln\,r + ln\,e^{j\theta} = ln\,r + j\theta + j2\pi k}$ **(k=integer)**
$\boldsymbol{e^{\pm j\pi} = -1}$
$\boldsymbol{e^{\pm j2\pi} = 1}$
$\boldsymbol{e^{j\pi/2} = j}$

$$e^{-j\pi/2} = -j$$

$$Re\left(e^{(\alpha+j\omega)t}\right) = Re\left(e^{\alpha t}e^{j\omega t}\right) = e^{\alpha t}\cos\omega t$$

$$Im\left(e^{(\alpha+j\omega)t}\right) = Im\left(e^{\alpha t}e^{j\omega t}\right) = e^{\alpha t}\sin\omega t$$

2.5.4 L'Hopital's Rule

If f(0) = 0 = h(0), then

$$\lim_{x\to 0}\frac{f(x)}{h(x)} = \lim_{x\to 0}\frac{f'(x)}{h'(x)}$$

2.5.5 Summation

$$\sum_{k=1}^{N} k = \frac{1}{2}N(N+1)$$

$$\sum_{k=1}^{N} k^2 = \frac{1}{6}N(N+1)(2N+1)$$

$$\sum_{k=1}^{N} k^3 = \frac{1}{4}N^2(N+1)^2$$

$$\sum_{k=0}^{N} a^k = \frac{a^{N+1}-1}{a-1} \quad a \neq 1$$

$$\sum_{k=M}^{N} a^k = \frac{a^{N+1}-a^M}{a-1} \quad a \neq 1$$

$$\sum_{k=0}^{N} \binom{N}{k} a^{N-k}b^k = (a+b)^N, \text{ where } \binom{N}{k} = \frac{N!}{(N-k)!\,k!}$$

2.5.6 Numerical Integration Approximations

$$TRAP(n) = \frac{b-a}{2n}[f(x_0) + 2f(x_1) + 2f(x_2) + \ldots + 2f(x_{n-1}) + f(x_n)]$$

$$SIMP(n) = \frac{b-a}{3n}[f(x_0) + 4f(x_1) + 2f(x_2) + 4f(x_3) + \ldots + 2f(x_{n-2}) + 4f(x_{n-1}) + f(x_n)]$$

2.6 Powers of the Trig Functions

1) Integrals of the form: $\int sin^m x\, cos^n x \;\; dx$ m or n odd
Strategy: If m is odd, save a sine factor and convert to the cosine
If n is odd, save a cosine factor and convert to sine

2) Integrals of the form: $\int sin^m x\, cos^n x \;\; dx$ m and n even and non-negative

Strategy: use ½ angle identities:

$cos^2 x = \frac{1+cos(2x)}{2}$; $sin^2 x = \frac{1-cos(2x)}{2}$

3) Integrals of the form: $\int sec^m x\, tan^n x \;\; dx$; if m is even
Strategy: Save a $sec^2 x$ and convert to the tangent

4) Integrals of the form: $\int sec^m x\, tan^n x \;\; dx$; if n is odd
Strategy: Save a secx tanx and convert to the secant

5) $\int tan^n x \;\; dx$ n any positive integer
Strategy: convert a tan2 x to sec2x-1 and distribute; repeat if necessary

6) $\int sec^m x \;\; dx$; m is odd
Strategy: Integrate by parts

2.7 Definition of the Fourier Transform

$$X(\omega) = \mathcal{F}[x(t)] = \int_{-\infty}^{\infty} x(t)\, e^{-j\omega t}\, dt$$

where F is the Fourier transform operator. and the inverse Fourier transform as

$$\mathrm{x(t)} = \mathcal{F}^{-1}[\mathrm{X}(\omega)] = \frac{1}{2\pi}\int_{-\infty}^{\infty} \mathrm{X}(\omega)\, \mathrm{e}^{j\omega t}\, d\omega$$

Example: Find the Fourier transform of the following functions: (a)$\delta(t)$, (b) $e^{j\omega_0 t}$, (c) $sin\, \omega_0\, t$, (d) $e^{-at}u(t)$.

Solution:

(a) For the impulse function,

$$X(\omega) = F[\delta(t)] = \int_{-\infty}^{\infty} \delta(t) e^{-j\omega t} dt = e^{-j\omega t}|(at\ t = 0) = 1$$

Thus,

$F[\delta(t)] = 1$ and $\delta(t) = F^{-1}[1]$

Using the inverse Fourier transform

$$\delta(t) = F^{-1}[1] = \frac{1}{2\pi}\int_{-\infty}^{\infty} 1 e^{j\omega t} d\omega$$

Or

$$\int_{-\infty}^{\infty} e^{j\omega t}\, d\omega = 2\pi\delta(t)$$

Interchanging variables t and ω results in

$$\int_{-\infty}^{\infty} e^{j\omega t}\, dt = 2\pi\delta(\omega)$$

Using this result, the Fourier transform of given function is

$$F[e^{j\omega t}] = \int_{-\infty}^{\infty} e^{j\omega_o t} e^{-j\omega t} dt = \int_{-\infty}^{\infty} e^{j(\omega_o - \omega)t} dt = 2\pi\delta(\omega_0 - \omega)$$

Since the impulse function is an even function, $\delta(\omega_0 - \omega) = \delta(\omega - \omega_0)$,

$$F[e^{j\omega_0 t}] = 2\pi\delta(\omega - \omega_0)$$

By simply changing the sign of ω_0, we readily obtain

$$F[e^{-j\omega_0 t}] = 2\pi\delta(\omega + \omega_0)$$

Also, by setting $\omega_0 = 0$,

$$F[1] = 2\pi\delta(\omega)$$

(c)

$$F[\sin \omega_0 t] = F\left[\frac{e^{j\omega_0 t} + e^{-j\omega_0 t}}{2j}\right]$$

$$= \frac{1}{2j} F[e^{j\omega_0 t}] - \frac{1}{2j} F[e^{-j\omega_0 t}]$$

$$= \frac{1}{2j} F[e^{j\omega_0 t}] - \delta(\omega - \omega_0)]$$

(d) Let $x(t) = e^{-at}u(t) = \begin{cases} e^{-at}, t > 0 \\ 0, t < 0 \end{cases}$

$$X(\omega) = \int_{-\infty}^{\infty} x(t) e^{-j\omega t} dt = \int_0^{\infty} e^{-at} e^{-j\omega t} dt = \int_0^{\infty} e^{-(a+j\omega)t} dt$$

$$L\left[e^{-at}u(t)\right] = X(\omega) = \frac{-1}{a+j\omega} e^{-(a+j\omega)t} \Big|_0^{\infty} = \frac{1}{a+j\omega}$$

Example: Obtain the Fourier transform of the signal given by

$$x(t) = \begin{cases} -A & -1 < t < 0 \\ A & 1 > t > 0 \end{cases}$$

Solution

$$X(\omega) = \int_{-\infty}^{\infty} x(t)e^{-j\omega t}dt = \int_{-1}^{0} (-A)\, e^{-j\omega t}dt + \int_{0}^{1} A\, e^{-j\omega t}dt$$

$$= \frac{A}{j\omega}e^{-j\omega t}\Big|_{-1}^{0} - \frac{A}{j\omega}e^{-j\omega t}\Big|_{0}^{1}$$

$$= \frac{-jA}{\omega}\left(1 - e^{j\omega} - e^{-j\omega} + 1\right)$$

$$= \frac{-j2A}{\omega}(1 - \cos\omega)$$

Example: Find the Fourier transform of the two-sided exponential is given by

$$x(t) = e^{-a|t|} = \begin{cases} e^{at}, t < 0 \\ e^{-at}. t > 0 \end{cases}$$

Solution

Let $x(t) = e^{-a|t|} = \begin{cases} e^{at}, t < 0 \\ e^{-at}. t > 0 \end{cases}$

The Fourier transform is

$$X(\omega) = \int_{-\infty}^{\infty} x(t)\, e^{-j\omega t}dt = \int_{-\infty}^{0} e^{at}\, e^{-j\omega t}dt + \int_{0}^{\infty} e^{-at}\, e^{-j\omega t}dt$$

$$= \frac{1}{a - j\omega} + \frac{1}{a + j\omega}$$

$$= \frac{2a}{a^2 + \omega^2}$$

2.8 Properties of Fourier Transform

2.8.1 Convolution

According to the convolution property, the Fourier transform maps convolution to multiplication.

The Fourier transform of the convolution of two time functions is the product of their corresponding Fourier transforms. This is particularly useful for analyzing linear, time-invariant systems because using the Fourier transform we can sometimes map the difficult problem of evaluating a convolution to a simpler algebraic operation, multiplication. If x(t) and h(t) are the convolution integral gives two signals, their convolution y(t)

$$y(t) = h(t) * x(t) = \int_{-\infty}^{\infty} h(\tau)x(t-\tau)d\tau$$

If X(ω), H(ω), and Y(ω) are the Fourier transforms of x(t), h(t), and y(t) respectively, then

$$Y(\omega) = \mathcal{F}\,[h(t) * x(t)] = H(\omega)X(\omega)$$

Table 2.1 Properties of the Fourier transform

No. Property	x(t)	X(ω)
1. Linearity	$a_1x_1(t) + a_2x_2(t)$	$a_1X_1(\omega) + a_2X_2(\omega)$
2. Scaling	x(at)	$\frac{1}{\lvert a\rvert}X\left(\frac{\omega}{a}\right)$
3. Time shift	x(t-a)	$e^{-j\omega a}X(\omega)$
4. Frequency shift	$e^{j\omega_0 t}x(t)$	$X(\omega-\omega_0)$
5. Modulation	cos(ω_0t)x(t)	$\frac{1}{2}[X(\omega+\omega_0) + X(\omega-\omega_0)]$
	sin(ω_0t)x(t)	$\frac{j}{2}[X(\omega+\omega_0) - X(\omega-\omega_0)]$
6. Time differentiation	$\frac{dx}{dt}$	jωX(ω)
	$\frac{d^nx}{dt^n}$	$(j\omega)^nX(\omega)$
7. Frequency differentiation	$(-jt)^nx(t)$	$\frac{d^n}{d\omega^n}X(\omega)$
8. Time integration	$\int_{-\infty}^{t} x(t)dt$	$\frac{X(\omega)}{j\omega} + \pi X(0)\delta(\omega)$
9. Time reversal	x(-t)	X(-ω) or X*(ω)
10. Duality	X(t)	2πx(-ω)
11. Time convolution	$x_1(t) * x_2(t)$	$X_1(\omega)X_2(\omega)$
12. Frequency convolution	$x_1(t)x_2(t)$	$\frac{1}{2\pi}X_1(\omega) * X_2(\omega)$
13. Parseval's relation	$\int_{-\infty}^{\infty} \lvert x(t)\rvert^2dt$	$\frac{1}{2\pi}\int_{-\infty}^{\infty} \lvert X(\omega)\rvert^2d\omega$

Table 2.2 Fourier transform pairs

No.	x(t)	X(ω)
1.	δ(t)	1
2.	1	$2\pi\delta(\omega)$
3.	u(t)	$\pi\delta(\omega) + \frac{1}{j\omega}$
4.	u(t+τ) – u(t-τ)	$2\frac{\sin \omega\tau}{\omega}$
5.	|t|	$\frac{-2}{\omega^2}$
6.	sgn(t)	$\frac{2}{j\omega}$
7.	$e^{-at}u(t)$	$\frac{1}{a+j\omega}$
8.	$e^{at}u(-t)$	$\frac{1}{a-j\omega}$
9.	$t^n e^{-at}u(t)$	$\frac{n!}{(a+j\omega)^{n+1}}$
10.	$e^{-a\|t\|}$	$\frac{2a}{a^2+\omega^2}$
11.	$\|t\|e^{-at}\, u(t)$	$\frac{4aj\omega}{a^2+\omega^2}$
12.	$e^{j\omega_0 t}$	$2\pi\delta(\omega - \omega_0)$
13.	$\sin \omega_0 t$	$j\pi[\delta(\omega + \omega_0) - \delta(\omega - \omega_0)]$
14.	$\cos \omega_0 t$	$\pi[\delta(\omega + \omega_0) - \delta(\omega - \omega_0)]$
15.	$(\sin \omega_0 t)u(t)$	$j\pi[\delta(\omega + \omega_0) - \delta(\omega - \omega_0)] + \frac{\omega_0}{\omega_0^2-\omega^2}$
16.	$(\cos \omega_0 t)u(t)$	$\pi[\delta(\omega + \omega_0) - \delta(\omega - \omega_0)] + \frac{j\omega}{\omega_0^2-\omega^2}$
17.	$\sin(\omega_0 t+\theta)$	$j\pi[e^{-j\theta}\delta(\omega + \omega_0) - e^{j\theta}\delta(\omega - \omega_0)]$
18.	$\cos(\omega_0 t+\theta)$	$\pi[e^{-j\theta}\delta(\omega + \omega_0) + e^{j\theta}\delta(\omega - \omega_0)]$
19.	$e^{-at}\sin \omega_o\, tu(t)$	$\frac{\omega_0}{(a+j\omega)^2+\omega_0^2}$
20.	$e^{-at}\cos \omega_o\, tu(t)$	$\frac{a+j\omega}{(a+j\omega)^2+\omega_0^2}$
21.	Π(t/τ)	$\tau \sin c\left(\frac{\omega\tau}{2}\right)$
22.	$\Lambda\left(\frac{t}{\tau}\right)$	$\tau \sin c^2\left(\frac{\omega\tau}{2}\right)$
23.	$\sum_{n=-\infty}^{\infty}\delta(t - nT)$	$\omega_0\sum_{n=-\infty}^{\infty}\delta(\omega - n\omega_0)$

Example: A signal x(t) has a Fourier transform given by

$$X(\omega) = \frac{5(1 + j\omega)}{8 - \omega^2 + 6j\omega}$$

Without finding f(t), find the Fourier transform of the following.

(a) x(t-3)

(b) x(4t)

(c) $e^{-j2t}x(t)$

(d) x(-2t)

Solution

We apply the relevant property for each case.

(a) $F[x(t-3)] = e^{-j\omega 3}X(\omega) = \frac{5(1+j\omega)e^{-j\omega 3}}{8-\omega^2+j6\omega}$

(b) $F[x(4t)] = \frac{1}{4}X\left(\frac{\omega}{4}\right) = \frac{\frac{5}{4}(1+j\omega/4)}{8-\omega^2/16+j6\omega/4} = \frac{5(4+j\omega)}{128-\omega^2+j24\omega}$

(c) $F[e^{-j2t}x(t)] = X(\omega + 2) = \frac{5[1+j(\omega+2)]}{8-(\omega+2)^3+6j(\omega+2)} = \frac{5(1+j\omega+j2)}{4-\omega^2-4\omega+6j\omega+j12}$

(d) $F[x(-2t)] = \frac{1}{2}X\left(\frac{\omega}{-2}\right) = \frac{\frac{5}{2}(1-j\omega/2)}{8-\frac{\omega^2}{4}-\frac{6j\omega}{2}} = \frac{5(2-j\omega)}{32-\omega^2-12j\omega}$

Inverse Fourier Transform

The equation defines as the inverse Fourier transform of X(ω).

$$x(t) = \frac{1}{2\pi}\int_{-\infty}^{\infty} X(\omega)e^{j\omega t}\,d\omega$$

Example: Find the inverse Fourier transform of:

(a) $G(\omega) = \dfrac{10j\omega}{(-j\omega+2)(j\omega+3)}$

(b) $Y(\omega) = \dfrac{\delta(\omega)}{(j\omega+1)(j\omega+2)}$

Solution:

(a) To avoid complex algebra, let s=jω. Using partial fraction,

$$G(s) = \frac{10s}{(2-s)(3+s)} = \frac{-10s}{(s-2)(s+3)} - \frac{A}{s-2} + \frac{B}{s+3}, s = j\omega$$

$$A = (s-2)G(s)|(at\ s=2) = \frac{-10(2)}{2+3} = -4$$

$$B = (s+3)G(s)|(at\ s = -3) = \frac{-10(-3)}{-3-2} = -6$$

$$G(\omega) = \frac{-4}{j\omega-2} - \frac{6}{j\omega+3}$$

Taking the inverse Fourier transform of each term,

$$g(t) = -4e^{2t}u(-t) - 6e^{-3t}u(t)$$

(b) Because of the delta function, we use eq. (5.7) to find the inverse.

$$y(t) = \frac{1}{2\pi}\int_{-\infty}^{\infty} \frac{\delta(\omega)e^{j\omega t}d\omega}{(2+j\omega)(j\omega+1)} = \frac{1}{2\pi}\frac{e^{j\omega t}}{(2+j\omega)(j\omega+1)}|(at\ \omega=0)$$

$$= \frac{1}{2\pi}\frac{1}{2} = \frac{1}{4\pi}$$

where the sifting property has been applied.

Multiple Choice Questions

1. Which of the following trigonometric statements is **not** identically true?

   A. $\boldsymbol{sin^2 x + cos^2 x = 1}$
   B. $\boldsymbol{sec^2 x - tan^2 x = 1}$
   C. $\boldsymbol{tan(-x) = tan\, x}$
   D. $\boldsymbol{cos(-x) = cos\, x}$

2. What is the equation of the tangent line to the curve $\boldsymbol{y = x^2 - 3}$ at the point (2,1)?

   A. $\boldsymbol{y = 2x - 3}$
   B. $\boldsymbol{y = 4x - 7}$

C. $y = 4x - 9$
D. $y = 4x - 5$

3. If $y = \tan x$, then $\frac{dy}{dx}$ equals?

A. $\cot^2 x$
B. $\sec^2 x$
C. $\sec x \tan x$
D. $\cos e\, cx$

4. Differentiate $f(x) = \frac{1}{2}e^{2x}$ concerning x.

A. $\frac{1}{4}e^{2x}$
B. e^x
C. $\frac{1}{4}e^x$
D. e^{2x}

5. The function $f(x) = x - \frac{x^3}{3!} + \frac{x^5}{5!} - \frac{x^7}{7!} + ...$ is the standard Taylor's series for

A. $\sin x$
B. $\cos x$
C. $\sin(-x)$
D. $\cos(-x)$

6. To find the derivative of the function $y = x \sin x$, the basic principle applied is

A. Trigonometry
B. Quotient principle
C. Parametric principle
D. Product principle

7. Integration is sometimes described as the_______________of differentiation.
(Fill in the blank with the appropriate word)

A. process
B. reverse
C. extreme
D. result

8. To work out the solution of $\int e^{2x} \sin x\, dx$, the commonest approach is to apply:

A. direct integration
B. substitution method
C. partial fractions
D. integration by parts

9. Express $\frac{x^3-1}{x^2-1}$ into partial fractions

A. $x + \frac{1}{x-1}$
B. $x - \frac{1}{x-1}$
C. $x + \frac{1}{x+1}$
D. $x - \frac{1}{x+1}$

10. Find the integral of $\frac{x^3-1}{x^2-1}$

A. $\frac{x^2}{2} + ln(x-1) + c$
B. $\frac{x^2}{2} - ln(x-1) + c$
C. $\frac{x^2}{2} - ln(x+1) + c$
D. $\frac{x^2}{2} + ln(x+1) + c$

ANSWER KEY

1. C	2. B	3. B	4. D.
5. A	6. D	7. B	8. D
9. C	10. D		

2.9 Differential Equations

A differential equation is a relation between a function and its derivatives. Differential equations form the language in which the basic laws of physical science are expressed.

The science tells us how a physical system changes from one instant to the next. The theory of differential equations then provides us with the tools and techniques to take this short term information and obtain the long-term overall behavior of the system.

Definition: Ordinary and partial differential equations

A differential equation (DE) involves derivatives of unknown function of one or more variables. If the unknown function depends on only one variable, the equation is called an ordinary differential equation (ODE). If the unknown function depends on more than one variable, the equation is called a partial differential equation (PDE).

Example: $\frac{dy}{dx} = 2x + y$ or $y' = 2x + y$ is an ordinary differential equation since the function $y = f(x)$ depends on only one variable x.In the function$y = f(x)$,x is the independent variable, and y is the dependent variable.

Example: $\frac{\partial y}{\partial x} = 2x + z$ is a partial differential equation since the function

$y = f(x, z)$ depends on two variables x and z.

Definition : Order and degree of a differential equation

A DE is the order of the highest ordered derivative involved in the expression. The degree of the an ordinary differential equation is the algebraic degree of its highest ordered–derivative.

Example: $\frac{dy}{dx} = 2x + y$ is first order, first degree ordinary DE

Example: $\frac{\partial y}{\partial x} = 2x + z$ is first order, first degree partial DE

Example: $\frac{d^2x}{dt^2} - 2\frac{dx}{\partial t} - 15x = 0$ is second order, first degree ordinary DE

2.10 Formation of a Differential Equation

Although the prime problem in the study of differential equations is finding the solution to a given differential equation, the converse problem is also interesting. That is the problem of finding a differential equation which satisfies by a given solution. The problem is solved by repeated differentiation and elimination of the arbitrary constants.

Example: Find the differential equation which has $y = c_1e^x + c_2e^{-x} + 3x$ as its general solution.

Solution: Differentiate the given expression twice

(1) $y = c_1e^x + c_2e^{-x} + 3x$

(2) $y^{/} = c_1e^x - c_2e^{-x} + 3$

(3) $y^{//} = c_1e^x + c_2e^{-x}$

Eliminate $c_1$and $c_2$by subtracting (3) from (1) to obtain $y - y^{//} = 3x$ gives which is the desired differential equation.Note that thje desired differential equation is free from the arbitrary constants.

2.11 Solutions of First Order Differential Equations

The systemic development of techniques for solving differential equations logically begins with the equations of the first order and first degree.

Equations of this type can in general, be written as $\frac{dy}{dx} = F(x, y)$,

where $F(x, y)$ is a given function. However, despite the apparent simplicity of this equation, analytic solutions are usually possible only when $F(x, y)$ has simple forms. Two such forms are introduced in this activity.

2.11.1 Variables Separable

If $F(x,y) = f(x)g(y)$

where $f(x)$ and $g(x)$ are respectively functions of $x$only and y only, then,

$$\frac{dy}{dx} = f(x)g(y) \tag{2.52}$$

Since the variables x and$y$are now separable , then

$$\int \frac{dy}{g(y)} = \int f(x)dx,$$

which expresses y implicitly in terms of x.

Example: Solve the equation $\frac{dy}{dx} = \frac{y+1}{x-1}$,

Solution:

$$\int \frac{dy}{y+1} = \int \frac{dx}{x-1},$$

or $log_e(y+1) = log_e(x-1) + log_e C$

where $C$is an arbitrary constant. Hence$\frac{y+1}{x-1} = C$, is the general solution.

2.11.2 Homogeneous Differential Equation

An expression of the nth degree in x and$y$is said to be homogeneous of degree n, if when$x$ and$y$are replaced by tx andty, the result will be the original expression multiplied by t^n; symbolically $f(tx,ty) = t^n f(x,y)$.

Example: Show that $x^2 + xy - y^2$is homogeneous and determines the degree.

Solution: Replace x and$y$ with tx and tyrespectively, to obtain $t^2x^2 + (tx)(ty) - t^2y^2$=$t^2(x^2 + xy - y^2)$. The degree is 2.

Consider the differential equation$M(x, y)dx + N(x, y)dy = 0$.

The equation is said to be homogeneous in x and$y$if M, and N are homogeneous functions of the same degree in x and$y$. The technique for solving this equation is to make the substitution$y = vx$ or $x = vy$ and is based upon the following theorem.

Theorem: *Any homogeneous differential of the first order and first degree can be reduced to the type of variables separable by substituting $y = vx$ or$x = vy$.*

Example: Find the general solution of the differential equation$2xydx - (x^2 - y^2)dy = 0$.

Solution: A simple check reveals that the equation is homogeneous(Refer to the first example in this section).

Let $y = vx$ and $dy = vdx + xdv$, and substitute into the differential equation to obtain

$$2x(vx)dx - (x^2 - v^2x^2)(vdx + xdv) = 0$$

$$x^2(v + v^3)dx + x^3(v^2 - 1)dv = 0.$$

Dividing by $x^3(v + v^3)$ separates the variables:

$$\frac{dx}{x}+\frac{(v^2-1)dv}{v(1+v^2)} = 0$$

Integration yields

$$ln\, x - ln\, v + ln(v^2 + 1) = ln\, c \text{ or } x(v^2 + 1) = cv$$

Rewriting the original variables by substituting $v = \frac{y}{x}$, we obtain $y^2 + x^2 = cy$ as the general solution.

Fill in the blanks for the fillowing differential equations:

	Differential equation	Ordinary or Partial	Order	Degree	Independent variable	Dependent variable
1						
2	$xy^{//} - 4y^{/} - 5y = e^{3x}$					
3	$\frac{\partial u}{\partial t} = 4\frac{\partial^2 u}{dx^2} + \frac{\partial u}{dy}$					
4	$\left(\frac{d^3 s}{dt^3}\right)^2 + \left(\frac{d^2 s}{dt^2}\right) = s - 3t$					

Solutions to Additional Learning Activities

	Differential equation	**Ordinary or partial**	**Order**	**Degree**	**Independent variable**	**Dependant variable**
1	$y^{/} = x^2 + 5y$	Ordinary	1	1	x	y
2	$xy^{//} - 4y^{/} - 5y = e^{3x}$	Ordinary	2	1	x	y
3	$\frac{\partial u}{\partial t} = 4\frac{\partial^2 u}{dx^2} + \frac{\partial u}{dy}$	Partial	2	1	x, y, t	u
4	$\left(\frac{d^3 s}{dt^3}\right)^2 + \left(\frac{d^2 s}{dt^2}\right) = s - 3t$	Ordinary	3	2	t	s

Key terms

Variable coefficients: Unlike differential equations with constant coefficients, there are also some differential equations with variable coefficients. Essentially, a variable coefficient is not constant, i.e., it is expressed in functional form.

Non-homogeneous equations: An homogeneous differential equation is one with the right-hand side equated to zero. A non-homogeneous differential equation is an equation with the right-hand side not equal to zero.

Undetermined coefficients: These are constants to be explicitly determined by solving the particular integral of a differential equation. The method that does this is called the method of undetermined coefficients.

Variation of parameters: This is a method of finding a particular linear differential equation solution when the general solution of the reduced equation(homogeneous equation) is known. (See notes below for details)

Inverse technique: This technique is applied in solving differential equations using the properties of a differential operator.

Definition: Linear equations

The first-order linear equation discussed in Unit 1 above is a special case of the general linear equation of order *n*

$$a_0(x)\frac{d^ny}{dx^n} + a_1(x)\frac{d^{n-1}y}{dx^{n-1}} + \cdots + a_{n-1}(x)\frac{dy}{dx} + a_n(x)y = f(x)$$

where $a_0(x), a_1(x) \dots a_n(x)$ and $f(x)$ are given functions of x or are constants.

Definition: Homogeneous and non-homogeneous equations

If$f(x) = 0$, it is called a **homogeneous** equation with variable or constant coefficients, depending on whether $a_0(x), a_1(x) \dots a_n(x)$

are functions of x, or are constants. It is also called the **reduced** linear differential equation.

Example: $x\frac{d^2y}{dx^2}+2\frac{dy}{dx}+3y=0$, is a second-order homogeneous equation with a variable coefficient.

If $f(x)\neq 0$, it is called non- homogeneous equation with variable or constant coefficients, depending on whether $a_0(x), a_1(x) \dots a_n(x)$ are functions of x or are constants.Example: $x\frac{d^2y}{dx^2}+2\frac{dy}{dx}+3y=\sin x$ is a second-order non-homogeneous equation with variable coefficient.

Definition: Solution of second-order differential equation

Suppose y_1 and y_2 are two independent solutions of the reduced equation, namely

$$a_0(x)\frac{d^ny}{dx^n}+a_1(x)\frac{d^{n-1}y}{dx^{n-1}}+\cdots+a_{n-1}(x)\frac{dy}{dx}+a_n(x)y=0$$

then the linear combination $y=c_1y_1+c_2y_2$ where c_1, c_2 are arbitrary constants, is also a solution.

Proof:

Substitute $y=c_1y_1+c_2y_2$, we have

$$\left[a_0(x)\frac{d^ny_1}{dx^n}+a_1(x)\frac{d^{n-1}y_1}{dx^{n-1}}+\ldots+a_{n-1}(x)\frac{dy_1}{dx}+a_n(x)y_1\right]$$
$$+\left[a_0(x)\frac{d^ny_2}{dx^n}+a_1(x)\frac{d^{n-1}y_2}{dx^{n-1}}+\ldots+a_{n-1}(x)\frac{dy_2}{dx}+a_n(x)y_2\right]\equiv 0$$

The equation vanishes identically to zero since each bracket is zero under y_1 and y_2 being solutions.

Theorem: If $y_1, y_2, \ldots, y_n$are *n* linearly independent functions of$x$ which satisfy a homogeneous, then the linear combination

$y_c=c_1y_1+c_2y_2+\ldots+c_ny_n$
where $c_1, c_2, \ldots, c_n$are arbitrary constants, is its solution. The solution for the homogeneous equation is called the complementary function.

Theorem: The general Solution of a complete non-homogeneous differential equation is equal to the sum of its *complementary function* and any *particular integral*. If P is a particular solution, then the general solution is

$$y = y_c + P = c_1y_1 + c_2y_2 + \ldots + c_n + P$$

Hence, for non-homogeneous equations:

General solution = Complementary function + Particular integral

2.11.3 Application of the theorem

An important and special case occurs when the coefficients. $a_0(x), a_1(x) \ldots a_n(x)$ are constants, the equation is called a constant coefficient equation. Consider a constant coefficient homogeneous equation

$$a_0\frac{d^ny}{dx^n} + a_1\frac{d^{n-1}y}{dx^{n-1}} + \ldots + a_{n-1}\frac{dy}{dx} + a_ny = 0$$

Denoting, $D_x^ny = \frac{d^ny}{dx^n}$, in equation (2.58) we have

$$a_0D_x^ny + a_1D_x^{n-1}y + \ldots + a_{n-1}D_xy + a_ny = 0$$

$$\Rightarrow (a_0D_x^n + a_1D_x^{n-1} + \cdots + a_{n-1}D_x + a_n)y = 0.$$

If we make a formal substitution $D_x = m$ in (2.59), we obtain a polynomial in m of degree n given:

$$g(m) = a_0m^n + a_1m^{n-1} + \ldots + a_{n-1}m + a_n = 0,$$

Theorem: If m_1 is a root of the auxiliary equation $a_0m_1^n + a_1m_1^{n-1} + \ldots + a_{n-1}m_1 + a_n = 0$, then $y = e^{mx}$ is a solution of the homogeneous linear differential equation
$(a_0D^n + a_1mD^{n-1} + \ldots + a_{n-1}D + a_n)y = 0$ where a_i is a constant.

Proof: By successive differentiation, we obtain

$$y = e^{m_1 x}$$
$$y' = m_1 e^{m_1 x}$$

$$y'' = {m_1}^2 e^{m_1 x}$$
$$y''' = {m_1}^3 e^{m_1 x}$$

$$y^{(n)} = {m_1}^n e^{m_1 x}$$

Substituting these derivatives into the differential equation, we obtain

$a_0 m_1^n e^{m_1 x} + a_1 m_1^{n-1} e^{m_1 x} + \ldots + a_{n-1} m_1 e^{m_1 x} + a_n e^{m_1 x} = 0$
or $(a_0 m_1^n + a_1 m_1^{n-1} + \ldots + a_{n-1} m_1 + a_n) e^{m_1 x} = 0$

and since m_1 is a root of the auxiliary equation, the expression in the parenthesis equals zero, and the equation is satisfied.

2.11.4 Summary of Solving a Homogeneous Differential Equation

The solution of the differential equation then reduces to solving the algebraic auxiliary equation for its n roots and forming a linearly independent combination $c_i e^{m_i x} (i = 1, \ldots, n)$ as the general solution *if the roots were all distinct.*

Example: Find the general solution of $y'' - y' - 6y = 0$

Solution: The auxiliary equation is $m^2 - m - 6 = 0$
$(m - 3)(m + 2) = 0$, which has the roots $m = 3$ or $m = -2$.
The Solution is
$$y = c_1 e^{3x} + c_2 e^{-2x}$$

Theorem: If the auxiliary equation of a homogeneous linear differential equation contains r as an
s-fold root, then $y = (c_0 + c_1 x + c_{2.} x^2 + \ldots + c_{s-1} x^{s-1}) e^{rx}$ is a solution of the differential equation

Example: If the $r = m$ (twice), the solution is $y = (c_0 + c_1 x) e^{mx}$

2.11.5 Auxiliary Equation with Complex Roots

If the auxiliary equation with real coefficients contains two complex roots $m_1 = a + bi$ and $m_2 = a - bi$ then $y = e^{ax}(C_1 \cos bx + C_2 \sin bx)$ is a solution of the differential equation, C_1, C_2, a, b are constants.

Problem: Write the auxiliary equations for the following differential equations and hence solve the equations:

(a) $y'' + 3y' + 2y = 0$ *Solution:* $y = Ae^{-x} + Be^{-2x}$]

(b) $\frac{d^2y}{dx^2} - 6\frac{dy}{dx} + 9y = 0$ [*Solution:* $y = (A + Bx)e^{3x}$].

Hint: Refer to Theorem 2.2.1 and the example following for some hints to this problem.

(c) $y''' - 3y'' + 7y' - 5y = 0$

2.11.6 Equations with Undetermined Coefficients

In the previous section, we learned that the solution to the complete linear differential equation is composed of the sum of the complementary function and the particular integral. Techniques for obtaining the complementary function $y_c$were developed in sections 2.1.4-2.1.6 with many working examples. What remains is only to provide techniques for finding a particular integral to obtain a complete solution. In this section, we shall discuss the method of undetermined coefficients.

Although the method of undetermined coefficients is not applicable in all cases, it may be used if the right-hand side $f(x)$, contains only terms which have a finite number of linearly independent derivatives, such as x^n,e^{mx},$\sin bx$, $\cos bx$ or products of these.

2.12 Procedure for the Techniques of Undetermined Coefficients

The general procedure in this technique is to assume the particular integral $y_p$of a form similar to the right member$f(x)$ .The necessary derivatives of $y_p$are then obtained and substituted into the given differential equation. This results in an identity for the independent variable, and consequently, the coefficients of the like terms are equated.The values of the undetermined coefficients are found from the resulting system of linear equations. The procedure is best illustrated with an example

Table 2.3: below summarizes a general rule for the formulation of the particular integral.

If $f(x)$ is of the form	Choose y_P to be
$c_0 + c_1x + c_{2.}x^2 + \ldots + c_nx^n$ $(c_0 + c_1x$ $+ c_{2.}x^2 + \ldots + c_nx^n)e^{rx}$ $c_0 \sin b\,x + c_1 \cos b\,x$	$C_0 + C_1x + C_{2.}x^2 + \ldots + C_nx^n$ $(C_0 + C_1x$ $+ C_{2.}x^2 + \ldots + C_nx^n)e^{rx}$ $C_0 \sin b\,x + C_1 \cos b\,x$

Example: Find the particular integral of

$y'' + 3y' + 2y = e^{2x}$

Solution: In the learning activity of section 2.1, you worked out the complementary function of this equation to be$y = Ae^{-x} + Be^{-2x}$.That is, the auxiliary equation is $m^2 + 3m + 2 = 0 \Rightarrow (m + 2)(m + 1) = 0$, giving $\Rightarrow m = -2$ or $\Rightarrow m = -1$.Hence the complementary function is $y_c = Ae^{-x} + Be^{-2x}$as before.

Looking at the right-hand side of the above differential equation example 2.2.1 and the general rule in table 2.2.1, the particular integral is

$Y = Ae^{2x}, Y' = 2Ae^{2x}, Y'' = 4Ae^{2x}$

Substituting in the original equation $(4A + 6A + 2A)e^{2x} = e^{2x}$
Comparing the coefficients on both sides $12A = 1 \Rightarrow A = \frac{1}{12}$

The general solution is then the complementary function + particular integral, which is

$y = Ae^{-x} + Be^{-2x} + \dfrac{1}{12}$

2.13 Method of Solution by Variation of parameters (VOP)

The method of undetermined coefficients discussed in the previous section is limited in its application. We need another technique with a wider application. The technique discussed in this section is called the method of variation of parameters.

The VOP procedure consists of replacing the constants in the complementary function by undetermined functions of the independent variable x and then determining these functions so that when the modified complementary function is substituted into the differential equation, $f(x)$ will be obtained on the left side. This places only one restriction on the n arbitrary functions. c_i,$(i = 1, \dots, n)$, and we have $(n-1)$ conditions at our disposal. We utilize this freedom in the following manner:

(a) As we differentiate y_c to find $D_x y_c = \frac{dy_c}{dx}$, there will now appear terms that contain $c_i'(x)$. We set this combination of terms to zero.

(b) As we differentiate again to find $D_x^2 y_c$,we again set the resulting combination of terms containing $c_i'(x)$ equal to zero.

(c) We continue this process through $D_x^{n-1} y_c$

(d) We then find $D_x^n y_c$ and substitute all these values into the given differential equation. Since y_c is the complementary function, the results of this substitution will contain only the terms of $D_x^n y_c$ which appears because the c_i are functions of$x$.

(e) The equations obtained from (d) and the $(n-1)$ conditions imposed by (a)-(c) will yield a system of n linear equations in n unknowns c_i',$(i = 1, \dots, n)$.This is solved and integrated to yield the n functions$c_i(x)$.

The procedure is not too difficult, provided the order of the differential equation is small. The following example illustrates the technique.

Example: Find the general solution of $y'' - y = x^2$

Solution: Auxiliary equation is $m^2 - 1 = 0 \Rightarrow m = 1$or $m = -1$
$y_c = k_{1.}e^x + k_2 e^{-x}$

Let c_i,$(i = 1,2)$ be functions of$x$:

$y_p = c_{1.}(x)e^x + c_2(x)e^{-x}$

Differentiate to obtain $y'_p = c_{1.}e^x \quad c_2e^{-x} + c_1'e^x + c_2'e^{-x}$

Impose the first condition i.e. $c_1'e^x + c_2'e^{-x} = 0$

$y_p'' = c_{1.}e^x + c_2e^{-x} + c_1'e^x - c_2'e^{-x}$

$c_{1.}e^x + c_2e^{-x} + c_1'e^x - c_2'e^{-x} - c_{1.}e^x - c_2e^{-x} = x^2,$
or $c_1'e^x - c_2'e^{-x} = x^2$

$2c_1{'}e^x = x^2$

or $dc_1 = \frac{1}{2}x^2e^{-x}dx$

$c_1 = \int \frac{1}{2}x^2e^{-x}dx$

Integration by parts yields

$c_1 = -(1 + x + \frac{1}{2}x^2)e^{-x}$

we have

$$c_2' = -c_1'e^{2x} = -\frac{1}{2}x^2e^x$$

Again integration by parts yields

$$c_2 = -(1 - x + \frac{1}{2}x^2)e^x$$

The general solution is then, as usual, the sum of the complementary function plus the particular integral.i.e. $y = k_{1.}e^x + k_2e^{-x} + c_1(x)e^x + c_2(x)e^{-x}$

$=k_{1.}e^x + k_2e^{-x} - [1 + x + (1/2)x^2] - [1 - x + (1/2)x^2]$

$=k_{1.}e^x + k_2e^{-x} - x^2 - 2$

2.14 Differential Operators

In this section the theory of differential operators is outlined. The application of the theory to the solution of linear differential equations is then discussed. Several examples are given. Together with these examples, there are also learning activities within the sections which you are to do before you proceed to the next sections.

To operate means to *produce an appropriate effect,* and an operator is that instrument or effect which *does* that . We have already used the notation

$$D_y^k y = \frac{d^k y}{dx^k}, k = 1,2,\ldots$$

to indicate the kth derivative of the function $y$with respect to $x.D^k_{\square}$denotes the derivative of the order k, concerning the appropriate independent variable. D^k is called the differential operator. Since it must produce an effect, it must operate on a function and behave according to the differentiation rules. The following properties are valid.

Property 1. If c is a constant $D^k\ (cy) = cD^k\ y$
Property 1. $D^k\ (y_1 + y_2) = D^k\ (y_1) + D^k\ (y_2)$
Property 3. Two operators, A and B, are equal if and only if$Ay = By$.
Property 4. If operators A, B, and C are any differential operators, they will satisfy the ordinary laws of algebra, which are:

- The commutative law of addition $A + B = B + A$;
- The associative law of addition $(A + B) + C = A + (B + C)$;
- The associative law of multiplication $(AB)\ C = A\ (BC)$;
- The distributive law of multiplication $A\ (B + C) = AB + AC$; and
- The commutative law of multiplication if the operators all have constant coefficients $AB = BA$.

Property 5. The exponential shift. If $P(D)$ is a polynomial in D with constant coefficients, then.

- $e^{rx}P(D)y = P(D-r)[e^{rx}y]$;
- $P(D)[e^{rx}y] = e^{rx}P(D+r)y$;
- $e^{-rx}P(D)[e^{rx}y] = P(D+r)y$

2.15 Inverse Operators

To complete the discussion of the differential operator, we now consider the meaning of $D^{-k}\ y$.To be consistent, we let

$D^{-1}\ y = \frac{1}{D}y = z$ be an expression such that $Dz = y$.

In other words, the net effect of the differential operator with a negative index is called integration. This operator is called the **inverse differential operator.**

Definition 1. The inverse differential operator $(D-c)^{-k}, k = 1,2,\ldots,$ is defined as the
integral $(D-c)^{-k}y(x) = \int_{x_0}^{x} \frac{(x-u)^{k-1}}{(k-1)!} e^{c(x-u)}y(u)du,$ where $x_0$is arbitrary but fixed
number.

Property 1 The following properties are relevant to the discussions in this section.

Property 2 $\frac{1}{P(D)}[e^{rx}] = \frac{e^{rx}}{P(r)},$ if $P(r) \neq 0$

Property 3 $\frac{1}{P(D)}e^{rx} = \frac{x^k e^{rx}}{k!\phi(r)}.$

Property 4 $\frac{1}{D^2+r^2}\sin r\,x = -\frac{x}{2r}\cos r\,x.$

Property 5 $\frac{1}{D^2+r^2}\cos r\,x = \frac{x}{2r}\sin r\,x$

Property 6 $\frac{1}{D^2+r^2}(c\sin b\,x) = \frac{c}{r^2-b^2}\sin b\,x, b \neq r$

Property 7 $\frac{1}{D^2+r^2}(c\cos b\,x) = \frac{c}{r^2-b^2}\cos b\,x, b \neq r$

Property 8 Illustrated by example: $\frac{1}{D(D+1)}[x^3] = \frac{x^4}{4} - 3x^2 + 6$

Proof: $\frac{1}{D(D+1)} = D^{-1}(1 - D^2 + D^4 - \ldots)$

$= D^{-1} - D^1 + D^3 - \ldots$

We have

$$\frac{1}{D(D+1)}[x^3] = D^{-1}[x^3] - D^1[x^3] + D^3[x^3] - \ldots$$

$$= \frac{x^4}{4} - 3x^2 + 6$$

Property 9 Exponential Shift.

$$\frac{1}{P(D)}[e^{rx}y] = e^{rx}\frac{1}{P(D+r)}[y].$$

2.16 Application of the Inverse Differential Operator to the Solution of Linear Differential Equations

The use of the theory of operators can save some labor in finding particular integrals for the complete linear differential equation:

$$P(D)y = f(x)$$

as a simple algebraic equation, you only then have to solve for $y$by a division $y = \frac{1}{P(D)}f(x)$

Example: Find a particular integral.

$$D(D-2)^3(D+1)y = e^{2x}$$

Solution: solve for$y$to obtain the particular integral

$$y_p = [D(D-2)^3(D+1)]^{-1}e^{2x}$$

with $r = 2$,$k = 3$, and $\phi(D) = D(D+1)$

so that

$\phi(r) = r(r+1)$ = (2) (3) = 6, and we have

$$y_p = \frac{x^3 e^{2x}}{3!\ 6} = \frac{1}{36} x^3 e^{2x}$$

2.17 Taylor's Series

From calculus, you learned that Taylor's series might represent a function.

$f(x) = f(x_0) + f'(x_0)(x - x_0) + \frac{f''}{2!}(x - x_0)^2 + \cdots,$

provided all the derivatives exit at $(x - x_0)$. We further say that the function is analytic at $x = x_0$ if $f(x)$ can be expanded in a power series valid about the same point.

Example: Find the solution of $xy'' + x^3y' - 3y = 0$ that satisfies $y = 0$ and $y' = 2$ at $x = 1$.

Solution

$$y'' = -x^2y' + 3x^{-1}y$$

$$y''' = -x^2y'' - (2x - 3x^{-1})y' - 3x^{-2}y,$$

$$y^{iv} = -x^2y''' - (4x - 3x^{-1})y'' - (2 + 6x^{-2})y' + 6x^{-3}y.$$

Evaluating these derivatives at $x = 1$

$$y''(1) = -2,$$

$$y'''(1) = 4,$$

$$y^{iv}(1) = -18.$$

Substituting in Taylor's series (3.2.3), the solution is

$$y(x) = 0 + 2(x - 1) - \frac{2(x - 1)^2}{2} + \frac{4(x - 1)^3}{6} + \frac{18(x - 1)^4}{24} + \ldots$$

$$= 2(x-1) - (x-1)^2 + \frac{2(x-1)^3}{3} + \frac{3(x-1)^4}{4} + \cdots$$

2.18 Solutions of Some Simple Partial Differential Equations

The class of partial differential equations containing partial derivatives with respect to a single variable may be solved by ordinary differential equations techniques.

To get some ideas concerning the nature of solutions of partial differential equations, let us consider the following problem for discussion.

Example: Obtain solutions for the PDE

$$\frac{\partial^2 U}{\partial x \partial y} = 6x + 12y^2$$

Here the dependent variable Udepends on two independent variables $x$and y.To find the solution, we seek to determine $U(x, y)$.i.e., Uin terms of $x$and y given by

$$\frac{\partial}{\partial x}\left(\frac{\partial U}{\partial y}\right) = 6x + 12y^2$$

We can integrate concerning x keeping y constant to find

$$\left(\frac{\partial U}{\partial y}\right) = 3x^2 + 12xy^2 + F(y)$$

where we have added the arbitrary "constant" of integration which can depend on y denoted by $F(y)$.We now integrate the above equation with respect to$y$keeping x constant, getting

$$U = 3x^2y + 4xy^3 + \int F(y)dy + G(x)$$

This time an arbitrary function of x,$G(x)$ is added. Since the integral of an arbitrary function of y is another arbitrary function of y,

$$U = 3x^2y + 4xy^3 + H(y) + G(x)$$

$H(y) = y^3$ and $G(x) = sin\, x$, is called *a particular solution.*

$U(1, y) = y^2 - 2y, U(x, 2) = 5x - 5$

Then from the general solution, we get

$U(1, y) = 3(1)^2y + 4(1)y^3 + H(y) + G(1) = y^2 - 2y$

or $H(y) = y^2 - 5y - 4y^3 - G(1)$

so that $U = 3x^2y + 4xy^3 + y^2 - 5y - 4y^3 - G(1) + G(x)$

If we now use the second condition in the general solution

$U(x, 2) = 3x^2(2) + 4x(2)^3 + (2)^2 - 5(2) - 4(2)^3 - G(1)+G(x)=5x - 5$

from which $G(x) = 33 - 27x - 6x^2 + G(1)$, the required solution.

$U = 3x^2y + 4xy^3 + y^2 - 5y - 4y^3 - G(1)+33 - 27x - 6x^2 + G(1)$

$U = 3x^2y + 4xy^3 + y^2 - 5y - 4y^3$

2.19 The Method of Separation of Variables

Example: Solve the boundary value problem

$\frac{\partial U}{\partial x} + 3\frac{\partial U}{\partial y} = 0 \qquad U(0, y) = 4e^{-2y} - 3e^{-6y}$

Solution: Here the dependent variables are x and y, so we substitute $U = XY$ in the given differential equation where X depends on only x and Y depends on only y.

$\frac{\partial(XY)}{\partial x} + 3\frac{\partial(XY)}{\partial y} = 0$ or $X'Y = -3XY'$

$\frac{X'}{3X} = -\frac{Y'}{Y}$

Since one side depends only on x and the other side depends only on y, and since $X$and $Y$are independent variables equation above can be true if and only if each side is equal to the same constant.

$$\frac{X'}{3X} = -\frac{Y'}{Y} = c$$

therefore

$$X' - 3cX = 0 \qquad Y' + cY = 0$$

From our knowledge of ordinary differential equations, we have solutions.

$$X = b_1 e^{3cx}, Y = b_2 e^{-cy}$$

Thus

$$U = XY = b_1 b_2 e^{c(3x-y)} = Be^{c(3x-y)}$$

where $B = b_1 b_2$

If we now use the condition we have

$$B\square e^{-cy} = 4e^{-2y} - 3e^{-6y}$$

$U_1 = b_1 e^{c_1(3x-y)}$ and $U_2 = b_2 e^{c_2(3x-y)}$

$$U = b_1 e^{c_1(3x-y)} + b_2 e^{c_2(3x-y)}$$

The boundary condition leads to

$$b_1 e^{-c_1 y)} + b_2 e^{-c_2 y} = 4e^{-2y} - 3e^{-6y}$$

which is satisfied if we choose $b_1 = 4, c_1 = 2, b_2 = -3, c_2 = 6$

This leads to the required solution

$$U = 4e^{2(3x-y)} - 3e^{6(3x-y)}$$

(a) $\frac{\partial u}{\partial t} = 4\frac{\partial^2 u}{\partial x^2}$ Solution: *linear; order 2; dependent var.u; independent var.x, t.]*

(b) $x^2\frac{\partial^3 R}{\partial y^3} = y^3\frac{\partial^2 R}{\partial x^2}$ Solution: *linear; order 3; dependent var.R; independent var.x, y.]*

(c) $W\frac{\partial^2 W}{\partial r^2} = rst$ [Solution: non*linear; order 2; dependent var.W; independent var.r, s, t.]*

(d) $\frac{\partial^2\phi}{\partial x^2} + \frac{\partial^2\phi}{\partial y^2} + \frac{\partial^2\phi}{\partial z^2} = 0$ [Solution: *linear; order 2; dependent var.ϕ; independent var.x, y, z..]*

(e) $\left(\frac{\partial z}{\partial u}\right)^2 + \left(\frac{\partial z}{\partial v}\right)^2 = 1$ [Solution: non*linear; order 1; dependent var.z; independent var.u, v.]*

Example: State the order and the degree of the following differential equations.

a) $\left(\frac{dy}{dx}\right)^2 = e^x$

b) $\frac{d^2y}{dx^2} + xy = sinx$

c) $\frac{d^2y}{dx^2} + xy\frac{dy}{dx} + y = 2$

d) $\left(y\frac{d^4y}{dx^4}\right)^2 + 3\cos x\frac{d^3y}{dx^3} + \frac{dy}{dx} = 0$

Solution:

	Order	Degree	Comments
(a)	1	2	Highest order = 1
(b)	2	1	Highest order = 2
(c)	2	1	Highest order = 2
(d)	4	2	Highest order = 4

Example: Obtain the differential equation associated with the following solution.

$\boldsymbol{y = Ax^2 + Bx + C}$

Solution:

$$\begin{aligned} y &= Ax^2 + Bx + C \\ y' &= 2Ax + B \\ y'' &= 2A \\ y''' &= 0 \end{aligned}$$

$\therefore$ The differential equation is $y''' = 0$

Example: Solve the initial value problem.

$\frac{dy}{dx} = x^2 + 3x^2y$, given that $y(0) = 2$

Solution:

$$\frac{dy}{dx} = x^2 + 3x^2y = x^2(1 + 3y)$$

Using the method of seperation of variables:

$$\frac{dy}{(1 + 3y)} = x^2 dx$$

Integrating both sides:

$$\frac{1}{3}\ln(1+3y) = \frac{x^3}{3} + \ln C$$

$$\ln(1+3y) = x^3 + K \qquad (K \text{ is a constant})$$

$$\therefore (1 + 3y) = Ae^{x^3} \qquad (A = e^K)$$

At $x = 0, y = 2$

$$(1+6) = A$$
$$A = 7$$

So, the required solution is:

$$(1 + 3y) = 7e^{x^3}$$

Example: Establish whether the following equations are exact or not.

a) $(3x^2 - 2y^2)dx + (1 - 4xy)dy = 0$

b) $(2x^3 + 3y)dx + (3x + y - 1)dy = 0$

Solution:

(a). The general solution $M(x, y)dx + N(x, y)dy = 0$
It is said to be exact if:

$$\frac{\partial M}{\partial y} = \frac{\partial N}{\partial x}$$

$$(3x^2 - 2y^2)dx + (1 - 4xy)dy = 0$$

$M = 3x^2 - 2y^2$ and $N = (1 - 4xy)$

$$\frac{\partial M}{\partial y} = -4y, \frac{\partial N}{\partial x} = -4y$$

Since, $\frac{\partial M}{\partial y} = \frac{\partial N}{\partial x} = -4y$

The given equation is exact.

(b). $(2x^3 + 3y)dx + (3x + y - 1)dy = 0$

$M = 2x^3 + 3y$ and $N = (3x + y - 1)$

$$\frac{\partial M}{\partial y} = 3, \frac{\partial N}{\partial x} = 3$$

Since, $\frac{\partial M}{\partial y} = \frac{\partial N}{\partial x} = 3$

The given equation is exact.

Example: Find the general solution of the homogeneous differential equation

$\frac{dy}{dx} = x^2 + 3x^2y,$

Solution:

$y'' + 4y' + 13y = 0$

Let $y = e^{mx}$

The auxiliary equation is:

$m^2 + 4m + 13 = 0$

$$m = \frac{-4 \pm \sqrt{16 - 52}}{2}$$
$$= \frac{-4 \pm 6i}{2}$$
$$= -2 \pm 3i \qquad \text{(complex roots)}$$

There are two roots: $-2 + 3i, -2 - 3i$

The general solution is:

$y = e^{-2x}[A\, cos\, 3\, x + B\, sin\, 3\, x]$

Example: Solve the different equation $y'' - 5y + 6y = 2e^x$ by the methods of variation of parameter.

Solution:

$y'' - 5y' + 6y = 2e^x$
$y'' - 5y' + 6y = 0$

The auxiliary equation is $m^2 - 5m + 6 = 0$

$\Rightarrow m = 2$ and $m = 3$

$\therefore$ The complementary function is $y_c = Ae^{2x} + Be^{3x}$

Let $y_1 = e^{2x}, y_2 = e^{3x}$

The particular integral y_p is given by:

$y_p = v_1y_1 + v_2y_2$

where $v_1(x) = \int \frac{-g(x)y_2}{y_1y_2' - y_2y_1'} dx$

$$v_2(x) = \int \frac{g(x)y_1}{y_1y_2' - y_2y_1'} dx$$

and $g(x) = 2e^x$

$$v_1 = \int \frac{-2e^x e^{3x}}{e^{5x}} dx$$
$$= -2\int e^{-x} dx$$
$$= 2e^{-x}$$
$$v_2 = \int \frac{2e^x e^{2x}}{e^{5x}} dx$$
$$= 2\int e^{-2x} dx$$
$$= -e^{-2x}$$

So, $y_p = 2e^x - e^x$

The general solution is:

$$y = y_c + y_p$$
$$= Ae^{2x} + Be^{3x} + e^{x}$$

Example: A cup of tea at 90^0C is placed on a dining table at a constant room temperature of 20^0C.From experience, it is known that it takes 10 minutes for the tea to cool from 90^0C to 70^0C. What will the temperature of the tea be after 30 minutes?

Solution:

Let *H*(*t*) be the temperature of the tea at time *t* and $H_0$the temperature of the room at time *t*.

Then:

$$\frac{dH}{dt} = -k(H - 20)$$
$$\Rightarrow (H - 20) = Ae^{-kt}$$

At $t = 0, H = 90, \Rightarrow A = 70$

Similarly at At $t = 10, H = 70, \Rightarrow k = \frac{1}{10} ln\left(\frac{7}{5}\right)$

$$\therefore H(t) = 70e^{-\frac{1}{10}ln\left(\frac{7}{5}\right)t}$$

After 15 minutes ($t = 15$),

$$H(t) = 70e^{-\frac{1}{10}ln\left(\frac{7}{5}\right)x15} = 51.6^0C$$

The temperature of tea after 15 minutes will be approximately 52^0C.

Example: Solve the differential equation $y'' + x^2y = 0$ by using the Power Series method.

Solution:

Let

$$y = \sum_{n=0}^{\infty} a_n x^n$$

Then

$$y' = \sum_{n=1}^{\infty} n a_n x^{n-1}$$

$$y'' = \sum_{n=2}^{\infty} n(n-1) a_n x^{n-2}$$

Substituting into $y'' + x^2 y = 0$

$$\sum_{n=2}^{\infty} (n(n-1) a_n x^{n-2} + x^2 \sum_{n=0}^{\infty} a_n x^n$$

Putting the series under the same powers of *x*:

$$\sum_{n=0}^{\infty} (n+2)(n+1) a_{n+2} x^n + \sum_{n=2}^{\infty} a_{n-2} x^n = 0$$

Summing all of the series for $n = 2$ gives:

$$2a_2 + 6a_3 x + \sum_{n=2}^{\infty} [(n+2)(n+1) a_{n+2} + a_{n-2}] x^n - 0$$

Equating coefficients of powers of x:

$x^0 : 2a_2 = 0 \Rightarrow a_2 = 0$
$x^1 : 6a_3 = 0 \Rightarrow a_3 = 0$

For x^nwe either obtain the recursive formula:

$$a_{n+2} = \frac{-1}{(n+2)(n+1)} a_{n-2}, n \geq 2$$

Thus for:

$$n = 2; \quad a_4 = \frac{-1}{12} a_0$$
$$n = 3; \quad a_5 = \frac{-1}{20} a_1$$
$$n = 4; \quad a_6 = \frac{-1}{30} a_2 = 0 \ \ (a_2 = 0)$$
$$n = 5; \quad a_7 = \frac{-1}{42} a_3 = 0 \ \ (a_3 = 0)$$
$$n = 6; \quad a_8 = \frac{-1}{56} a_4 = \frac{1}{56 \times 12} a_0$$

and

$$y = \sum_{n=0}^{\infty} a_n x^n$$
$$= a_0 (1 - \frac{1}{12} x^4 + \frac{1}{672} x^8 - \ldots) + a_1 (x - \frac{1}{20} x^5 + \ldots)$$

Chapter 3

Concepts of Probability and Statistics

Probability and statistics are two branches of mathematics widely used in many fields, including engineering. Probability is concerned with the study of random events and the likelihood of their occurrence, while statistics is concerned with the analysis and interpretation of data.

Probability theory provides a framework for understanding the likelihood of different events occurring. It is based on the notion of probability, which measures the likelihood of a particular event occurring. Probability can range from 0 (indicating that an event is impossible) to 1 (indicating that an event is certain to occur).

Some key concepts in probability theory include:

Sample space: the set of all possible outcomes of a random experiment.
Event: a subset of the sample space.
Probability distribution: a function that assigns probabilities to each possible outcome in a sample space.
Random variable: a variable that takes on different values with certain probabilities.
Statistics is concerned with the analysis and interpretation of data. It involves collecting, organizing, and summarizing data and drawing conclusions from the data. Statistics can be divided into two main branches: descriptive and inferential.
Descriptive statistics involves describing and summarizing data using mean, median, mode, variance, and standard deviation measures. These measures can provide a general overview of the data and identify patterns or trends.
Inferential statistics involves concluding a population based on a sample of data. This involves using probability theory to make inferences about the population based on the data collected from the sample. Inferential statistics can be used to test hypotheses, make predictions, and estimate the parameters of a population.

Some key concepts in statistics include:

Population: the entire group of individuals or objects the study is concerned with.
Sample: a subset of the population that is used to collect data.
Parameter: a numerical value that describes a characteristic of a population.
Statistic: a numerical value that describes a characteristic of a sample.
Probability and statistics are closely related and often used together. Probability theory provides the foundation for statistical inference, which is the process of concluding a population based on a sample of data.

This chapter contains an overvoiew on:

✓Experiments, outcomes, sample space, and events,
✓Union, Intersection, complement, disjoint Events.
✓Probability.
✓Axioms of Probability.
✓Properties of Probability.

3.1 Probability

The probability of an event, *A*, denoted as P(*A*), is a quantity to describe how likely event *A* occurs.

Ex. $P(A) = 0 \Rightarrow$ Event A will never occur

Axiom of probability
1. The probability of any event must lie between 0 and 1.
That is, for any event A,

$$0 \leq P(A) \leq 1$$

2. The total probability assigned to the sample space of an experiment must be 1.

That is, $P(S) = 1$

3.1.1 Probability Formulas

Table 3.1 Probability formulas

#	Description	Formula
1	Probability of an event A	$P(A) = \frac{f}{n}$ where f = frequency of occurrence of an event n = sample size
2	Probability of the complement of event A	$P(not\ A) = 1 - P(A)$
3	Multiplication rule for independent events	$P(A\ and\ B) = P(A).P(B)$
4	General multiplication rules	$P(A\ and\ B) = P(A).P(B\ given\ A)$ $P(A\ and\ B) = P(B).P(A\ given\ B)$
5	Addition rule for mutually exclusive events	$P(A\ or\ B) = P(A) + P(B)$
6	General addition rule	$P(A\ or\ B) = P(A) + P(B) - P(A\ and\ B)$
7	Permutation rule	$P_{n,r} = \frac{n!}{(n-r)!}$
8	Combination rule	$C_{n,r} = \frac{n!}{r!\ (n-r)!}$
9	The mean of a discrete probability distribution	$\mu = \sum x\ P(x)$
10	The standard deviation of a discrete probability distribution	$\sigma = \sqrt{\sum (x-\mu)^2\ P(x)}$

3.1.2 Experiments, Outcomes, Sample Space, and Events

Table 3.2 Experiments, outcomes, sample space, and events

Experiment	Possible Outcomes
Toss a dice	1, 2, 3, 4, 5, 6
Toss two dices	2, 3, 4, 5, 6,7.8.9.10.11.12
Toss three dices	3, 4, 5,

	6,7,8,9,10,11,12,13,14,15,16,17,18
Flip a coin	H, T
Flip two coins	HH, HT, TH, TT
Flip three coins	HHH, HHT, HTH, THH, HTT, THT, TTH, TTT
Examine one fuse in sequence (fail or pass)	P, F
Examine two fuses in sequence (fail or pass)	PP, PF, FP, FF
Examine three fuses in sequence (fail or pass)	PPP, PPF, FPP, PFP, PFF, FPF, FFP, FFF

The sample space, *S*, of an experiment, is the set (collection) of all possible outcomes from an experiment. An event, *A*, is a subset of the sample space S.

Example: Three fuses are examined in sequences; each receives a pass (P) or fail (F) rating because of the inspection.
(1) S = sample space = {PPP, PPF, FPP, PFP, PFF, FPF, FFP, FFF}
(2) Let *A* denote the event that exactly one fuse fails inspection. How would *A* be defined?
A = {PPF, FPP, PFP}

3.1.3 Union, Intersection, Complement Disjoint Events

Sometimes it is useful to visualize the relationships between events using a Venn diagram.

(1) $A \cup B$, the union of events A and B. It reads as "A *union* B" or "A *or* B" (The area covered by either A or B)
(2) $A \cap B$, the Intersection of events A and B. It reads as "A *intersect* B" or "A *and* B" (The area covered by both A and B)

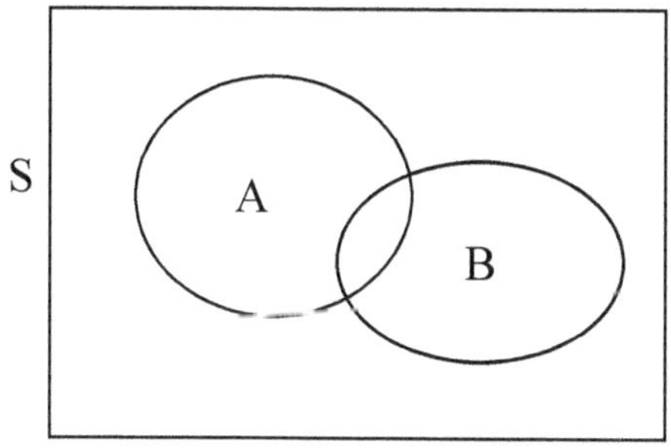

(3) A', the complement of event A. It reads as "A *compliment*" or "*not* A" (The area inside S but not covered by A)

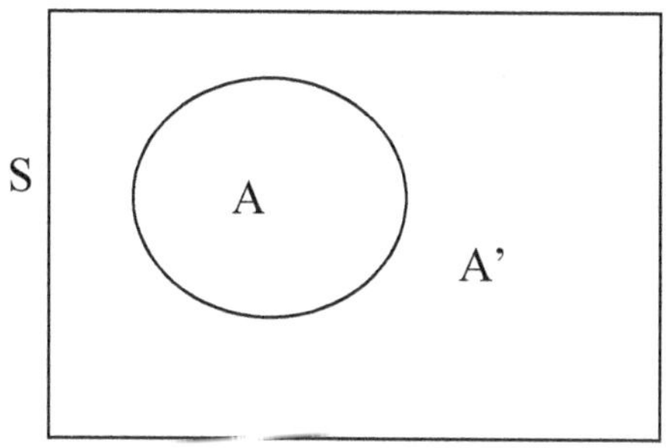

(4) A and B are disjoint. That is, $A \cap B = \Phi$ (A and B do not have common part)

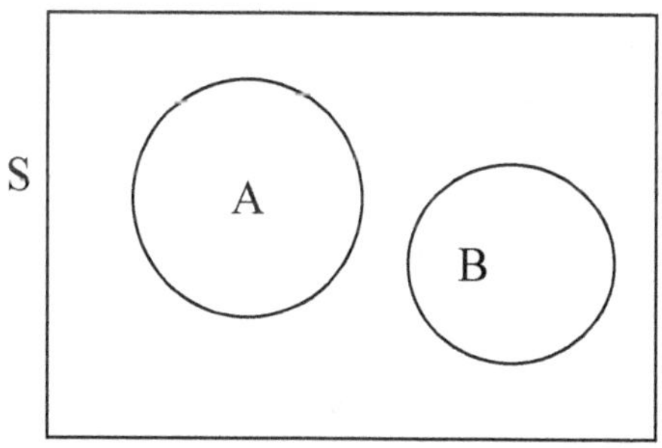

Example: Three fuses are examined in sequences; each receives a pass (P) or fail (F) rating because of the inspection. Let A denote the event that, at most, one fuse fails inspection. What $A \cup B$ and $A \cap B$?

Solution:
A = { PPP, PPF, FPP, PFP}, B = {PPF, FPP, PFP}
$A \cup B = \{PPP, PPF, FPP, PFP\} = A, A \cap B$ ={PPF, FPP, PFP}=B

3.1.4 Properties of Probability

1. The addition rule: for any two events, *A* and *B*,

$$P(A \cup B) = P(A) + P(B) - P(A \cap B)$$

(this should be clear if we view P(A) as the area covered by A in the sample space S)

2. If *A* and *B* are disjoint, then $P(A \cap B) = 0$
As a result, the addition rule *for disjoint events* can be simplified as $P(A \cup B) = P(A) + P(B)$ (only true if A and B are disjoint)

3. The complement rule: for any event *A*,

$$P(A') = 1 - P(A)$$

Proof: $S = A \cup A'$, A and A' are disjoint. So $1 = P(S) = P(A) + P(A')$

Example: What is the probability of picking an orange ball out of a bag containing 12 orange balls, 14 red balls, and eight green balls?

Solution:
There are a total of 34 balls, and 12 of them are orange, so $P = 12/34$.

Example: A bag contains 120 balls numbered from 1 to 120. An even-numbered ball is removed and set aside. What is the probability that the next ball removed is an odd number greater than 40?

Solution:
Of the 99 remainings, 60 are odd and greater than 40, so $P = 60/119$.

Example: Two dice are rolled. What is the probability that the sum is 7?

Solution:
There are 36 outcomes are the set of [(1,1),(1,2),...,(6,6)]
The sum of 7 come from [(1,6),(2.5),(3,4),(4,3),(5,2),(6,1)] So $P = 6/36 = 1/6$.

Notes

1. The probability of each pattern for throwing two dice.

Table 3.3 The probability of each pattern for throwing two dice

Dices	1	2	3	4	5	6
1	1/36	1/36	1/36	1/36	1/36	1/36
2	1/36	1/36	1/36	1/36	1/36	1/36
3	1/36	1/36	1/36	1/36	1/36	1/36
4	1/36	1/36	1/36	1/36	1/36	1/36
5	1/36	1/36	1/36	1/36	1/36	1/36
6	1/36	1/36	1/36	1/36	1/36	1/36

2. The probability for thrown two dice

Table 3.4 The probability for thrown two dice

Sum	1	2	3	4	5	6	7	8	9	10	11	12
P(A)	0/36	1/36	2/36	3/36	4/36	5/36	6/36	5/36	4/36	3/36	2/36	1/36

Example: Seven dice are rolled. What is the probability that at least one of them is a four?

Solution:
The probability that none of them are four is (5/6) chance on each roll, and since there are five rolls, we multiply $(5/6) * (5/6) * (5/6) * (5/6) * (5/6) * (5/6) * (5/6) = (5/6)^7$. Then the complement of this is that at least one of them is a six or $1 - (5/6)^7 = 0.720918$.

Example: Three dice are rolled. What is the probability that the sum is less than 18?

Solution:
The probability of getting an 18 is $(1/6)^3$ $so\ P(X < 18) =$ $1 - P(X = 18) = 1 - (1/6)^3 = 0.99537,$

Example: If a fair coin is tossed 30 times, what is the probability of the coin landing on tails exactly eight times?

Solution:
Let X is the number of tails in 30 tosses, then X is binomial with $n = 30\ and\ p = q = 0.5.$

Therefore, for any possible outcome $y = 8$

$$P(X = y) = C(n, y)p^y q^{n-y}$$

$$P(X = 5) = C(30,8)0.5^8 0.5^{30-8} = \frac{30!}{8!\,(30-8)!} \times 0.5^8 0.5^{22}$$
$$= 0.0054509$$

Example: A student is randomly selected from a class where 45% of the class is left-handed and 40% are sophomores. We further know that 3% of the class consists of left-handed sophomores.

a)What is the probability of selecting a student as either left-handed OR a sophomore?
b) What is the probability of selecting a right-handed sophomore?
c) Are *selecting a left-handed student* and *selecting a sophomore* considered disjoint?

Solution:
Define A = event that a randomly selected student is left-handed
B = event that a randomly selected student is a sophomore

a) P(A) = 0.45, P(B) = 0.4, and $P(A \cap B) = 0.03$
$P(A \cup B) = P(A) + P(B) - P(A \cap B) = 0.45 + 0.4 - 0.03 = 0.82$

b) $P(B) = P(A' \cap B) + P(A \cap B)$ (since $(A' \cap B) and (A \cap B)$ are disjoint).
$P(A' \cap B) = P(B) - P(A \cap B) = 0.4 - 0.05 = 0.35$

c) $A \cap B = \varphi$

Example: Find the probability of rolling a die, which is an even number.

Solution:
$Possible\ outcome\ = \{1, 2, 3, 4, 5, 6\}$

$Favourable\ outcome\ = \{2, 4, 6\}$

$P\ (even\ number) = \frac{No.of\ Possible\ outcome}{Total\ no,\ \ of\ Favourable\ outcome} = \frac{1}{2}$

Example: Six-sided dice are rolled. Find the probability of getting:
a. $P(8)$
b. $P(\text{an odd number})$
c. $P(\text{multiple of 2})$
d. $P(\text{prime number})$
e. $P(\text{A number less than 5})$
f. $P(8)$

Solution:
Sample space: S = {1, 2, 3, 4, 5, 6}
a. $P(3) =\ \ Favourable\ outcome\ = \{\ 3\ \} = 1/6$
b. $P\ (odd\ number) = Favourable\ outcome\ = \{1, 3, 5\} = 3/6 = 1/2$
c. $P(multiple\ of\ 2) =\ Favourable\ outcome\ = \{2, 4, 6\} = 3/6 = 1/2$
d. $P(prime\ number)\ =\ Favourable\ outcome\ = \{2, 3, 5\} = 3/6 = 1/2$
e. $P(number\ less\ than\ 5\ =\ Favourable\ outcome\ = \{1, 2, 3, 4\} = 4/6 = 2/3$
f. $P(8)\ = 0/6\ \ = 0$
Impossible to obtain an 8 with dice

Example: Two coins are tossed together:

a. List the sample space.
b. List the probability of getting:
1. 2 Heads
2. 1 Head and 1 Tail
3. At least one head

Solution:
a. $S = \{HH, HT, TH, TT\}$
b. 1. $P\ (2\ Heads)$ =1/4
2. $P\ (1\ Head\ and\ 1\ tail) = 1/2$
3. $P\ (at\ least\ 1\ head)\ =\ 3/4$

Example: A teacher chooses a student at random from a class of 25 girls. What is the probability that the student chosen is a girl?

Solution:
Since all the students in the class are girls, the teacher is certain to choose a girl.

$P(girl) = 25/25 = 1$

Example: A card player has a total of 8 cards in her hand. How many different ways can she put them in order?

Solution:

$$P(n, r) = \frac{n!}{(n-r)!}\quad [r \leq n]$$

n permutations took n at a time:

$$P(n, n) = \frac{n!}{(n-n)!} = \frac{n!}{(0)!} = n! \ \rightarrow\ P(8,8) = \frac{8!}{(8-8)!} = \frac{8!}{(0)!}$$
$$= 40320$$

Example: A " Do List" has a total of 15 tasks on it. Assuming each task isindependent of the other tasks and that because of time constraints, only eight tasks can be performed. How many different ways can they order?

Solution:

$$P(n,r) = \frac{n!}{(n-r)!}\ [r \le n] \rightarrow P(15,8) = \frac{15!}{(15-8)!} = \frac{15!}{(7)!}$$
$$= 259459200$$

Example: How many ways can a study group of 4 students be formed from a class of 20 students without regard to the groups' order?

Solution

$$P(n,r) = \frac{n!}{(n-r)!}\ [r \le n] \rightarrow P(20,4) = \frac{20!}{(20-4)!} = \frac{20!}{(16)!}$$
$$= 116280$$

Example: Six power books and four electronic books will be placed on a shelf. How many ways are there to arrange the books if books of the same subject must be kept together?

Solution
The equivalent of separately arranging the power books is

$$P(6,6) = \frac{6!}{(6-6)!} = \frac{6!}{(0)!} = 720$$

ways, and separately arranging the electronic books

$$P(4,4) = \frac{4!}{(4-4)!} = \frac{4!}{(0)!} = 24$$

ways. So there are $720 \times 6 = 4320$ ways.

Example: If the set is $S = \{a, b, c, d\}$, find the total number of possible permutations of 3 letters taken from S.

Solution:
Looking at all subsets of size $r = 3$ that can be selected from the set of $n = 4$ elements.

Using the permutations formula,

$$P(n,r) = \frac{n!}{(n-r)!} \ [r \leq n] \rightarrow P(4,3) = \frac{4!}{(4-3)!} = 24$$

24 Possible "Orderings."
a,b,c a,c,b b, a,c c, a,b b,c, a c,b, a
a,b,d a,d,b b,a,d d,a,b b,d,a d,b,a
d,b,c d,c,b b,d,c c,d,b b,c,d c,b,d
a,d,c a,c,d d,a,c c,a,d d,c,a c,d,a

Example: A simple language consists of only two symbols, A and B, produced in a long continuous sequence. BAAA BBBA AAAA BBBA AAAB, find the probability of occurrence A and B.

Solution:

$$P(A) = 12/20 \ , P(B) = 8/20$$

Joint Probability: If we have two experiments A & B, experiment A has $A_1, A_2, A_3 \ldots A_n$ Outcomes and experiment B has $B_1, B_2, B_3 \ldots B_n$ outcomes then

P(A_i , B_j) joint probability that the event A_i occurs from experiment A and event B_j from experiment B. Note that:

$$\sum_{i=1}^{n}\sum_{j=1}^{m} p\left(A_i, B_j\right) = 1$$

$p\left(A_i, B_j\right)$ *is written in matrix form as follows.*

$$p\left(A_i, B_j\right)=\begin{bmatrix} B_1 & B_2 & B_3 & & & B_m \\ \cdots & \cdots & \cdots & \cdots & \cdots & \cdots \\ \cdots & & & & & \\ \cdots & & & & & \\ \cdots & & & & & \\ \cdots & & & & & \\ \cdots & & & & & \end{bmatrix} \begin{matrix} A_1 \\ A_2 \\ \cdots \\ A_n \end{matrix}$$

Note that

$$\sum_{i=1}^{n} \mathrm{p}\left(\mathrm{A_i}, \mathrm{B_j}\right)=\mathrm{p}\left(\mathrm{B_j}\right) \quad and \quad \sum_{j=1}^{m} \mathrm{p}\left(\mathrm{A_i}, \mathrm{B_j}\right)=\mathrm{p}\left(\mathrm{A_i}\right)$$

Example: A simple language consists of only two symbols A and B produced in a long continuous sequence. BAAA BBBA AAAA BBBA AAAB A, assume that.
Find $\mathrm{p(A, A)}$, $\mathrm{p(A, B)}$, $\mathrm{p(B, A)}$, and $\mathrm{p(B, B)}$.

Solution
There are 20 pairs, so
$\mathrm{p(A, A)} = 9/20$, $\mathrm{p(A, B)} = 3/20$, $\mathrm{p(B, A)} = 3/20$, and $\mathrm{p(B, B)} = 5/20$. Or in matrix form

$$\mathrm{p}\left(\mathrm{A_i}, \mathrm{B_j}\right)=\begin{bmatrix} \frac{9}{20} & \frac{3}{20} \\ \frac{3}{20} & \frac{5}{20} \end{bmatrix}$$

Conditional Probability: Two experiments, A & B, with their outcomes, affect each other. $P(A_i/B_j)$ *= Conditional probability of* A_i *given that* B_j *has already occurred in experiment* $P(B_j/A_i)$ *= Conditional probability of* B_j *given that* A_i *has already occurred in the experiment. Note that:* $P(A_i, B_j) = P(A_i) \,.\, P(B_j/A_i)$

$$P(A_i, B_j) = P(B_j) \,.\, P(Ai\,/Bj\,)$$

Both $P(A_i/B_j)$ *and* $P(B_j/A_i)$ *can be written in matrix form*

$$p\,(A/B) = \begin{bmatrix} B_1 & B_2 & B_3 & & & B_m \\ \cdots & \cdots & \cdots & \cdots & \cdots & \cdots \\ \cdots & & & & & \\ \cdots & & & & & \\ \cdots & & & & & \\ \cdots & & & & & \\ \cdots & & & & & \end{bmatrix} \begin{matrix} A_1 \\ A_2 \\ \cdots \\ A_n \end{matrix}$$

Note that

$$\sum_{i=1}^{n} p\left(A_i/B_j\right) = 1 \quad and \quad \sum_{j=1}^{m} p\left(B_j/A_i\right) = 1$$

Example: From the previous example, $\mathrm{p(A,A)} = 9/20, \mathrm{p(A,B)} = 3/20, \mathrm{p(B,A)} = 3/20,$ and $\mathrm{p(B,B)} = 5/20,$

We have on matrix form

$$p(x/y) = \begin{bmatrix} \frac{9}{12} & \frac{3}{8} \\ \frac{3}{12} & \frac{5}{8} \end{bmatrix}$$

Statistical Independence: If Ai does not affect the probability of B_j, then they are called independent, and $P(A_i/B_j) = P(A_i)$ and $P(B_j/A_i) = P(B_j)$ and $\mathrm{P(Ai/Bj)} = \mathrm{P(Ai).P(Bj)}$

Example: The two experiments, A & B, have the joint probability matrix:

$$\mathrm{p(A,B)} = \begin{bmatrix} 0.1 & 0.25 \\ 0 & 0.2 \\ 0.25 & 0.2 \end{bmatrix}$$

find $\mathrm{p(A), p(B), p(A/B), p(B/A)}$

Solution:

$$p(A_1) = \sum_{j=1}^{2} P(A_1, B_j) = 0.1 + 0.25 = 0.35$$

$$p(A_2) = \sum_{j=1}^{2} P(A_2,\ B_j) = 0 + 0.2 = 0.2$$

$$p(A_3) = \sum_{j=1}^{2} P(A_3,\ B_j) = 0.25 + 0.2 = 0.45$$

So $p(A) = [0.35\ \ 0.2\ \ 0.45]$

Note that: $p(A_1) + p(A_2) + p(A_3) = 1$

To find p(B) we have:

$$p(B_1) = \sum_{i=1}^{3} P(A_i, B_1) = 0.1 + 0 + 0.25 = 0.35$$

$$p(B_2) = \sum_{i-1}^{3} P(A_i, B_2) = 0.25 + 0.2 + 0.2 = 0.65$$

So $p(B) = [0.35\ \ 0.65]$

Also note that: $p(B_1) + p(B_2) = 1$

To find $p(A/B)$ matrix, we have:

$$p\left(\frac{A_i}{B_j}\right) = \frac{p(A_i, B_j)}{p(B_j)} = \begin{bmatrix} \frac{0.1}{0.35} & \frac{0.25}{0.65} \\ 0 & \frac{0.2}{0.65} \\ \frac{0.25}{0.45} & \frac{0.2}{0.65} \end{bmatrix} = \begin{bmatrix} \frac{2}{7} & \frac{5}{13} \\ 0 & \frac{4}{13} \\ \frac{5}{7} & \frac{4}{13} \end{bmatrix}$$

$$p\left(\frac{B_j}{A_i}\right) = \frac{p(A_i, B_j)}{p(A_j)} = \begin{bmatrix} \frac{0.1}{0.35} & \frac{0.25}{0.35} \\ 0 & \frac{0.2}{0.2} \\ \frac{0.25}{0.45} & \frac{0.2}{0.45} \end{bmatrix} = \begin{bmatrix} \frac{2}{7} & \frac{5}{7} \\ 0 & 1 \\ \frac{5}{9} & \frac{4}{9} \end{bmatrix}$$

3.2 Statistics

3.2.1 Statistics Formulas

Table 3.5 Statistics formulas

#		
1	Sample mean (n = sample size; N = population size)	$\bar{x} = \frac{\sum x}{n}$
2	Population mean	$\mu = \frac{\sum x}{N}$
3	Sample standard deviation	$s = \sqrt{\frac{\sum(x-\bar{x})^2}{n-1}}$
4	Population standard deviation	$\sigma = \sqrt{\frac{\sum(x-\mu)^2}{N}}$
5	The sample means a frequency distribution.	$\bar{x} = \frac{\sum xf}{n}$
6	Sample standard deviation for a frequency distribution	$s = \sqrt{\frac{\sum(x-\bar{x})^2 f}{n-1}}$
7	Sample coefficient of variation	$CV = \frac{s}{\bar{x}}$
8	$Range = Largest\ data\ value - smallest\ data\ value$	
9	Standard z value	$z = \frac{x-\mu}{\sigma}$
10	Original x value	$x = \mu + z\sigma$

11	Central limit theorem	$z = \frac{\bar{x} - \mu}{\sigma/\sqrt{n}}, n \geq 30$

3.2.2 Binomial Distribution Formulas

Table 3.6 Binomial distribution formulas

#	Description	Formula
1	Binomial probability distribution	$P(r) = \frac{n!}{r!\,(n-r)!}\, p^r\, q^{n-r}$ where r = number of successes; p = probability of success; q = 1 - p
2	Mean for a binomial distribution.	$\mu = np$
3	The standard deviation for a binomial distribution	$\sigma = \sqrt{npq}$

Arithmetic mean

$$\bar{Y} = \frac{\sum_{i=1}^{n} y_i}{n}$$

where $y_i = value\ of\ the\ ith\ observation$

Example: What is the mean of the following data: 10,15,15,13,23,45,25

$$\overline{Y_w} = \frac{\sum_{i=1}^{n} y_i}{n} = \frac{10 + 15 + 15 + 13 + 23 + 45 + 25}{7} = 20.85$$

Weighted arithmetic mean

$$\overline{Y_w} = \frac{\sum_{i=1}^{n} w_i y_i}{\sum_{i=1}^{n} w_i}$$

where: $y_i = value\ of\ the\ ith\ observation,\ w_i\ weight\ applied\ to\ y_i.$

Example: What is the weighted arithmetic mean of the following data?

Data	Weighting factor
70	1
65	2
49	3
89	4
95	5

Solution

$$\overline{Y_w} = \frac{\sum_{i=1}^{n} w_i y_i}{\sum_{i=1}^{n} w_i} = \frac{1 \times 70 + 2 \times 65 + 3 \times 49 + 4 \times 89 + 5 \times 95}{1 + 2 + 3 + 4 + 5}$$
$$= 78.533$$

3.2.3 The z-Table

The following z-Table indicates the area to the **right** of the vertical center-line of the z-curve (or standard normal curve) for different standard deviations.

Example: The green shaded area in the diagram below represents 1.45 standard deviations from the mean (which is 0). The area of this shaded portion is 0.4265 (or 42.65% of the total area under the curve).

To get this area, we read down the left side of the table for the standard deviation's first two digits (the whole number and the first number after the decimal point, in this case, 1.4), then we read across the table for the "0.05" part (the top row represents the 2nd decimal place of the standard deviation that we are interested in.)

Table 3.7 Part of z-Table

z	**0.00**	**0.01**	**0.02**	**0.03**	**0.04**	**0.05**	**0.06**
1.4	0.4192	0.4207	0.4222	0.4236	0.4251	**0.4265**	0.4279

We have:
(left column) 1.4 + (top row) 0.05 = 1.45 standard deviations

The area represented by 1.45 standard deviations to the right of the mean is shaded in green in the following standard normal curve.

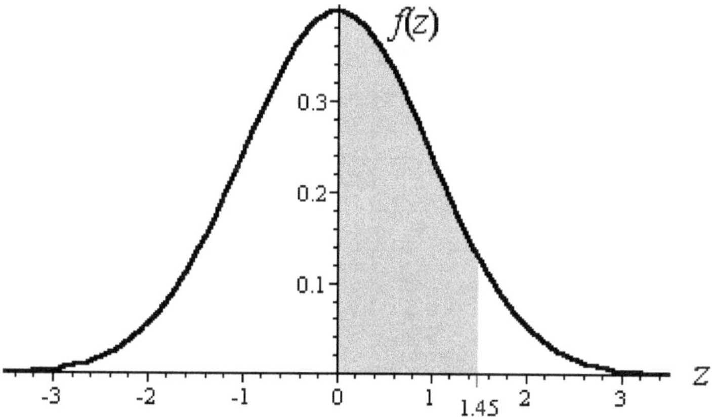

You can find the appropriate value in the full *z*-table below.

Table 3.8 z-Table

z	0	0.01	0.02	0.03	0.04	0.05	0.06	0.07	0.08	0.09
0	0	0.0040	0.0080	0.0120	0.0160	0.0199	0.0239	0.0279	0.0319	0.0359
0.1	0.0398	0.0438	0.0478	0.0517	0.0557	0.0596	0.0636	0.0675	0.0714	0.0753
0.2	0.0793	0.0832	0.0871	0.0910	0.0948	0.0987	0.1026	0.1064	0.1103	0.1141
0.3	0.1179	0.1217	0.1255	0.1293	0.1331	0.1368	0.1406	0.1443	0.1480	0.1517
0.4	0.1554	0.1591	0.1628	0.1664	0.1700	0.1736	0.1772	0.1808	0.1844	0.1879
0.5	0.1915	0.1950	0.1985	0.2019	0.2054	0.2088	0.2123	0.2157	0.2190	0.2224
0.6	0.2257	0.2291	0.2324	0.2357	0.2389	0.2422	0.2454	0.2486	0.2517	0.2549
0.7	0.2580	0.2611	0.2642	0.2673	0.2704	0.2734	0.2764	0.2794	0.2823	0.2852
0.8	0.2881	0.2910	0.2939	0.2967	0.2995	0.3023	0.3051	0.3078	0.3106	0.3133
0.9	0.3159	0.3186	0.3212	0.3238	0.3264	0.3289	0.3315	0.3304	0.3365	0.3389
1.0	0.3413	0.3438	0.3461	0.3485	0.3508	0.3531	0.3554	0.3577	0.3599	0.3621
1.1	0.3643	0.3665	0.3686	0.3708	0.3729	0.3749	0.3770	0.3790	0.3810	0.3830
1.2	0.3849	0.3869	0.3888	0.3907	0.3925	0.3944	0.3962	0.3980	0.3997	0.4015
1.3	0.4032	0.4049	0.4066	0.4082	0.4099	0.4115	0.4131	0.4147	0.4162	0.4177
1.4	0.4192	0.4207	0.4222	0.4236	0.4251	0.4265	0.4279	0.4292	0.4306	0.4319

1.5	0.4332	0.4345	0.4357	0.4370	0.4382	0.4394	0.4406	0.4418	0.4429	0.4441
1.6	0.4452	0.4463	0.4474	0.4484	0.4495	0.4505	0.4515	0.4525	0.4535	0.4545
1.7	0.4554	0.4564	0.4573	0.4582	0.4591	0.4599	0.4608	0.4616	0.4625	0.4633
1.8	0.4641	0.4649	0.4656	0.4664	0.4671	0.4678	0.4686	0.4693	0.4699	0.4706
1.9	0.4713	0.4719	0.4726	0.4732	0.4738	0.4744	0.4750	0.4756	0.4761	0.4767
2.0	0.4772	0.4778	0.4783	0.4788	0.4793	0.4798	0.4803	0.4808	0.4812	0.4817
2.1	0.4821	0.4826	0.483	0.4834	0.4838	0.4842	0.4846	0.4850	0.4854	0.4857
2.2	0.4861	0.4864	0.4868	0.4871	0.4875	0.4878	0.4881	0.4884	0.4887	0.4890
2.3	0.4893	0.4896	0.4898	0.4901	0.4904	0.4906	0.4909	0.4911	0.4913	0.4916
2.4	0.4918	0.4920	0.4922	0.4925	0.4927	0.4929	0.4931	0.4932	0.4934	0.4936
2.5	0.4938	0.4940	0.4941	0.4943	0.4945	0.4946	0.4948	0.4949	0.4951	0.4952
2.6	0.4953	0.4955	0.4956	0.4957	0.4959	0.4960	0.4961	0.4962	0.4963	0.4964
2.7	0.4965	0.4966	0.4967	0.4968	0.4969	0.4970	0.4971	0.4972	0.4973	0.4974
2.8	0.4974	0.4975	0.4976	0.4977	0.4977	0.4978	0.4979	0.4979	0.4980	0.4981
2.9	0.4981	0.4982	0.4982	0.4983	0.4984	0.4984	0.4985	0.4985	0.4986	0.4986
3.0	0.4987	0.4987	0.4987	0.4988	0.4988	0.4989	0.4989	0.4989	0.4990	0.4990
3.1	0.4990	0.4991	0.4991	0.4991	0.4992	0.4992	0.4992	0.4992	0.4993	0.4993
3.2	0.4993	0.4993	0.4994	0.4994	0.4994	0.4994	0.4994	0.4995	0.4995	0.4995
3.3	0.4995	0.4995	0.4995	0.4996	0.4996	0.4996	0.4996	0.4996	0.4996	0.4997
3.4	0.4997	0.4997	0.4997	0.4997	0.4997	0.4997	0.4997	0.4997	0.4997	0.4998
3.5	0.4998	0.4998	0.4998	0.4998	0.4998	0.4998	0.4998	0.4998	0.4998	0.4998
3.6	0.4998	0.4998	0.4999	0.4999	0.4999	0.4999	0.4999	0.4999	0.4999	0.4999
3.7	0.4999	0.4999	0.4999	0.4999	0.4999	0.4999	0.4999	0.4999	0.4999	0.4999
3.8	0.4999	0.4999	0.4999	0.4999	0.4999	0.4999	0.4999	0.4999	0.4999	0.4999

Chapter 4

Engineering Economics

Engineering economics is a subset of economics for application to engineering projects. Engineers try to find solutions to problems, and the economic viability of each potential solution is normally considered along with the technical aspects. Engineering economics is the application of economic techniques to the evaluation of design and engineering alternatives. The role of engineering economics is to assess the appropriateness of a given project, estimate its value, and justify it from an engineering standpoint.

The time value of money is central to most engineering economic analyses. Cash flows are *discounted* using an interest rate, i, except in basic economic studies. For each problem, there are usually many possible *alternatives*. One alternative that must be considered in each study, often the *choice*, is the *do-nothing alternative*. The *opportunity cost* of making one choice over another must also be considered. There are also non-economic factors to be considered, like color, style, public image, etc.; such factors are termed *attributes*.

For each alternative, costs, and revenues are considered for an *analysis period* that is either a fixed number of years or the estimated life of the project. The *salvage value* is often forgotten but is important and is either the net cost or revenue for decommissioning the project.

Since engineering is an essential part of the manufacturing sector of the economy, industrial engineering economics is an important part of industrial or business economics.

4.1 Factor Table

Table 4.1 shows a factor table for economics; for example, where F/P is the column selector for the Factor Tables, the interpretation is:
Calculate "F" / Given "P"
A = Annual Amount
F = Future Worth P = Present worth

G = Uniform Gradient Amount
n = number of compounding periods or the life of asseti = effective rate per period

Table 4.1 Factor table equations

Economic Process	Changes	Shorthand	Equation
Single Payment Compound Amount	Calculate *F* given *P*	$(F/P, i\%, n)$	$(1+i)^n$
Single Payment Present Worth	Calculate *P* given *F*	$(P/F, i\%, n)$	$(1+i)^{-n}$
Uniform Series Sinking Fund	Calculate *A* given *F*	$(A/F, i\%, n)$	$\frac{i}{(1+i)^n-1}$
Uniform Series Present Worth	Calculate P given A	$(P/A, i\%, n)$	$\frac{(1+i)^n-1}{i(1+i)^n}$
Capital Recovery	Calculate *A* given *P*	$(A/P, i\%, n)$	$\frac{i(1+i)^n}{(1+i)^n-1}$
Uniform Series Compound Amount	Calculate *F* given *A*	$(F/A, i\%, n)$	$\frac{(1+i)^n-1}{i}$
Uniform Gradient Present Worth	Calculate *P* given *G*	$(P/G, i\%, n)$	$\frac{(1+i)^n-1}{i^2(1+i)^n}-\frac{n}{i(1+i)^n}$
Uniform Gradient Future Worth	Calculate *F* given *G*	$(F/G, i\%, n)$	$\frac{(1+i)^n-1}{i^2}-\frac{n}{i}$
Uniform Gradient Uniform Series	Calculate *A* given *G*	$(A/G, i\%, n)$	$\frac{1}{i}-\frac{n}{(1+i)^n-1}$

Note the mathematical relationships among the equations:

$$F/G = (F/A - n)/i = (F/A) \times (A/G)$$

4.2 Time Value of Money

The **time value** is the money figured in a given amount of interest earned over time. The time value of money is the central concept in **finance theory.**

Time Value of Money (TVM) is an important concept in financial management. It can be used to compare investment alternatives and to solve problems involving loans, mortgages, leases, savings, and annuities.

TVM is based on the concept that a dollar you have today is worth more than the promise or expectation that you will receive a dollar in the future. The money you hold today is worth more because you can invest it and earn Interest. After all, you should receive some compensation for preceding spending. For instance, you can invest your dollar for one year at a 6% annual interest rate and accumulate $1.06 at the end of the year. You can say that the **dollar's future value** is $1.06, given a 6% **interest rate** and one year. The **present value** of the $1.06 you expect to receive in one year is only $1.

A key concept of TVM is that a single sum of money or a series of equal, evenly spaced payments or receipts promised in the future can be converted to an equivalent value today. Conversely, you can determine the value to which a single sum or a series of future payments will grow at some future date.

You can calculate the fifth value if given four: Interest Rate, Number of Periods, Payments, Present Value, and Future Value. Each of these factors is very briefly defined in the right-hand column below. The left column references more detailed explanations, formulas, and examples.

For example, $100 of today's money invested for one year and earning 5% interest will be worth $105 after one year. Therefore, $100 paid now or $105 paid exactly one year from now both have the same value to the recipient who assumes 5% interest; using **time value of money terminology**, $100 invested for one year at 5% interest has a *future value* of $105. This notion dates to Martín de Azpilcueta (1491–1586) of the School of Salamanca.

The method also allows the valuation of a likely income stream in the future so that the annual incomes are discounted and then added together, thus providing a lump-sum "present value" of the entire income stream.

All of the standard calculations for the time value of money are derived from the most basic algebraic expression for the present value of a future sum, "discounted" to the present by an amount equal to the time value of money. For Example, a sum of *FV* to be received in one year is discounted (at the rate of interest **r**) to give a sum of *PV* at present: $PV = FV - r \cdot PV = FV/(1+r)$.

Some standard calculations based on the time value of money are:

Present Value. The current worth of a future sum of money or stream of cash flows given a specified rate of return. Future cash flows are discounted at the discount rate, and the higher the discount rate, the lower the present value of the future cash flows. Determining the appropriate discount rate is the key to properly valuing future cash flows, whether earnings or obligations.

Present value of an annuity. An annuity is a series of equal payments or receipts at evenly spaced intervals. Leases and rental payments are examples. The payments or receipts occur at the end of each period for an ordinary annuity, while they occur at the beginning of each period for an annuity due.

The present value of a perpetuity is an infinite and constant stream of identical cash flows.

Future value is the value of an asset or cash at a specified future date that is equivalent to a specified sum today.

The future value of an annuity (FVA) is the future value of a stream of payments (annuity), assuming the payments are invested at a given interest rate.

Number of Periods	Periods are evenly-spaced intervals of time. They are intentionally not stated in years since each interval must correspond to a compounding period for a single amount or a payment period for an annuity.
Payments	Payments are a series of equal, evenly-spaced cash flows. In TVM applications, payments must represent all outflows (negative amount) or all inflows (positive amount).
Present Value Single Amount Annuity	**Present value** is an amount today equivalent to a future payment or series of payments that an appropriate interest rate has discounted. The future amount can be a single sum received at the end of the last period, as a series of equally-spaced payments (an annuity), or both. Since money has time value, the present value of a promised future amount is less the longer you wait to receive it.
Future Value Single Amount Annuity	**Future value** is the money invested with a fixed, compounded interest rate that will grow by some future date. The investment can be a single sum deposited at the beginning of the first period, a series of equally-spaced payments (an annuity), or both. Since money has time value, we naturally expect the future value to be greater than the present value. The difference between the two depends on the number of compounding periods involved and the going interest rate.
Loan	A method for repaying a loan in equal

Amortization	installments. Part of each payment goes toward interest; any remainder is used to reduce the principal. As the loan balance is gradually reduced, a progressively larger portion of each payment goes toward reducing the principal.
Cash Flow Diagram	A cash flow diagram is a picture of a financial problem that shows all cash inflows and outflows along a timeline. It can help you to visualize a problem and to determine if TVM methods can solve it.

4.3 Interest

Interest is a charge for borrowing money, usually stated as a percentage of the amount borrowed over a specific time.

Interest is the cost of borrowing money. An interest rate is a cost stated as a percent of the amount borrowed per time, usually one year. The prevailing market rate is composed of the following:

1. The **Real Rate of Interest** compensates lenders for postponing their spending during the loan term.

2. An **Inflation Premium** to offset the possibility that inflation may erode the value of the money during the term of the loan. A unit of money (dollar, peso, etc.) will purchase progressively fewer goods and services during inflation, so the lender must increase the interest rate to compensate for that loss.
3. Various **Risk Premiums to** compensate the lender for risky loans such as unsecured, made to borrowers with questionable credit ratings, or illiquid loans that the lender may not readily resell.

The first two components of the interest rate listed above, the real rate of interest and an inflation premium, collectively are referred to as the nominal risk-free rate. The rate of US Treasury bills can approximate the nominal risk-free rate in the USA since they are generally considered to have a very small risk.

4.3.1 Simple Interest

It is computed only on the original amount borrowed. It is the return on that principal for one period.

Simple interest is calculated on the original principal only. Accumulated interest from prior periods is not used in calculations for the following periods. Simple interest is normally used for less than a year, such as 30 or 60 days.

$$\textbf{\textit{Simple Interest}} = \boldsymbol{p \times i \times n}$$

where:

p = principal (original amount borrowed or loaned)
i = interest rate for one period
n = number of periods

Example: You borrow $20,000 for five years at 7% simple annual interest.

$$interest = p \times i \times n = 20{,}000 \times 7\% \times 5 = 7000$$

Example: You borrow $1,000 for 60 days at 7% simple interest per year (assume a 365-day year).

$$interest = p \times i \times n = 1{,}000 \times 7\% \times (60/365) = 11.5$$

4.3.2 Compound Interest

It is calculated each period on the original amount borrowed **plus** all unpaid interest accumulated. Compound interest is always assumed in TVM problems.

Compound interest is calculated each period on the original principal and all interest accumulated during past periods. Although the interest may be stated as a yearly rate, the compounding periods can be yearly, semi-annually, quarterly, or even continuous.

Example: Tom borrowed $20,000 for three years at 4% annual interest compounded annually:

$$interest\ 1\ year = p \times i \times n = 20{,}000 \times 0.04 \times 1 = 800$$

$$interest\ 2\ year = (p_2 = p_1 + i_1) \times i \times n$$
$$= (20{,}000 + 800) \times 0.04 \times 1 = 832$$
$$interest\ 3\ year = (p_3 = p_2 + i_2) \times i \times n$$
$$= (20{,}800 + 832) \times 0.04 \times 1 = 865.28$$
$$Total\ interest\ earned\ over\ the\ three\ years$$
$$= 800 + 832 + 865.28 = 2{,}497.28.$$

4.3.3 Rate of Return

When we know the Present Value (amount today), Future Value (amount to which the investment will grow), and Number of Periods, we can calculate the rate of return with this formula:

$$i = (FV / PV)^{(1/n)} - 1$$

Example: Find the Simple, Compounded Yearly, and Compounded Quarterly for a one-time investment of $10,000 for 30 years using 12% simple interest and 12% interest compounded yearly and quarterly. Also, find the interest rate i.

Solution

Simple 46,000.00
Compounded Yearly 299,599.22
Compounded Quarterly 347,109.87

The Present Value is $10,000, the Future Value is $299,599.22, and there are 30 periods. Confirm that the annual compound interest rate is 12%, so

$$FV = 299{,}599.22$$
$$PV = 10{,}000$$
$$n = 30$$
$$i = (299{,}599.22 / 10{,}000)\ 1/30 - 1 = 29.959922\ .0333 - 1$$
$$= .12$$

4.3.4 Effective Rate (Effective Yield)

The effective rate is the actual rate you earn on an investment or pay on a loan after the effects of compounding frequency are considered.
To convert a nominal rate to an equivalent effective rate:

$$Effective\ Rate\ =\ (1\ +\ \ (i\ /\ n))^{n}\ -\ 1$$

Where:

$$i\ =\ Nominal\ or\ stated\ interest\ rate$$
$$n\ =\ Number\ of\ compounding\ periods\ per\ year$$

Example: What effective rate will a stated annual 8% yield when compounded semi-annually?

Solution

$$n = 2, i = 0.08\ \ so\ , Effective\ Rate\ =\ (1\ +\ 0.08\ /\ 2)^{2} - 1$$
$$=\ 0.0816$$

4.3.5 Interest Rate in Calculations

• Time Value of Money calculations always use compound interest.
• You must adjust the interest rate and the number of periods to be consistent with compounding periods.
For Example, a 6% interest rate compounded semi-annually for five years should be entered as 3% (6 / 2) for 10 (5 * 2) periods.
• A calculator expects a 6% interest rate to be entered as the whole number 6, whereas formulas typically use a decimal value 0.06.

4.4 Number of Periods

The variable **n** in the Time Value of Money formulas represents the number of **periods**. It is intentionally not stated in years since each interval must correspond to a compounding period for a single amount or a payment period for an annuity.

The interest rate and the number of periods must be adjusted to reflect the number of compounding periods per year before using them in TVM formulas. For Example, if you borrow $1,000 for two years at 12% interest compounded **quarterly**, you must **divide** the interest rate by 4 to obtain the interest rate per period (i = 3%). You must **multiply** the number of years by 4 to obtain the total number of periods (n = 8).

You can determine the number of periods required for an initial investment to grow to a specified amount with this formula:

$$\boldsymbol{number\ of\ periods = natural\ log\ [(FV * i) / (PV * i)] / natural\ log\ (1 + i)}$$

Where:

PV = Present Value, the amount you invested.

FV = future value, the amount your investment will grow to

i = interest per period

Example: You put $10,000 into a savings account at a 9.05% interest rate compounded annually. How long will it take to double your investment?
LN [(20,000 * .0905) / (10,000 * .0905)] / *LN* (1 .0905) =
LN (2) / *LN* (1.0905) =.69314 / .08663 = 8 years

4.5 Annuity Payments

Payments in Time Value of Money formulas are a series of equal, evenly-spaced cash flows of an annuity, such as payments for a mortgage or monthly receipts from a retirement account.
Payments must:

- be the same amount each period
- occur at evenly spaced intervals
- occur exactly at the beginning or end of each period
- be all inflows or all outflows (payments or receipts)
- represent the payment during one compounding (or discount) period

To Calculate Payments When Future Value is Known: The Future Value is an amount you wish to have after several periods have passed. For Example, you may need to accumulate $20,000 in ten years to pay for college tuition. When you know the future value, interest rate, and the number of periods of an ordinary annuity, you can solve for the payment with this formula:

$$\boldsymbol{payment = FVoa\ /\ [((1 + i)^n - 1)\ /\ i]}$$

Where:

FVoa = Future Value of an ordinary annuity (payments are made at the end of each period)

i = interest per period

n = number of periods

Example: In 10 years, you will need $50,000 to pay for college tuition. Your savings account pays 5% interest compounded monthly. How much should you save each month to reach your goal?

FVoa = 50,000, the future savings goal

i = 0.004167 interest per month (0.05 / 12)

n = 120 periods (12 payments per year for ten years)

payment = 50,000 / [(1.004167^{120} - 1) / .004167] = 321.99

4.6 Present Value

4.6.1 Present Value of a Single Amount

Present value is today's amount equivalent to a future payment, or series of payments, discounted by an appropriate interest rate. Since money has time value, the present value of a promised future amount is less the longer you wait to receive it. The difference between the two depends on the number of compounding periods involved and the interest (discount) rate.

The relationship between the present value and future value can be expressed as:

$$PV = FV\,[\,1\,/\,(1 + i)^n\,]$$

Where:

PV = Present Value

FV = Future Value

i = Interest Rate Per Period

n = Number of Compounding Periods

Example: You want to buy a house five years from now for $150,000. Assuming a 6% interest rate compounded **annually**, how much should you invest today to yield $150,000 in 5 years?

FV=150,000
i=0.06
n = 5

PV = 150,000 [1 / (1 + .06)^5] = 150,000 (1 / 1.3382255776) = 112,088.73

End of Year	**1**	**2**	**3**	**4**	**5**
Principal	112,088.73	118,814.05	125,942.89	133,499.46	141,509.43
Interest	6,725.32	7,128.84	7556.57	8,009.97	8,490.57
Total	**118,814.05**	**125,942.89**	**133,499.46**	**141,509.43**	**150,000.00**

Example: You find another financial institution that offers an interest rate of 6% compounded **semi-annually**. How much less can you deposit today to yield $150,000 in five years?

Interest is compounded twice per year, so you must **divide the annual interest rate by two** to obtain a rate per period of 3%. Since there are two compounding periods per year, you must **multiply the years by two** to obtain the total number of periods.
FV = 150,000
i = .06 / 2 = .03
n = 5 * 2 = 10

PV = 150,000 [1 / (1 + .03)^{10}] = 150,000 (1 / 1.343916379) = 111,614.09

4.6.2 Present Value of Annuities

An **annuity** is a series of equal payments or receipts at evenly spaced intervals. Leases and rental payments are examples. The payments or receipts occur at the **end** of each period for an **ordinary annuity,** while they occur at the **beginning** of each period for an **annuity due**.

4.6.3 Present Value of an Ordinary Annuity

The **Present Value of an Ordinary Annuity** (PVoa) is the value of a stream of expected or promised future payments discounted to a single equivalent value today. It is extremely useful for comparing two separate cash flows that differ in some way.

PVoa can also be considered the amount you must invest today at a specific interest rate so that when you withdraw an equal amount each period, the original principal and all accumulated interest will be completely exhausted at the end of the annuity.

The Present Value of an Ordinary Annuity could be solved by calculating the present value of each payment in the series using the present value formula and then summing the results. A more direct formula is:

$PVoa = PMT\,[(1 - (1 / (1 + i)^n)) / i]$
Where:

PVoa = Present Value of an Ordinary Annuity
PMT = amount of each payment
i = Discount Rate Per Period
n = Number of Periods

Example: What amount must you invest today at 6% compounded **annually** so you can withdraw $5,000 each year's end for the next five years?
PMT = 5,000
i = 0.06
n = 5
$PVoa = 5{,}000\,[(1 - (1/(1 + .06)^5)) / .06] = 5{,}000\,(4.212364) = 21{,}061.82$

Year	1	2	3	4	5
Begin	21,061.82	17,325.53	13,365.06	9,166.96	4,716.98
Inter-est	1,263.71	1,039.53	801.90	550.02	283.02
With-draw	-5,000	-5,000	-5,000	-5,000	-5,000
End	**17,325.53**	**13,365.06**	**9,166.96**	**4,716.98**	**.00**

Example: In practical problems, you may need to calculate both the present value of an annuity (a stream of future periodic payments) and the present value of a single future amount:

For Example, a computer dealer offers to lease a system to you for \$50 per month for two years. You can buy the system for \$500 at the end of two years. You will pay at the **end** of each month. He will sell the same system to you for \$1,200 cash. If the interest rate is 12%, which is the better offer?

You can treat this as the sum of two separate calculations:

1. The present value of an ordinary annuity of 24 payments at \$50 per monthly period **Plus**
2. the present value of \$500 paid as a single amount in two years.

PMT = 50 per period

i = .12 /12 = 0.01

Interest per period (12% annual rate / 12 payments per year)

n = 24 number of periods

PVoa = 50 [(1 - (1/(1.01)^{24})) / .01] = 50 [(1- (1 / 1.26973)) /.01] = **1,062.17**

FV = 500 Future value (the lease buyout)

i = .01 interest per period

n = 24 Number of periods

PV = FV [1 / (1 + i)^{n}] = 500 (1 / 1.26973) = **393.78**

The lease's present value (cost) is \$1,455.95 (1,062.17 + 393.78). So if taxes are not considered, you would be $255.95 better off paying cash right now if you have it.

4.6.4 Present Value of an Annuity Due (PVad)

The **Present Value of an Annuity Due** is identical to an ordinary annuity except that each payment occurs at the beginning of a period rather than at the end. Since each payment occurs one period earlier, we can calculate the present value of an **ordinary annuity** and then multiply the result by (1 + i).

$$PVad \;=\; PVoa\,(1 + i)$$

Where:
PV-ad = Present Value of an Annuity Due
PV-oa = Present Value of an Ordinary Annuity
i = Discount Rate Per Period

Example: What amount must you invest today at a 6% interest rate compounded annually so you can withdraw $5,000 at the **beginning** of each year for the next five years?
PMT = 5,000
i = 0.06
n = 5
PVoa = 21,061.82 (1.06) = 22,325.53

Year	1	2	3	4	5
Begin	22,325.53	18,365.06	14,166.96	9,716.98	5,000.00
Inter-est	1,039.53	801.90	550.02	283.02	
With-draw	-5,000.00	-5,000.00	-5,000.00	-5,000.00	-5,000.00
End	**18,365.06**	**14,166.96**	**9,716.98**	**5,000.00**	**.00**

Combined Formula

You can also combine these formulas and the present value of a single amount formula into one.

4.7 Future Value

4.7.1 Future Value of a Single Amount

Future value is the amount of money an investment made today (the present value) will grow by some future date. Since money has time value, we naturally expect the future value to be greater than the present value. The difference between the two depends on the number of compounding periods involved and the going interest rate.
The relationship between the future value and present value can be expressed as:

$$FV = PV\,(1 + i)^n$$

Where:
FV = Future Value
PV = Present Value
i = Interest Rate Per Period
n = Number of Compounding Periods

Example: You can afford to put $10,000 in a savings account today that pays 6% interest compounded annually. If you make no withdrawals, how much will you have five years from now?
PV = 10,000
i = 0.06
n = 5

FV = 10,000 (1 + .06)^5 = 10,000 (1.3382255776) = 13,382.26

End of Year	**1**	**2**	**3**	**4**	**5**
Principal	10,000.00	10,600.00	11,236.00	11,910.16	12,624.77

Interest	600.00	636.00	674.16	714.61	757.49
Total	**10,600.00**	**11,236.00**	**11,910.16**	**12,624.77**	**13,382.26**

Example: Another financial institution offers to pay 6% compounded **semi-annually**. How much will your $10,000 grow in five years at this rate?

Interest is compounded twice per year, so you must **divide the annual interest rate by two** to obtain a rate per period of 3%. Since there are two compounding periods per year, you must multiply **the years by two** to obtain the total number of periods.

PV = 10,000

i = .06 / 2 = .03

n = 5 * 2 = 10

FV = 10,000 (1 + .03)^{10} = 10,000 (1.343916379) = 13,439.16

4.7.2 Future Value of Annuities

An **annuity** is a series of equal payments or receipts at evenly spaced intervals. Leases and rental payments are examples. The payments or receipts occur at the **end** of each period for an **ordinary annuity,** while they occur at the **beginning** of each period for an **annuity due**.

4.7.3 Future Value of an Ordinary Annuity

The **Future Value of an Ordinary Annuity** (FVoa) is the value that a stream of expected or promised future payments will grow to after a given number of periods at a specific compounded interest.

The Future Value of an Ordinary Annuity could be solved by calculating the future value of each payment in the series using the future value formula and then summing the results. A more direct formula is:

$$\boldsymbol{FVoa = PMT\,[((1 + i)^n - 1)\,/\,i]}$$

Where:

FVoa = Future Value of an Ordinary Annuity

PMT = amount of each payment
i = Interest Rate Per Period
n = Number of Periods

Example: What amount will accumulate if we deposit $5,000 at the **end** of each year for the next five years? Assume an interest of 6% compounded annually.
PV = 5,000
i = 0.06
n = 5
FVoa = 5,000 [(1.3382255776 - 1) /.06] = 5,000 (5.637092) = 28,185.46

Year	**1**	**2**	**3**	**4**	**5**
Begin	0	5,000.00	10,300.00	15,918.00	21,873.08
Inter-est	0	300.00	618.00	955.08	1,312.38
De-posit	5,000.00	5,000.00	5,000.00	5,000.00	5,000.00
End	**5,000.00**	**10,300.00**	**15,918.00**	**21,873.08**	**28,185.46**

Example: In practical problems, you may need to calculate both the future value of an annuity (a stream of future periodic payments) and the future value of a single amount that you have today:

For example, you are 40 years old and have accumulated $50,000 in your savings account. You can add $100 to your account at the end of each month, which pays an interest rate of 6% annually. Will you have enough money to retire in 20 years?

You can treat this as the sum of two separate calculations:

1. the future value of 240 monthly payments of $100 plus
2. the future value of the $50,000 now in your account.

PMT = $100 per period
i = .06 / 12 = 0 .005
Interest per period (6% annual rate / 12 payments per year)
n = 240 periods

FVoa = 100 [(3.3102 - 1) /0.005] = 46,204
PV = 50,000 Present value (the amount you have today)
i = 0.005 interest per period
n = 240 Number of periods
FV = PV $(1+i)^{240}$ = 50,000 $(1.005)^{240}$ = 165,510.22
After 20 years, you will have accumulated $211,714.22 (46,204.00 + 165,510.22).

4.7.4 Future Value of an Annuity Due (FVad)

The **Future Value of an Annuity Due** is identical to an ordinary annuity except that each payment occurs at the beginning of a period rather than at the end. Since each payment occurs one period earlier, we can calculate the present value of an **ordinary annuity** and then multiply the result by (1 + i).

$$FVad = FVoa\,(1 + i)$$

Where:
FVad = Future Value of an Annuity Due
FVoa = Future Value of an Ordinary Annuity
i = Interest Rate Per Period

Example: What amount will accumulate if we deposit $5,000 at the **beginning** of each year for the next five years? Assume an interest of 6% compounded annually.
PV = 5,000
i = .06
n = 5
FVoa = 28,185.46 (1.06) = 29,876.59

Year	**1**	**2**	**3**	**4**	**5**
Begin	0	5,300.00	10,918.00	16,873.08	23,185.46
Deposit	5,000.00	5,000.00	5,000.00	5,000.00	5,000.00
Interest	300.00	618.00	955.08	1,312.38	1,691.13
End	**5,300.00**	**10,918.00**	**16,873.08**	**23,185.46**	**29,876.59**

Combined Formula

You can also combine these formulas and the future value of a single amount formula into one.

4.8 Loan Amortization

4.8.1 Amortization

Amortization is a method for repaying a loan in equal installments. Part of each payment goes toward interest due for the period, and the remainder is used to reduce the principal (the loan balance). As the loan balance is gradually reduced, a progressively larger portion of each payment goes toward reducing the principal.

For example, the 15 and 30-year fixed-rate mortgages common in the US are fully amortized loans. To pay off a $100,000, 15-year, 7%, fixed-rate mortgage, a person must pay $898.83 monthly for 180 months (with a small adjustment at the end to account for rounding). $583.33 of the first payment goes toward interest, and $315.50 is used to reduce the principal. But by payment 179, only $10.40 is needed for interest, and $888.43 is used to reduce principal.

4.8.2 Amortization Schedule

An amortization schedule is a table with a row for each payment period of an amortized loan. Each row shows the payment amount needed to pay interest, the amount used to reduce the principal, and the loan balance remaining at the end of the period.

The first and last five months of an amortization schedule for a $100,000, 15-year, 7%, fixed-rate mortgage will look like this:

Amortization Schedule			
Month	**Principal**	**Interest**	**Balance**
1	-315.50	-583.33	99,684.51
2	-317.34	-581.49	99,367.17

Amortization Schedule			
Month	**Principal**	**Interest**	**Balance**
3	-319.19	-579.64	99,047.98
4	-321.05	-577.78	98,726.93
5	-322.92	-575.91	98,404.01
Rows 6-175 omitted			
176	-873.07	-25.76	3,543.48
177	-878.16	-20.67	2,665.32
178	-883.28	-15.55	1,782.04
179	-888.43	-10.40	893.61
180	-893.62	-5.21	-0.01

4.8.3 Negative Amortization

Negative amortization occurs when the payment is insufficient to cover the interest due for a period. This will cause the loan balance to increase after each payment - a situation that should certainly be avoided. This might occur, for instance, if the rate of an adjustable-rate loan increase but the payment does not.

4.9 Cash Flow Diagrams

A cash flow diagram is a picture of a financial problem that shows all cash inflows and outflows plotted along a horizontal timeline. It can help you to visualize a financial problem and to determine if it can be solved using TVM methods.

Engineering economics is the application of economic techniques to the evaluation of design and engineering alternatives.

The role of engineeringeconomics is to assess the appropriateness of a given project, estimate itsvalue, and justify the benefits to the client from an engineering standpoint.

Cash flow is the stream of monetary (dollar) values— costs (inputs) andbenefits (outputs)—resulting from a project investment.

It is often difficult to solve a problem if you cannot see it. The easiest way toapproach problems in economic analysis is to draw a picture.

The picture should show three things:

- A time interval divided into an appropriate number of equal periods
- All cash outflows (deposits, expenditures, etc.) in each period
- All cash inflows (withdrawals, income, etc.) for each period
- Unless otherwise indicated, all cash flows are considered to occur *at theend of their respective periods.* The figure below is a cash-flow diagram showing an outflow or disbursement of $1000 at the beginning of year one and an inflow or return of $2,000 at the end of year 5.

4.10 Time Value of Money

The following are reasons why $1000 today is "worth" more than $1000one year from today:

- Inflation
- Risk
- Cost of money

Of these, the cost of money is the most predictable and essential component of economic analysis. The cost of money is represented by the following:

- Money paid for the use of borrowed money, or
- Return on investment.

An interest rate determines the cost of money. *The time value of money* is defined as the time-dependent value stemming from changes in money's purchasing power (inflation or deflation) and the real earning potential of alternative investments over time.

4.11 Fixed costs (FC)

These include costs such as buildings, insurance, fixedoverhead, some minimum level of labor, equipment capital recovery, and information systems.

4.12 Variable costs (VC)

These include direct labor, materials,indirect costs, contractors, marketing, advertisement, and warranty.

The fixed-cost component is essentially constant for all variable values, so it does not vary for a large range of operating parameters, suchas production level or workforce size. Even if no units are produced, fixed costs are incurred at some threshold level.

In this situation, manufacturingcannot last long before the plant must shut down to reduce fixed costs.

Fixed costs are reduced through improved equipment, information systems,workforce utilization, less costly fringe benefit packages, subcontracting specific functions, etc. Variable costs change with production level, workforce size, and other parameters. It is usually possible to decrease variable costs through better product design, manufacturing efficiency, improved quality and safety, and higher sales volume.

Example: Handheld fiber-optic meters with white light polarization interferometry are useful for measuring temperature, pressure, and strain inelectrically noisy environments. The fixed costs associated with manufacturing are $800,000 per year. If a base unit sells for $2,950 and itsvariable cost is $2,075, the number of units that must be sold each year to breakeven is most nearly:

a. 885
b. 914
c. 953
d. 998

And, the profit for sales of 3000 units per year is most nearly: a. 1,250,000
b. 1,422,000
c. 1,825,000
d. 1,985,000

Quantity breakeven = 800,000 ÷ (2950 – 2075)
= 914 units per year

Profit = (2950 – 2075) (3000) – 800,000
= $1,825,000 per year

Example: In linear breakeven analysis, if a company expects tooper-ate at a point *above* the breakeven point, it should select the alter-native:

a. With the lower fixed cost
b. With the higher fixed cost
c. With the lower variable cost
d. With the higher variable cost,

Answer: c

Example: A die-cast manufacturer plans todevelop a new product that will sell for $2.00 per unit. The manufacturing parameters are found in the table below:

Description	***Allocation***
Cost of manufacturing equipment	$200,000
Net Salvage Value for the manufacturing equip-ment	$10,000
The economic life of the equipment	10-years
Number of units per year the equipment can man-ufacture	1,000,000-units
Direct Labor Cost	$0.30 per unit
Direct Material cost	$0.65 per unit
Variable administrative and selling expense	$0.20 per unit
Fixed overhead costs per year (not including	$150,000-per year

depreciation)	

Demand Units/ Year	*Probabilityof Sell*
500,000	0.1
600,000	0.2
700,000	0.4
800,000	0.2
900,000	0.1

If the demand follows the projected probability distribution shown in the table and straight-line depreciation is used, the expected average net profitper year, before taxes, is most nearly:

a. $445,000
b. $426,000
c. $226,000
d. $45,000

Solution: Analysis; since there is no provided information related to the time value ofmoney, thegiven parameters are as given and solve for the present worth costs.

Step 1- Determine the cost for the lifespan of the manufacturing equipment based on per year:

Equipment Cost = Total Equipment Cost – Salvage ValueEquipment Cost = $200,000 – 10,000
Equipment Cost = $190,000-per 10-years Equipment Cost = $19,000-per-year

Step 2 - Determine overhead costs based on a per-year basis

Overhead Cost per year = Overhead/year + Equipment cost per year
Overhead Cost per year = $150,000 + $19,000
Overhead Cost per year = $169,000

Step 3 - Determine the project production capacity per year using the probability distribution.

Demand Units/ Year	*Probability of Sell*	*Units Produced per year*	
500,000	0.1	50,000	
600,000	0.2	120,000	
700,000	0.4	280,000	
800,000	0.2	160,000	
900,000	0.1	90,000	
	Total 100%	**700,000**	**< 1,000,000 Therefore, checks**

Step 4 - Determine the direct/ indirect cost per unit

Direct Indirect Cost = Labor + Material + IndirectDirect/Indirect Cost = 0.30 + 0.65 +0.20 Direct/Indirect Cost = $1.15-per unit

Step 5 - Determine the project sales revenue per year

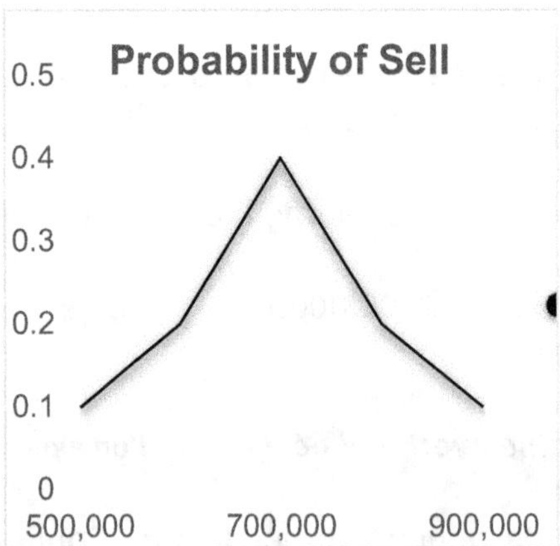

Revenue = Sell Price x Annual Production UnitsRevenue = $2.00 x 700,000
Revenue = $1,400,000

Step 6 - Determine the projected cost per year

Total Cost = Number of Units x Cost per unit + Fixed CostsTotal Cost = (700,000 x $1.15) + $169,000
Total Cost = $974,000

Step 7 - Determine the net profit per year

Net Profit = Revenue – Cost
Net Profit = $1,400,000 - $974,000Net Profit = $426,000

Answer is b

Example: An engineer deposits an "X" amount every six months for three years, so she'll have $10,000 at the end of this period. The interest rate is 5% annually, which is rebalanced every six months. The amount deposited is most nearly:

a. $1,565.50
b. $1,585.50
c. $1,595.50
d. $1,600.50

Solution: Note that the word "rebalance" changes the calculations for thequestion. ***"Rebalance every six months"*** requires adjusting the annual interest rate. The interest rate is divided by two since the 5%-per-yearrate is rebalanced every six months or 2.5% per 6-month period.
Determine the components and apply the equation.:
n = 6 deposits i = 2.5% per 6-month period
F = $10,000(**A/F**, 2.5%, 6) = 0.15655 A = $1,565.50 (answer is a)

Example: A credit card advertises a nominal rate of 18% compounded monthly. If no monthly payments are made, the effective annual interest rate is most nearly:

a. 18.00%
b. 18.25%

c. 18.75%
d. 19.56%

The actual rate is (18% ÷ 12) = 1.5% per month. The effective annual rateis:

$i = (1 + .015)^{12} - 1 = 0.1956$ or 19.56% if you do not pay anything monthly. (the answer is d)

Example: How long will the owner have to maintain a business with an income of $5,000 per year and expenses of $1,500 per year if the owner's initial investment was $28,000 and the MARR is 10% per year?

Solution: Determine the annual worth and establish a zero-balance equation to arrive at the Answer.

-28,000(A/P,10%, n) + 5000 – 1500 = 0
(A/P,10%, n) = 3500 / 28,000 = 0.125

Find 0.125 in the A/P column of the 10% Factor Tables and interpolate the amount
n = 16.9 years

Example: A city engineer decides which alternatives toselect for a public works project. If the bond rate is 7.5% per year, calculate the Benefit to Cost ratio of the alternative.

Alterna-tive	***Capital Cost***	***AnnualBen-efit***	***Salvage Value***	***Disposal Cost***	***Useful Life***
A	$480,000	$75,000	$25,500	$16,500	16 yrs
B	$400,000	$75,000	$0	$12,500	12 yrs

Alternative A:
PW of Benefits = $75,000 (P/A, 7.5%, 16)

= $75,000 (9.1490)
= $686,175

PW of Costs = $480,000 + $16,500 (P/F, 7.5%, 16) - $25,500 (P/F, 7.5%, 16)
=$477,162

B/C = $686,175 ÷ $477,162 = 1.438

Alternative B:
PW of Benefits = $75,000 (P/A, 7.5%, 12)
= $75,000 (7.7394)
= $580,455

PW of Costs = $400,000 + $12,500 (P/F, 7.5%, 12) – 0 (P/F, 7.5%, 12)
= $405,256.88

B/C = $580,455 ÷ $405,256 = 1.432
Since the B/C ratio of Alternative A is more than Alternative B, select Alternative A.

Example: Two different methods are under consideration for building a bypass road. Material J will cost $100,000 per mile and last for ten years. Its annual maintenance cost will be $10,000 per year per mile. Material K will cost $30,000 per mile and last for five years. Calculate the annual maintenance cost for material K at an interest rate of 6% per year, which will make the two methods cost the same.

Solution: Set up an equivalence equation between both methods and find thevariable x.

-100,000(A/P,6%,10) -10,000 = -30,000(A/P,6%,5) – x
-100,000 (0.13587) -10,000 = -30,000 (0.23740) – x
x = 16,465

Chapter 5

Electrical Materials

Electrical materials have electrical properties and are used to construct electrical components and devices. Some common electrical materials include:

Conductors: These materials allow the electric current to flow through them with minimal resistance. Examples include copper, aluminum, gold, and silver.

Insulators: These materials do not conduct electricity and are used to prevent current from flowing between conductors. Examples include rubber, plastic, glass, and ceramic.

Semiconductors: These materials have intermediate electrical conductivity between conductors and insulators. They are used in electronic devices such as transistors, diodes, and integrated circuits. Examples include silicon, germanium, and gallium arsenide.

Superconductors: These materials have zero electrical resistance when cooled to extremely low temperatures. They are used in applications with critical low energy loss, such as in power transmission lines and medical imaging devices. Examples include niobium-titanium and yttrium-barium-copper oxide.

Magnetic materials: These are materials that exhibit magnetic properties and are used in constructing magnetic components such as transformers, motors, and generators. Examples include iron, cobalt, nickel, and neodymium.

Dielectric materials: These have high electrical resistance and are used as insulators in capacitors, transformers, and other electrical components. Examples include mica, ceramic, and polypropylene.
Piezo-electric materials: These are materials that generate an electrical charge when subjected to mechanical stress. They are used in sensors, actuators, and other devices. Examples include quartz, lithium niobate, and lead zirconate titanate.

The selection of electrical materials depends on the specific application and requirements of the component or device. Factors such as electrical conductivity, thermal conductivity, mechanical strength, and cost are considered when choosing materials for a particular application.

5.1 Properties of Electrical Materials

Semiconductor materials can be categorized into two main classes: intrinsic and extrinsic. Intrinsic semiconductors **are the pure form of semiconductor materials,** while extrinsic semiconductor materials are formed by adding some impurity to the pure semiconductors.

5.1.1 n-type semiconductor

An extrinsic semiconductor doped with pentavalent impurities like P, As, Sb, Bi, etc., in whichnegatively charged electrons work as charge carriers, is called an n-type semiconductor. The pentavalent foreign atom becomes positively charged when it donates an extra electron.

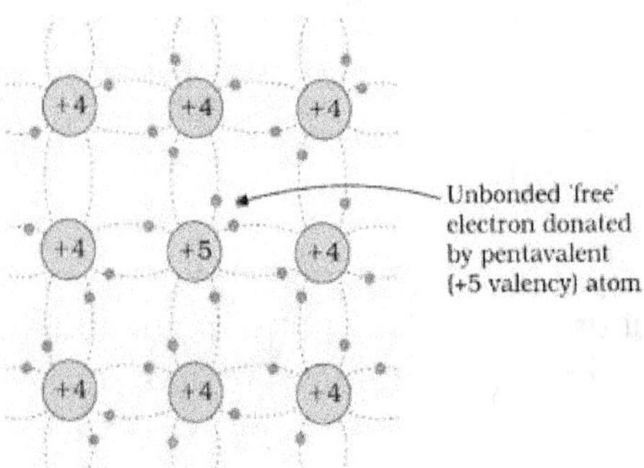

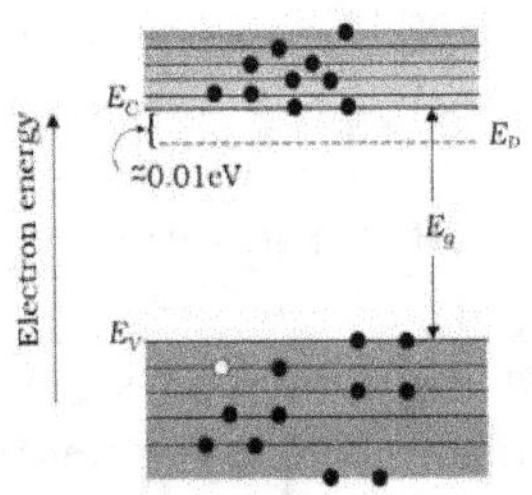

one thermally generated electron-hole pair + 9 electrons from donor atoms
Energy bands of n-type semiconductor at $T > 0K$

Table 5.1 Difference between intrinsic and extrinsic semiconductor

#	Intrinsic Semiconductors	Extrinsic Semiconductors
1	It is a pure semi-conducting material, and no impurity atoms are added.	It is prepared by doping several impurity atoms into the pure semi-conducting material.
2	The number of free electrons in the conduction band and the no. of holes in the valence band are equal and very small.	The number of free electrons and holes is never equal. There is an excess of electrons in n-type semiconductors and holes in p-type semiconductors.
3	Its electrical conductivity is low.	Its electrical conductivity is high.
4	Its electrical conductivity is a function of temperature alone.	Its electrical conductivity depends upon the temperature and the number of impurity atoms doped in the structure.
5	Examples: crystalline forms of pure silicon and germanium.	Examples: silicon "Si" and germanium "Ge" crystals with impurity atoms of As, Sb, P, etc., or In B, Aℓ, etc.
6	An intrinsic semiconductor will behave like an insulator at T = 0 K. At higher temperatures (T > 0K), some electrons are excited and move from the valence band to the conduction band.	At 0K, an **extrinsic semiconductor** behaves as an insulator, but as the **temperature** increases(0-300K), conductivity starts increasing because covalent bonds start breaking and majority/minority carriers increase, which means resistivity starts decreasing.
7	In intrinsic semiconductors, the number of free electrons, n_e is equal to the number of holes, n_h. $$n_e = n_h = n_i$$ Where, n_i is called intrinsic carrier concentration. Both electrons and holes move to conduct.	In extrinsic semiconductors, the number of free electrons, n_e is not equal to the number of holes, n_h. $$n_i^2 = n_e \times n_h$$ Where, n_i is called intrinsic carrier concentration.

N-type semiconductor, the total number of conduction electrons ne is due to the electrons contributed by donors and those generated intrinsically, while the total number of holes nh isonly due to the holes from the intrinsic source. But the rate of recombination of holes would increase due to the increase in the number of electrons. As a result, electrons are the majoritycarriers, and holes are the minority carriers.

5.1.2 p-type semiconductor

This is obtained when Si or Ge is doped with a trivalent impurity like Al, B, In, etc. The dopant has one valence electron less than Si or Ge. This atom can form covalent bonds with neighboring three Si atoms but has no electron to offer to the fourth Si atom. The bond between the fourth neighbor and the trivalent atom has a vacancy or hole. Since the neighboring Si atom in the lattice needs an electron in place of a hole, anelectron in the outer orbit of an atom in the neighborhood may jump to fill this vacancy, leaving a vacancy or hole at its site, and the moves increase the conductivity.

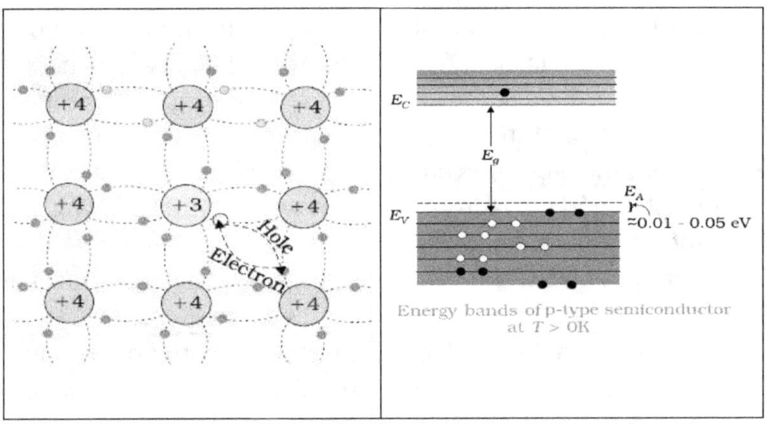

The trivalent foreign atom becomes negatively charged when it shares the fourth electron with the neighboring Si atom. Therefore, the dopant atom of p-type materialcan be treated as a core of one negative charge along with its associated hole.

These holes are in addition to the intrinsically generated holes, while the source of conduction electrons is only intrinsic generation. Thus, for such a material, the holesare the majority carriers, and electrons are the minority carriers

Table 5.2 The difference between a p-type and n-type semiconductor is in the tabulated form below.

Basis of difference	p-type semiconductor	n-type semiconductor
Group of Doping Element	A P-type semiconductor adds the III group element as a doping element.	An n-type semiconductor adds a V group element as a doping element.
Nature of Doping Element	The impurity added creates a vacancy of electrons (holes) called an acceptor atom.	The added impurity provides extra electrons and is known as the donor atom.
Type of impurity added.	Al, Ga, In, etc.	P, As, Sb, Bi, etc.
Majority Carriers	Holes	Electrons
Minority Carriers	Electrons	Holes
The density of Electrons and Holes	The hole density is much greater than the electron density. nh >> ne	The electron density is much greater than the hole density. ne >> nh
Energy level	The acceptor energy level is close to the valence band and away	The donor energy level is close to the conduction band and away from the valence band.

Basis of difference	p-type semiconductor	n-type semiconductor
	from the conduction band.	
Fermi level	Fermi level lies between the acceptor energy level and the valence band.	The Fermi level lies between the donor energy level and the conduction band.
Movement of Majority carriers	The majority of carriers move from higher to lower potential.	The majority of carriers move from a lower to a higher potential.

5.1.3 p-n junction

An arrangement consisting of a p-type semiconductor brought into close contact withan-type semiconductor is called a p -n junction.

Diffusion of charge and diffusion current

During the formation of the p-n junction, and due to the concentration gradient across p-and n-sides, holes diffuse from p-side to n-side (p→n), and electrons diffuse from n-side to p-side (n→p), and a diffusion current is set up across the junction.

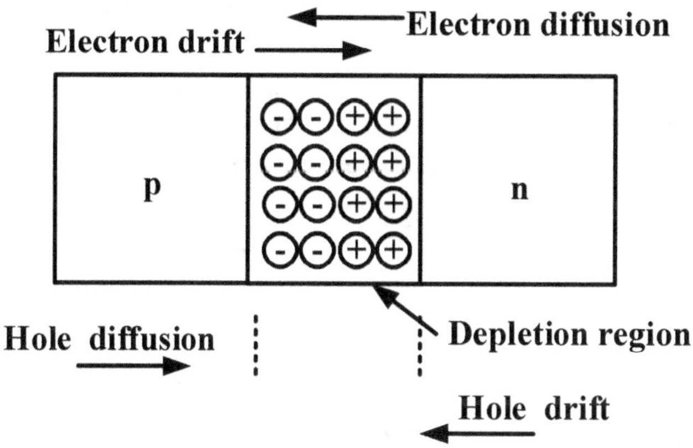

Depletion region

Due to the diffusion of holes p→n and electrons n→p, a –ve region is created on the p-side, anda +ve region is created on the n-side.

This space-charge region on either side of the junction is known as the depletionregion, and the net charge in this region is 0.

Due to the positive space-charge region on the n-side of the junction and the negative space-charge region on the p-side of the junction, an electric field directed from the positive chargetowards the negative charge develops.

Due to this field, an electron on the p-side of the junction moves to the n-side, and a hole on n-teside of the junction moves to the p-side.

Drift of charge carriers and drift current

The motion of charge carriers due to the electric field is called drift, and the current is calleddrift current.

Initially, the diffusion current is large, and the drift current is small. As the diffusion continues, the space-charge regions on either side of thejunction extend, thus increasing the electric field strength and drift current. This process continues until the diffusion current equals the drift current, forming the p-n junction.

In a p-n junction under equilibrium, there is no net current.

The loss of electrons from the n-region and the gain of electrons by the p-region; causesa difference in potential across the junction of the two regions. The polarity of this potential is to oppose the further flow of carriers so that acondition of equilibrium exists. Potential barrier across a p-n junction. The potential difference across the depletion layer is called the potential barrier (Vo).

5.2 Semiconductor diode

A semiconductor diode is a p-n junction with metallic contacts provided at the ends to apply an external voltage. It is a two-terminal device.

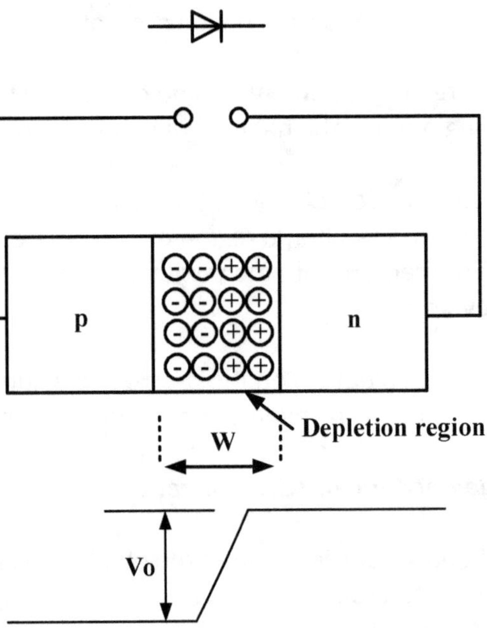

Diode under equilibrium (V=0)

Cathode

The electrode that receives electrons. Positive ions are discharged, and negative ions are formed. The cathode is protected from corrosion. The cathode is the negatively charged electrode. The cathode attracts cations or positive charges. The cathode is the source of electrons or an electron donor. It may accept a positive charge, because the cathode may generate electrons. Typically the electrical species doing the actual movement, it may be said that cathodes generate charge or that current moves from the cathode to the anode. This can be confusing because the direction of the current would be defined by how a positive charge would move. Just remember, any movement of charged particles is current.

Anode

The electrode where galvanic reaction(s) generate electrons. Negative ions are discharged, and positive ions are formed. Corrosion occurs at the anode. The anode is the positively charged electrode. The anode attracts electrons or anions.

Electrolyte

The conductor through which current is carried. Electrolytes include aqueous solutions or other liquids.

Return Current Path

The metallic pathway connects the anode to the cathode. It is often the underlying metal substrate.

Potential Difference

A voltage difference must exist between the anode and the cathode. All five elements (anode, cathode, electrolyte, and return current path and potentialdifference.) are necessary for corrosion to occur. The Galvanic Series of Metals (right) lists metals and alloys in decreasing order ofelectrical activity. Metals nearer the top of the table are less noble metals and tend to lose electrons more than those found lower on the list.

5.3 Oxidation and reduction

Reduction-Oxidation is a Redox reaction. It is an electrochemical process involving electron transfer to or from a molecule orion, changing its oxidation state. This reaction occurs with external voltage or through the release of chemical energy. Oxidation state change occurs in an electrochemical reaction's atoms, ions, or molecules. An atom or ion that gives up an electron to another atom or ion has its oxidation. The recipient of the negatively charged electron has its oxidation state decrease.

5.4 Corrosion

For corrosion to occur, there must be an anode and a cathode in electrical contact in thepresence of an electrolyte. When dissimilar metals are in contact, the more electropositive one becomes the anodein a corrosion cell. Different regions of carbon steel can also result in a corrosion reaction: e.g., cold-worked regions are anodic to noncold-worked. Different oxygen concentrations can cause oxygen-deficient regions to becomecathodic to oxygen-rich regions. Grain boundary regions are anodic to bulk grain. In multiphase alloys, various phases may not have the same galvanic potential.

5.5 Standard Oxidation Potentials for Corrosion Reactions

Zinc is a common “sacrificial anode” employed by attaching zinc strips to the topsof car fenders or coating a metal sheet in zinc (galvanizing the metal). It is lower on the list than steel by 0.323 volts. Note that uncoated aluminum parts that come into contact with uncoated steel parts will create a galvanic cell (when an electrolyte like water is present), causing the aluminum todecay. Al is 1.222 volts lower than Fe. Copper (Cu), a component of brass, is much higher on the table than iron (Fe). So, if you have copper pipes or brass fittings atop your cast iron water heater, the iron waterheater will corrode through at the top fittings.

Remedy this by installing a plastic-lined pipe (a dielectric fitting), which effectively creates distance through the electrolyte between the materialsthough the metal outsides of the fittings are still in contact. Cu is materials, though the metal outsides of the fittings are still in contact. Cu is 0.781 lower than Fe.

5.6 Types of Corrosion

Corrosion is a natural process that converts a refined metal to its more Stable Oxide. It is thegradual destruction of materials (usually metals) by chemical reactions with their environment. Corrosion cannot be eliminated, but it can be controlled. Corrosion is a complicated issue, depending on the combinations of materials and fluids, the fluid temperatures, the surrounding environment, and the galvanic currents in the constructions. Corrosion cannot be eliminated, but it can be controlled.

Table 5.3 Types of corrosion

Types of Corrosion		
1	Uniform	Electrochemical corrosion that occurs with equivalent intensity over entire surface
2	Galvanic	Occurs when two metals or alloys having different compositions are electrically coupled while exposed to an electrolyte
3	Crevice	Occurs when concentration differences of ions or dissolved gases exist in an electrolyte system with corrosion occurring preferentially at areas of low concentration
4	Pitting	A very localized form of corrosion similar to crevice corrosion, in which small holes or pits form
5	Intergranular	Occurs preferentially along grain boundaries of some alloys in certain environments
6	Selective Leaching	Occurs when one element is preferentially removed by a corrosion process from a solid solution alloy
7	Erosion-Corrosion	Occurs as a combined consequence of chemical attack and mechanical abrasion due to fluid motion
8	Stress-Corrosion	Occurs due to the combined influence of an applied tensile stress and a corrosive environment

5.6.1 Uniform corrosion

Uniform corrosion is characterized by a corrosive attack proceeding evenly over the entire surfacearea or a large fraction of the total area. General thinning takes place until failure. Based on tonnage wasted, this is the most important form of corrosion.

Uniform corrosion occurs in very aggressive environments, typically chemical production or use, pulp, paper industries, etc. It depends on the kind and concentration of acid or base and the solution temperature. Organic acids are generally less corrosive than mineral acids, suchas hydrochloric and sulfuric acids. As the molecular weight of organic acids increases, their corrosivity decreases, the whole surface of the steel is attacked, and the corrosion is expressed as corrosion rate in mm/year (usually less than 0.1 mm/year is acceptable for such cases).

5.6.2 Galvanic corrosion

Galvanic corrosion is an electrochemical process that erodes the anode. Corrosion damage occurs when two dissimilar materials are coupled in a corrosive electrolyte. It happens when two or more dissimilar metals are brought into electrical contact under water or metal ions dissolved in water.

Example:
When a galvanic couple forms, one of the metals in the couple becomes the cathode and corrodes slower than it would alone, and the other becomes the anode and corrodes faster than it would on its own. When in seawater, either metal in the couple may corrode.

5.6.3 Crevice corrosion

Crevice corrosion is localized in a stagnant solution in a small (micro) crevice. Local chemistry changes in crevices (shielded areas), such asthose formed under insulation material, gaskets, disbanded coatings, threads, fastener heads, surface deposits, washers, lap joints, and clamps, can cause crevice corrosion.

The factors that influence localized (Crevice and Pitting) corrosion:

- Chloride ion concentration (for example, in services (under gaskets), vapor spaces due toevaporation and condensation.)
- Increasing temperature increases susceptibility.
- Increasing acidity increases susceptibility.
- Stagnant conditions increase susceptibility.
- The presence of oxidizing species, such as ferric and cupric ions
- Crevice Corrosion Pitting Corrosion

5.6.4 Pitting Corrosion

Pitting corrosion is a localized form of corrosion by which pin holes or pits are created in the material. Pitting is seen to be more dangerous than uniform corrosion damage as it is more difficult to predict and, therefore, designed against. Corrosion products often cover the pits, which often makes detection extremely difficult. A narrow, small pit with minimal overall metal loss can fail a whole engineering system.

5.6.5 Intragranular corrosion

Intergranular corrosion (IGC), also known as an intergranular attack (IGA), is a form of corrosion where the boundaries of crystallites of the material are more susceptible to corrosion than their insides.

5.6.6 Selective Leaching

One material was removed from the alloy. Selective leaching is a corrosion type in some solid solution alloys when a component of the alloys is preferentially leached from the material in suitable conditions.

The most common example is the selective leaching of zinc from brass alloys containing more than15% zinc (dezincification) in the presence of oxygen and moisture.

5.6.7 Erosion Corrosion

The relative movement between a corrosive fluid and a metal surface causes erosion-corrosion. The mechanical aspect of the movement is important, and friction and wear phenomenacan be involved. This process leads to the formation of grooves, valleys, wavy surfaces, holes, etc., with a characteristic directional appearance

5.6.8 Stress Corrosion

Internal or external stresses. Stress corrosion cracking (SCC) is the growth of crack formation ina corrosive environment. It can lead to

unexpected sudden failure of normally ductile metal alloys subjected to tensile stress, especially at elevated temperatures. SCC is highly chemicallyspecific in that certain alloys are likely to undergo SCC only when exposed to several chemical environments.
SCC results from a combination of three factors – a susceptible material, exposure to a corrosive environment, and tensile stresses above a threshold. If any one of these factors is eliminated, SCC initiation becomes impossible.

5.6.9 Hydrogen Embrittlement Corrosion

Hydrogen penetrates steel to form methane. Hydrogen embrittlement (HE), also known as hydrogen-assisted cracking (HAC) and hydrogen-induced cracking (HIC), describes the embrittling of metal after being exposed to hydrogen.

Metals can exchange electrons with each other. For plastic/polymer/wood, there is no exchangeof electrons. Hence, no corrosion occurs. The exchange of electrons occurs through oxidation & reduction reactions, where oxidation reaction (loss of electron) occurs at anodes while reduction reaction (gain of the electron) occurs at cathodes.

5.7 Electrical Passive Elements

5.7.1 Capacitance

A capacitor is a device that stores electric charge. A capacitor is constructed as two conducting surfaces separated by an insulator. Capacitance is the charge-carrying capacity of an insulating material.
A parallel plate capacitor is a simple capacitor constructed as two parallel plates. If the plates are connected across voltage potential, charges of opposite polarity will build up on the plates.

Charge held by a capacitor:

$$q = CV$$

where
q = charge

C = capacitance
V = voltage

The capacitance of a parallel plate capacitor:

$$C = \frac{\mathbf{EA}}{d}$$

where
E = permittivity of the material
A = cross-sectional area of the plates
d = distance between the plates

Standard Units of Capacitance

Microfarad (μF)
1μF = 1/1,000,000 = 0.000001 = 10^{-6} F

Nanofarad (nF)
1nF = 1/1,000,000,000 = 0.000000001 = 10^{-9} F

Picofarad (pF)
1pF = 1/1,000,000,000,000 = 0.000000000001 = 10^{-12} F

Permittivity

Permittivity, E, is also expressed as the product of the dielectric constant k and the permittivity of vacuum (free space), E_{o}.:

$$\epsilon_0 = 8.854 \, x \, 10^{-12} \, F/m$$

where, F = Farad, m = meter

The capacitance can be written as,

$$C = \epsilon \frac{A}{d}$$

Example: A 0.15 cm thick insulator separates two square parallel plates with 0.04 m x 0.04 m withadielectric constant of 4.3. What is the nearest capacitance?

(A) 4.05 x10
(B) 6.05 x10
(C) 4.05 x10
(D) 6.05 x10

Ans: A

Example: A capacitor is constructed from two conductive metal plates 40cm x 40cm, spaced 8mm apart, and uses dry air as its only dielectric material. Calculate the capacitance of the capacitor.

(A) 0.180 n
(B) 0.210nF
(C) 0.03nF
(D) 0.8 nF

Solution:

$$C = \epsilon \frac{A}{d}$$

where the permittivity of vacuum (free space) is $8.854\ x\ 10^{-12}\ F/m$ Conductive plate area= $40\ cm\ \times\ 40\ cm$= $(0.04\ m)^2$ Distance between the plates = 8mm= $8x10^{-3}$ m

Then the value of the capacitor consisting of two plates separated by air is calculated as 0.177nFor 177 pF

(A) is the correct answer

5.7.2 Resistivity

Resistivity is the material property that determines the resistance of a resistor. Resistance is the property of a circuit or element to oppose the current flow. A circuit with zero resistance is a short circuit, whereas an open circuit has infinite resistance.

Materials such as copper and aluminum are known for their low levels of resistivity, thus allowingelectrical current to easily flow through them, making these materials ideal for making electricalwires and

cables. Silver and gold have much low resistivity values but, for obvious reasons, are more expensive to turn into electrical wires.
The resistivity of a material within a resistor:

$$R = \rho \frac{L}{A}$$

where p = resistivity of the material
R = resistance of the resistor, in ohms
A = cross-sectional area of the resistor, in square meters
L = length of the resistor, in meters

5.7.3 Conductivity

Conductivity (σ) is the reciprocal of resistivity. That is 1/ρ and is measured in Siemens permeter (S/m). 1 siemens (S) equals $1\Omega^{-1}$ electrical conductivity σ = 1/ρ,
R can be written as:

$$R = \frac{L}{\sigma A}$$

Example: Calculate the total D.C. resistance of a 100-meter roll of 2.5mm^2 copper wire if the resistivity ofcopper at 20°C is 1.72×10^{-8} Ω m.

(A) 0.688 Ω
(B) 0.990 Ω
(C) 0.590 Ω
(D) 0.790 Ω

Solution: resistivity of copper at 20°C is 1.72×10^{-8}coil length L = 100m
The cross-sectional area of the conductor is 2.5 mm^{2}A = 2.5×10^{-6} m^2

$$R = \rho \frac{L}{A} = 0.688\ \Omega$$

(A) is the correct answer

Example: A 15 m cable length has a cross-sectional area of 1mm^2 and a resistance of 5 ohms. Calculatethe conductivity of the cable.

(A) 3.5 X10^6 s/m

(B) 3X10^4 s/m

(C) 1.3 X10 s/m

(D) 3X10^6 s/m

Solution:

Given, R = 5 ohms cable length, L = 15 m

cross-sectional area of the conductor is 1mm^{2}A = 1 x 10^{-6} m^2

$$R = \rho \frac{L}{A}$$

That is three mega-siemens per meter in length.

(A) is the correct answer

Example: A long hollow tube with I.D. 2 cm and O.D. 4 cm, K is 0.58 w/m.K. Inner face maintained at

70 ^{0} C and outer at the surface at a temperature of 100^0 C. Find the value of the heat flow rate permeter length

(A) -160 w/m

(B) -170 w/m

(C) -140 w/m

(D) 170 w/m

(A) is the correct answer

Multiple Choice Questions

Q.1. A semiconductor is formed by .... bonds.

A. covalent

B. electrovalent

C. co-ordinate

D. none of the above

Answer: A

Q.2. A semiconductor has...... temperature coefficient of resistance.

A. positive

B. zero
C. negative
D. none of the above
Answer: C

Q.3. The most commonly used semiconductor is ....
A. germanium
B. silicon
C. carbon
D. sulfur
Answer: B

Q.4. A semiconductor has .... valence electrons.
A. 6
B. 4
C. 3
D. 2
Answer: B

Q.5. The resistivity of pure germanium under standard conditions is about ....
A. 6×10^4 Ω cm
B. 60 Ω cm
C. 3×10^6 Ω cm
D. 6×10^{-4} Ω cm
Answer: B

Q.6. The resistivity of pure silicon is about ....
A. 100 Ω cm
B. 6000 Ω cm
C. 3×10^5 Ω cm
D. 1.6×10^{-8} Ω cm
Answer: B

Q.7. When a pure semiconductor is heated, its resistance ....
A. goes up
B. goes down
C. remains the same
D. cannot say

Answer: B

Q.8. The strength of a semiconductor crystal comes from ....

A. forces between nuclei

B. forces between protons

C. electron-pair bonds

D. none of the above

Answer: C

Q.9. When a pentavalent impurity is added to a pure semiconductor, it becomes ....

A. an insulator

B. an intrinsic semiconductor

C. *p*-type semiconductor

D. *n*-type semiconductor

Answer: *D*

Q.10. Addition of pentavalent impurity to a semiconductor creates many ....

A. free electrons

B. holes

C. valence electrons

D. bound electrons

Answer: A

Q.11. A pentavalent impurity has .... valence electrons.

A. 3

B. 5

C. 4

D. 6

Answer: B

Q.12. An *n*-type semiconductor is ....

A. positively charged

B. negatively charged

C. electrically neutral

D. none of the above

Answer: C

Q.13. A trivalent impurity has .... valence electrons.

A. 4

B. 5

C. 6

D. 3

Answer: D

Q.14. Addition of trivalent impurity to a semiconductor creates many ....

A. holes

B. free electrons

C. valence electrons

D. bound electrons

Answer: A

Q.15. A hole in a semiconductor is defined as ....

A. a free electron

B. the incomplete part of an electron pair bond

C. a free proton

D. a free neutron

Answer: B

Q.16. The impurity level in an extrinsic semiconductor is about .... of pure semiconductor.

A. Ten atoms for 10^8 atoms

B. One atom for 10^8 atoms

C. One atom for 10^4 atoms

D. One atom for 100 atoms

Answer: B

Q.17. As the doping to a pure semiconductor increases, the bulk resistance of the semiconductor ....

A. remains the same

B. increases

C. decreases

D. none of the above

Answer: C

Q.18. A hole and electron close would tend to ....

A. repel each other
B. attract each other
C. not affect each other
D. none of the above

Answer: B

Q.19. In a semiconductor, current conduction is due ....
A. only to holes
B. only to free electrons
C. to holes and free electrons
D. none of the above

Answer: C

Q.20. The random motion of holes and free electrons due to thermal agitation is called ....
A. diffusion
B. pressure
C. ionization
D. none of the above

Answer: A

Q.21. A forward-biased *pn* junction has a resistance of the ....
A. order of Ω
B. order of kΩ
C. order of MΩ
D. none of the above

Answer: A

Q.22. The battery connections required to forward bias a *pn* junction are ....
A. +ve terminal to *p* and −ve terminal to *n*
B. −ve terminal to *p* and +ve terminal to *n*
C. −ve terminal to *p* and −ve terminal to *n*
D. none of the above

Answer: A

Q.23. The barrier voltage at a *pn* junction for germanium is about ....
A. 3.5 V

B. 3V
C. zero
D. 0.3 V

Answer: D

Q.24. In the depletion region of a *pn* junction, there is a shortage of ....

A. acceptor ions
B. holes and electrons
C. donor ions
D. none of the above

Answer: B

Q.25. A reverse biased *pn* junction has ....

A. a very narrow depletion layer
B. almost no current
C. very low resistance
D. large current flow

Answer: B

Q.26. A *pn* junction acts as a ....

A. controlled switch
B. bidirectional switch
C. unidirectional switch
D. none of the above

Answer: C

Q.27. A reverse biased *pn* junction has a resistance of the........

A. order of Ω
B. order of kΩ
C. order of MΩ
D. none of the above

Answer: C

Q.28. The leakage current across a *pn* junction is due to ....

A. minority carriers
B. majority carriers
C. junction capacitance
D. none of the above

Answer: A

Q.29. When the temperature of an extrinsic semiconductor is increased, the pronounced effect is on ....

A. junction capacitance
B. minority carriers
C. majority carriers
D. none of the above

Answer: B

Q.30. With a forward bias to a *pn* junction, the width of the depletion layer ....

A. decreases
B. increases
C. remains the same
D. none of the above

Answer: A

Q.31. The leakage current in a *pn* junction is of the order of ....

A. A
B. mA
C. kA
D. μA

Answer: D

Q.32. In an intrinsic semiconductor, the number of free electrons ....

A. equals the number of holes
B. is greater than the number of holes
C. is less than the number of holes
D. none of the above

Answer: A

Q.33. At room temperature, an intrinsic semiconductor has ....

A. many holes only
B. a few free electrons and holes
C. many free electrons only
D. no holes or free electrons

Answer: B

Q.34. At absolute temperature, an intrinsic semiconductor has........
A. a few free electrons
B. many holes
C. many free electrons
D. no holes or free electrons
Answer: D

Q.35. At room temperature, an intrinsic silicon crystal acts approximately as ....
A. a battery
B. a conductor
C. an insulator
D. a piece of copper wire
Answer: C

Q.36. Electrical conductivity of insulators is the range ___________.
(a) 10-10(Ω-mm)-1
(b) 10-10(Ω-cm)-1
(c) 10-10(Ω-m)-1
(d) 10-8(Ω-m)-1

Further Questions

1. Units for electric field strength
(a) A/cm^2
(b) mho/meter
(c) cm2/V.s
(d) V/cm

2. Energy band gap size for semiconductors is in the range _______eV.
(a)1-2
(b) 2-3
(c) 3-4
(d) > 4

3. Energy band gap size for insulators is in the range _______ eV.

(a) 1-2
(b) 2-3
(c) 3-4
(d) > 4

4. Flow of electrons is affected by the following: (a) Thermal vibrations, (b) Impurity atoms, (c) Crystal defects, (d) all

5. Not a superconductive metallic element: (a) Fe (b) Al (c) Ti (d) W

6. Fermi energy level for intrinsic semiconductors lies (a) At the middle of the band gap, (b) Close to the conduction band, (c) Close to the valence band, (d) None

7. Fermi energy level for p-type extrinsic semiconductors lies (a) At the middle of the band gap, (b) Close to the conduction band, (c) Close to the valence band, (d) None

8. Fermi energy level for n-type extrinsic semiconductors lies (a) At the middle of the band gap, (b) Close to the conduction band, (c) Close to the valence band, (d) None

9. Not an example for intrinsic semiconductor: (a) Si (b) Al (c) Ge (d) Sn

10. In intrinsic semiconductors, many electrons __________ many holes. (a) Equal (b) Greater than (c) Less than (d) Can not define

11. In n-type semiconductors, many holes __________ number of electrons. (a) Equal (b) Greater than (c) Less than (d) Can not define

12. In p-type semiconductors, the number of holes __________ number of electrons. (a) Equal (b) Greater than (c) Less than (d) Twice

13. Mobility of holes is __________ mobility of electrons in intrinsic semiconductors. (a) Equal (b) Greater than (c) Less than (d) Can not define

14. Fermi level for extrinsic semiconductors depends on: (a) Donor element, (b) Impurity concentration, (c) Temperature, (d) All

15. Value of dielectric constant for a material __________. (a) Equal to 1 (b) Greater than 1 (c) Less than 1 (d) Zero.

16. High dielectric constant material is a must for __________. (a) Insulation of wires (b) Generators (c) Switch bases (d) Generators.

17. Dielectric constant for most polymers lies in the range of ______. (a) 1-3 (b) 2-5 (c) 4-7 (d) 6-10.

18. Example for piezo-electric material: (a) Rochelle salt (b) Lead zirconate (c) Potassium niobate (d) Barium Titanium oxide

19. Example for ferroelectric material: (a) Potassium niobate (b) Lead titanate (c) Lead zirconate (d) quartz

Thermal properties

1. Gas thermometers are more sensitive than liquid thermometers as the gases
   a) have a low specific heat
   b) have a high specific heat
   c) have a large coefficient of expansion
   d) are lighter

Answer: (c) have a large coefficient of expansion

2. The earth intercepts approximately one billionth of the power radiated by the sun. if the surface temperature of the sun were to drop by a factor of 2, the average radiant energy incident on earth per second would reduce by a factor of
   a) 2

b) 4
c) 8
d) 16

Answer: (d) 16

3. Thermal equilibrium implies equality of:
   a) energy
   b) internal energy
   c) K.E.
   d) temperature

Answer: (d) temperature

4. A surface at temperature T0°K receives power P by radiation from a small sphere at temperature T >> T0 and at a distance d. if both T and d are doubled, the power received by the surface will become approximately
   a) P
   b) 2P
   c) 4P
   d) 16P

Answer: (c) 4P

5. Two blocks of ice, when pressed together, join to form one block. This happens because
   a) melting point rises with pressure
   b) melting point falls with pressure
   c) heat is rejected outside
   d) heat is absorbed from outside

Answer: (b) melting point falls with pressure

6. A black body at a hot temperature of 227°C radiates heat at a rate of 5 cal/cm^2s. at a temperature of 727°C, the rate of heat radiated per unit area will be
   a) 50 cal/cm^2s
   b) 250 cal/cm^2s
   c) 80 cal/cm^2s
   d) 100 cal/cm^2s

Answer: (c) 80 cal/cm^{2}s

7. Two bars of copper having the same length but an equal diameter are heated to the same temperature. The change in length will be:
    a) more in thicker bar
    b) more in thinner bar
    c) same for both
    d) determined by the ratio of the length and diameter of the bars

Answer: (c) same for both

8. The top of the lake is frozen as the atmospheric temperature is -10°C. The temperature at the bottom of the lake is most likely to be
    a) 0°C
    b) -4°C
    c) 4°C
    d) -10°C

Answer: (c) 4°C

9. A bimetallic strip is made of aluminum and steel ($\alpha_{Al} > \alpha_{steel}$). On heating, the strip will
    a) remain straight
    b) get twisted
    c) will bend with aluminum on the concave side
    d) will bend with steel on the concave side

Answer: The correct answer is d) will bend with steel on the concave side

10. The earth receives at its surface radiation from the sun at 1400 W/m^2. The distance of the sun's center from the earth's surface is 1.5 1011 m, and the sun's radius is 7.0 108 m. treating the sun as a black body, it follows from the above data that its surface temperature is,
    a) 5801 K
    b) 106 K

c) 50.1 K
d) 2801 C

Answer: (a) 5801 K

11. As the temperature is increased, the period of a pendulum
    a) increases as its effective length increase even though its center of mass remains at the center of the bob
    b) decreases as its effective length increase even though its center of mass remains at the center of the bob
    c) increases as its effective length increases due to shifting of the center of mass below the center of the bob
    d) decreases as its effective length remain the same, but the center of mass shifts above the center of the bob

Answer: The correct answer is a) increase as its effective length increases even though its center of mass remains at the center of the bob

12. Heat is associated with
    a) the kinetic energy of random motion of molecules
    b) the kinetic energy of the orderly motion of molecules
    c) the total kinetic energy of random and orderly motion of molecules
    d) the kinetic energy of random motion in some cases and the kinetic energy of orderly motion in other

Answer: The correct answer is a) kinetic energy of random motion of molecules

13. Emissivity of the perfect black body is
    a) 1
    b) 2
    c) 5
    d) 0

Answer: (a) 1

14. The running of the fan makes us comfortable during summer because of it:
  a) increases the conductivity of the air
  b) reduces the thermal radiation
  c) cools the air
  d) enhances the rate of evaporation of perspiration

Answer: (d) enhances the rate of evaporation of perspiration

15. The temperature determines the direction of net change of:
  a) gross potential energy
  b) intermolecular potential energy
  c) gross kinetic energy
  d) intermolecular kinetic energy

Answer: (d) intermolecular kinetic energy

16. The good absorber of heat is
  a) Non-emitter
  b) Poor-emitter
  c) Good-emitter
  d) Highly polished

Answer: (c) Good-emitter

17. Unit of Stefan constant is given by
  a) $W/ m K^2$
  b) $W/ m^2 K^2$
  c) $W^2/ m^2 K4$
  d) W/ mK

Answer: (b) $W/ m^2 K^2$

18. A metallic ball has a spherical cavity at its center. If the ball is heated, what happens to the cavity?
  a) Its volume decreases
  b) Its volume increases
  c) Its volume remains unchanged
  d) Its volume may increase or decrease depending on the nature of the metal

Answer: (b) Its volume increases

19. The temperature of a piece of metal is raised from 27°C to 51.2°C. The rate at which metal radiates energy increases nearly
    a) 1.36 times
    b) 2.36 times
    c) 3.36 times
    d) 4.36 times

Answer: (a) 1.36 times

20. Co-efficient of reflection, coefficient of absorption, and coefficient transmission are related as
    a) $a + r + 1 = 1$
    b) $a + r + t^1 = 1$
    c) $a + r = -t$
    d) $a^1 r + 1$

Answer: (a) $a + r + 1 = 1$

Chapter 6

Electrical Circuits

An electrical circuit is a path that allows electricity to flow through a system. It consists of components connected in a closed loop, allowing current to flow continuously. Electrical circuits are used in various applications, including power generation, transmission, and distribution, and in electronic devices such as computers, smartphones, and televisions.

There are two types of electrical circuits: series and parallel. In a series circuit, the components are connected end-to-end, forming a single path for the current to flow. The circuit's total resistance is the sum of the individual resistances of each component. The current is the same throughout the circuit, while the voltage is divided among the components based on their resistance.

In a parallel circuit, the components are connected side-by-side, forming multiple paths for the current to flow. The voltage is the same across all components, while the current is divided among the components based on their resistance. The circuit's total resistance is less than the resistance of the individual components, making parallel circuits more efficient than series circuits.

6.1 DC Circuits

Table 6.1 Quantity of DC circuits.

Quantity	Variable	Unit and Unit Symbol
Voltage (Potential Difference)	V, Joule per Coulomb	Volt (V)
Current	I, Coulombs per second	Amperes/Amps (A)
Resistance	R	Ohms (Ω)
Power	P	Watts (W)

Calculation of the charge passing through a point in a circuit Q = It
Current (I in amps) is the rate of flow of electrical charge around a circuit.

The greater the flow of charge each time, the greater the current. The rate of flow of electric charge is measured in coulombs/second. You can calculate the charge passing a point in a circuit in a given time from the formula

$$Q = It$$

where Q = electric charge in coulombs (C) - the unit of electric charge
I = current flow in amperes (A), and t = time (s)
rearrangements from $Q = It$, $I = Q/t$ and $t = Q/I$

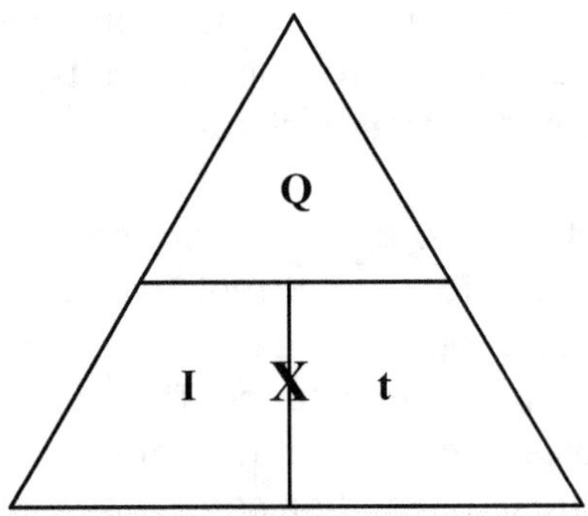

Figure 6.1 Current triangle.

A current flow of 1 A equals a rate of flow of charge of 1 C/s.

Example: If a current of 3.0 A passes through an appliance for 1 hour and 30 minutes, how much electrical charge is transferred in the process?

$$Q = It,\ Q = 3.0 \times 1.5 \times 60 \times 60 = 16\,200\ C = 1.62 \times 10^4\ C$$

Example: What is the current flow if 9000 C of charge passes a point in an electrical circuit in 12.0 minutes?

I = Q/t = 9000/(12 x 60) = 9000/720 = 12.5 A

Example: How long will it take, in minutes and seconds, for an electrical circuit current of 20.0 A to transfer 5000 C of charge?

t = Q/I = 5000/20 = 250 seconds = 4 mins and 10 seconds

Example: A laptop computer battery charger passes a current of 1.20 A for 30 minutes with an output p.d. of 15.0 V.

(a) Calculate how much charge is transferred to the computer battery.

(b) What is the resistance of the battery charger?

(c) When the laptop battery is fully charged, it stores 3000 C.

How long will it take to fully charge a flat battery?

a) Q = It = 1.2 x 30 x 60 = 2160 C
b) V = IR, R = V/I = 15 / 1.2 = 12.5 Ω
c) Q = It, t = Q / I = 3000 / 1.2 = 2500 s (41 min 40 secs)

Potential difference and energy transfer

Introduction to Electrical Energy Transfer

Energy transfer per unit of charge = potential difference (p.d.) and calculations based on $E = QV$

Potential difference (p.d. in volts, V) is the energy transferred per unit charge as the electric charge moves from one point to another in an electric circuit. It is measured with a voltmeter, always wired parallel across a circuit component. An electrical current transfers energy. Just think of all the electrical appliances you use - all need supplying energy to work! A power supply does work on a charge and transfers energy to it. Work must be done on the charge to increase its potential energy.

Electric charge is measured in coulombs (C). It transfers energy to components as they pass through by working against the component's resistance. If work is done, then energy is transferred. If an electrical charge experiences a potential difference, that charge will flow, transferring energy. The energy is supplied from the energy store of the power supply - battery, mains electricity, etc.

When a charge passes through any p.d. fall, it releases energy (from a higher to a lower potential energy level). E.g., in thin wire resistance, heat is released. The potential difference between the two points equals the work done per unit charge.

Potential difference (V) = work done (energy transferred in J) ÷ charge (C) i.e. 1 Volt corresponds to 1 Joule/coulomb or V = J/C

The bigger the fall in p.d., the greater the energy transferred because the charge starts with greater potential energy. Therefore, a power supply with a bigger source p.d. (V) can supply more energy to the circuit per unit of electric charge (the coulomb, C). The bigger the p.d., the more energy the same quantity of electric charge can carry.

3b. Another equation to calculate electrical energy transfers

The quantity of energy carried can be calculated from the equation:

$$Energy\ transferred\ =\ charge\ x\ potential\ difference.$$

$$E\ =\ QV\ \ (so\ Q\ =\ E/V\ and\ V\ =\ E/Q)$$

E = energy transferred in joules (J)
Q = quantity of electric charge in coulombs (C)
V = potential difference (V)

Noting that: V = E/Q = energy transferred per unit charge (J/C)

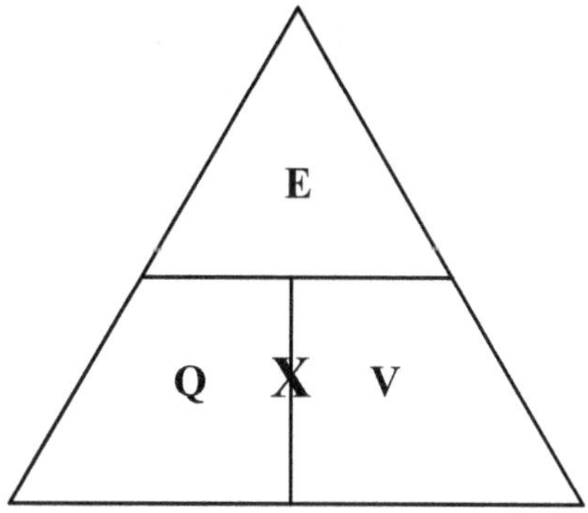

Figure 6.2 Voltage triangle.

Just in passing and some reminders:

The more energy transferred in a given time, the greater the power of a device or electrical appliance.
The p.d. V tells you how much energy each unit of electrical charge transfers,

so V = E/Q (units J/C), see E = QV calculations below).

The current (I) tells you how much charge passes a given point in a circuit per unit of time (coulombs/second, C/s).

This means both p.d. V and current I affect the rate at which energy is transferred to an appliance from the electrical energy store to other energy stores.

And some mathematical connections based on section 2. Q = It, and here in section 3, E = QV

From

Q = It and E = QV, substituting gives E = ITV,

so

i. E = IVt (I in A, t in s, V in volts)

Rearranging E = IVt gives IV = E/t

This connects with the equations for power

(ii) Power = energy transferred / time taken = E/t (J/s), and

(iii) Power = current x voltage = P (W) = I (A) x V (V), P = IV

From (ii) and (iii)

$$E/t = IV, \text{ so } E = IVt, \text{ which is equation (i)}$$

Example: An electric motor of a model car is powered by a 1.5 V battery. If 120 C of charge passes through the motor circuit in the moving car,

(a) how much energy is transferred?
(b) Describe the likely energy store changes when the car runs.

a)

$$E = QV = 120 \times 1.5 = 180 \text{ J}$$

b) The chemical potential energy stored in the battery decreases and becomes electrical energy. The kinetic energy stored in the car increases with some wasted heat from friction and sound energy transferred to the thermal energy stored in the surroundings.

Example: What quantity of charge is needed to transfer 500 J of energy if the p.d. of a circuit is 24.0 V?

$$E = QV, Q = E/V = 500/24 = 20.8 \text{ C}$$

Example: What potential difference is required in a circuit to transfer 2000 J energy with a charge of 50 coulombs?

$$E = QV, V = E/Q = 2000/50 = 40 \text{ V}$$

Example: A 12.0 V battery passes a current of 2.0 A through a lamp for 5 minutes.

(a) Calculate how much charge passed through the lamp.
(b) Calculate how much the lamp transferred electrical energy.

a) $Q = It = 2 \times 5 \times 60 = 600C$
b) $E = QV = 600 \times 12 = 7200$ J or $E = IVt = 2 \times 12 \times 5 \times 60 = 7200$ J

Example: An appliance has a power of 1.5 kW and works with a 230 V mains supply. How much charge has flowed through the circuit if the appliance is used for 15 minutes?

$$1.5 \text{ kW} \equiv 1500 \text{ W} \equiv 1500 \text{ J/s}$$

Total energy transferred = power x time

$$= 1500 \times 15 \times 60 = 1\,350\,000 \text{ J}$$

$$E = QV,$$

So, $$Q = E / V$$

$$= 1\,350\,000 / 230 = 5870 \text{ C (3 sf)}$$

Another route can calculate the answer

$P = IV$, $I = P / V = 1500 / 230 = 6.522$ A

$Q = It = 6.522 \times 15 \times 60 = 5870$ C (3 sf)

Kirchhoff's Current Law states that the current entering a point in a circuit equals the summation of the currents exiting.

Kirchhoff's Voltage Law states that the summation of all voltage drops in a closed loop must equal zero, resulting from the conservative electrostatic field.

(Conventional) current flowing through the cell has a positive voltage (gains energy). Current going through a resistor has a negative voltage (loses energy)

Series & Parallel Circuits

Below is a configuration of a circuit that has three resistors connected in series:

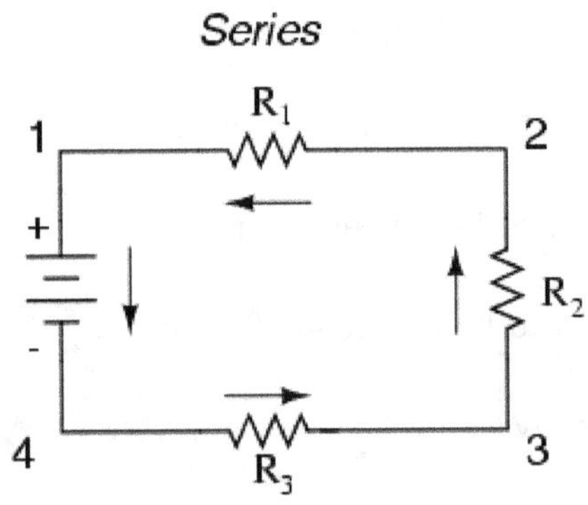

Now let's use what we know about electricity to develop the mathematical properties of a series circuit. First, there is only one pathway for the current to travel, and we know that no charge will accumulate on the resistors. This implies that the current through each resistor is of the same value. Using this property and applying Ohm's law to each resistor, we come up with the following equations for the above circuit diagram:

$$V_{total} = V_1 + V_2 + V_3$$

$$I_1 = I_2 = I_3$$

$$R_{total} = R_1 + R_2 + R_3$$

It is important to understand the quantity R_{total}. This quantity is commonly called the equivalent resistance. Let's say you wanted to reduce the number of resistors in the circuit from three resistors to only one resistor. You would need a resistor whose value equals the equivalent resistance (R_{total}).

Electrical circuits may be connected in parallel as well as in series. In this configuration, there are multiple paths for current to travel, and therefore there are different mathematical properties. We will start with a picture of a parallel configuration as shown below:

Parallel

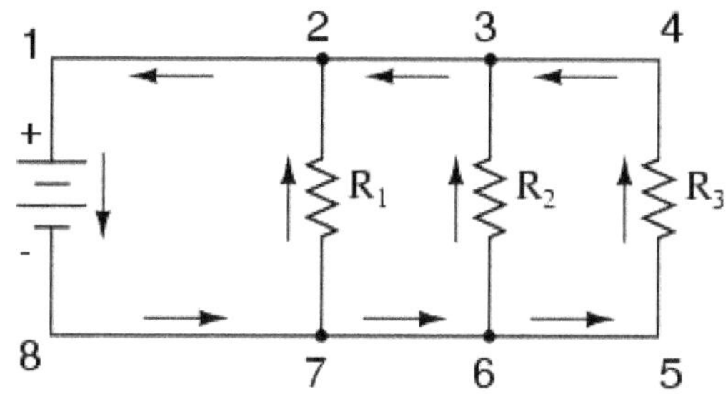

Can you see why this circuit is called parallel? If you said because the circuit elements (in this case, the circuit elements are resistors) are connected parallel to each other, you are correct. Unlike a series circuit, a parallel circuit allows current to travel in multiple paths. You can think of a parallel circuit as a "current-splitter." Again, we would like to derive mathematical formulas representing the circuit's total voltage, current, and resistance. By looking at the parallel configuration, can you say something about the voltage drop across each resistor? You are correct if you said they are the same in each element. Let us use this knowledge and Ohm's law to derive some mathematical laws for parallel circuits:

$$V_1 = V_2 - V_3 = V$$

$$I_{total} = I_1 + I_2 + I_3$$

$$I_{total} = \frac{V_1}{R_1} + \frac{V_2}{R_2} + \frac{V_3}{R_3} = V\left(\frac{1}{R_1} + \frac{1}{R_2} + \frac{1}{R_3}\right)$$

$$\frac{1}{R_{total}} = \frac{1}{R_1} + \frac{1}{R_2} + \frac{1}{R_3}$$

How is all this useful? Let's start to answer this by considering the light bulb experiment of last week. What would happen if we used two connected in series instead of one bulb? What if they were connected in parallel? Today we'll find those answers along with others as we examine light bulbs in parallel and series circuits.

Questions

1. Determine the equivalent (total) resistance for each of the following circuits below.

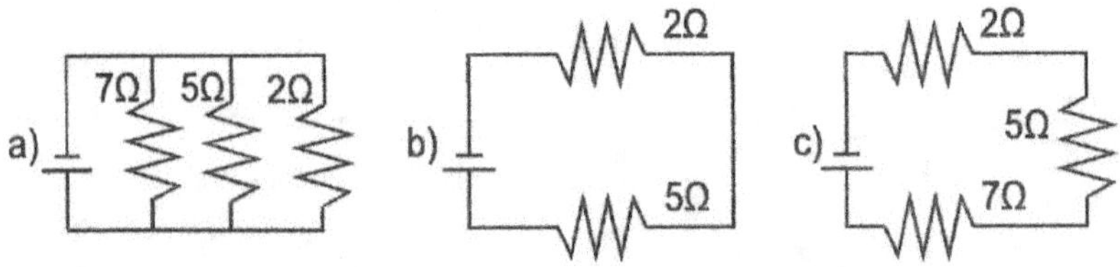

2. Determine the total voltage (electric potential) for each of the following circuits below.

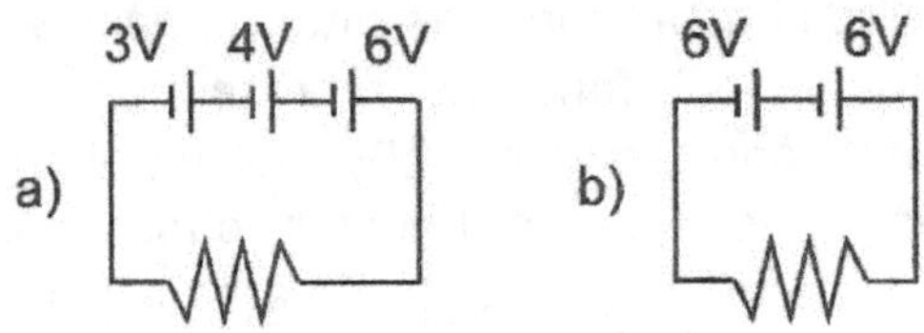

3. Fill out the table for the circuit diagram below.

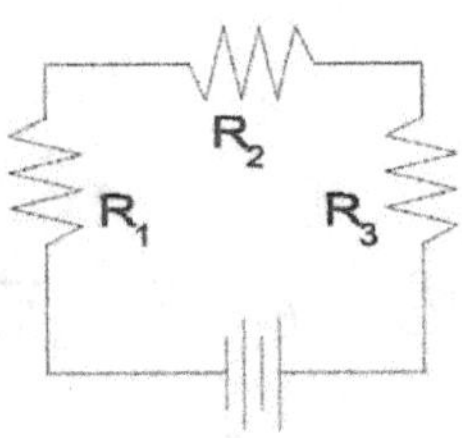

Circuit Position	Voltage (V)	Current (A)	Resistance (Ω)
1			10.0
2			20.0
3			30.0

Total	6.00		

4. Fill out the table for the circuit diagram below.

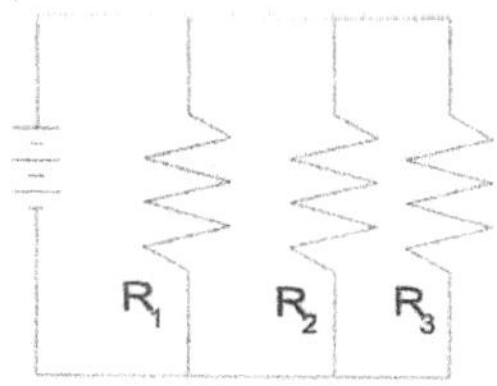

Circuit Position	Voltage (V)	Current (A)	Resistance (Ω)
1			10.0
2			20.0
3			30.0
Total	6.00		

5. Fill out the table for the circuit diagram below.

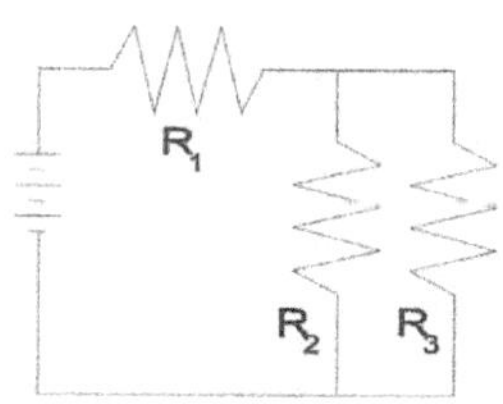

Circuit Position	Voltage (V)	Current (A)	Resistance (Ω)
1			10.0
2			20.0
3			30.0
Total	6.00		

For questions 6 and 7, refer to the following:

The diagram below represents an electric circuit consisting of four resistors and a 12-volt battery.

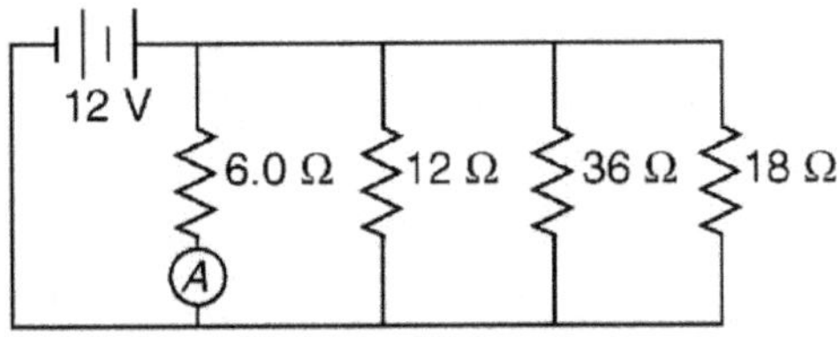

6. What is the equivalent resistance of the circuit shown?

7. What current is measured by ammeter *A* shown in the diagram?

8. A 6.0-ohm lamp requires 0.25 ampere of current to operate. Which circuit below would the lamp operate correctly when switch *S* is closed?

A) 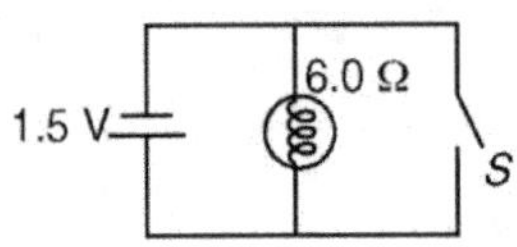

B) 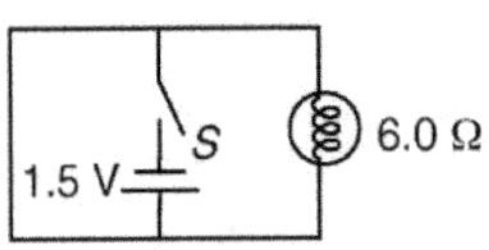

C) 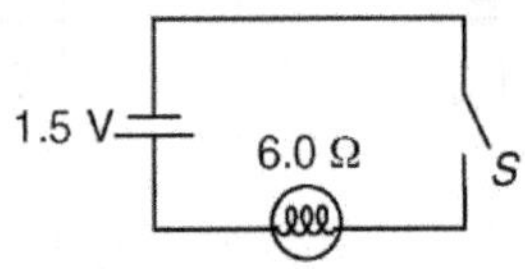

D) 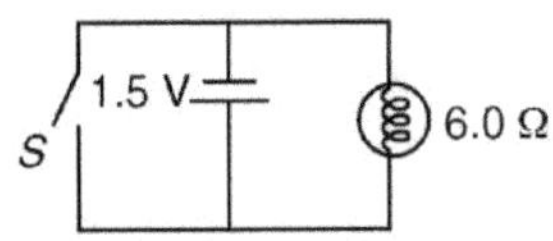

For questions 9 and 10, refer to the following:

A 50-ohm resistor, an unknown resistor *R*, a 120-volt source, and an ammeter are connected in a complete circuit. The ammeter reads 0.50 ampere.

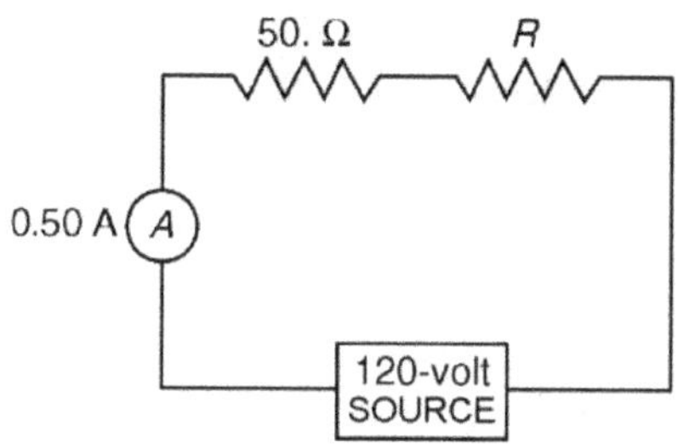

9. Calculate the equivalent resistance of the circuit shown.
10. Determine the resistance of resistor R shown in the diagram.

For questions 11 through 13, refer to the following:

A 3.0-ohm resistor, an unknown resistor, R, and two ammeters, A_1 and A_2, are connected below with a 12-volt source. Ammeter A_2 reads a current of 5.0 amperes.

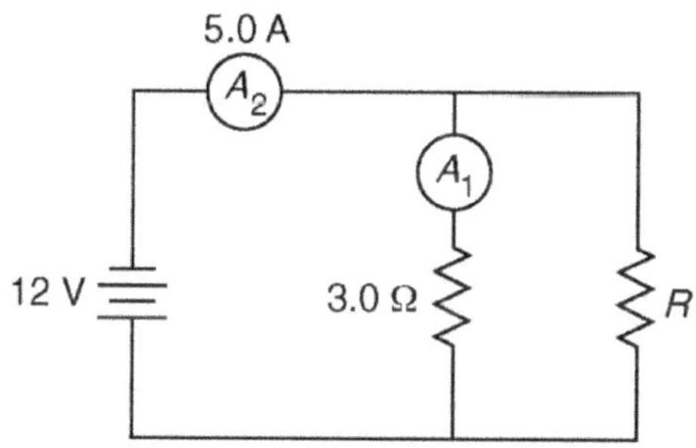

11. Determine the equivalent resistance of the circuit shown.

12. Calculate the current measured by ammeter A_1 in the diagram shown.

13. Calculate the resistance of the unknown resistor, R, in the diagram shown.

14. The load across a 50.0-V battery consists of a series combination of two lamps with resistances of 125 Ω and 225 Ω.

a. Find the total resistance of the circuit.

b. Find the current in the circuit.
c. Find the potential difference across the 125Ω lamp.

15. The load across a 12-V battery consists of a combination of three resistances, 15 Ω, 21 Ω, and 24 Ω, respectively.

a. Draw the circuit diagram.
b. What is the total resistance of the load?
c. What is the magnitude of the circuit current?
16. The load across a 40-V battery consists of three resistances R_1, R_2, and R_3. R_1 is 240Ω and R_3 is 120 Ω. The potential difference across R_1 is 24 V.

a. Find the current in the circuit.
b. Find the equivalent resistance of the circuit.
c. Find the resistance of R_2.

17. The load across a 12-V battery consists of three resistances R_1, R_2, and R_3. R_1 is 210 Ω, R_2 is 350 Ω, and R_3 is 120 Ω.

a. Find the equivalent resistance of the circuit.
b. Find the current in the circuit.
c. Find the potential difference across R_3.

18. Two resistances, one 12Ω and the other 18 Ω, are connected in parallel. What is the equivalent resistance of the parallel combination?

19. Three resistances of 12Ω each is connected in parallel. What is the equivalent resistance?

20. Two resistances, one 62Ω and the other 88 Ω, are connected in parallel. The resistors are then connected to a 12-V battery.

a. What is the equivalent resistance of the parallel combination?
b. What is the current through each resistor?

21. A 110-V household circuit that contains an 1800-W microwave, a 1000-W toaster, and 800-W coffeemaker is connected to a 20-A

fuse. Determine the current. Will the fuse melt if the microwave and the coffeemaker are both on?

22. A 35Ω, 55Ω, and 85Ω, resistors are connected in parallel. The resistors are then connected to a 35 V battery.

a. What is the equivalent resistance of the parallel combination?
b. What is the current through each resistor?

23. Resistors R_1, R_2, and R_3 have resistances of 15Ω, 9Ω, and 8Ω respectively. R_1 and R_2 are connected in series, and their combination is in parallel with R_3 to form a load across a 6.0-V battery.

a. Draw the circuit diagram.
b. What is the total resistance of the load?
c. What is the current in R_3?
d. What is the potential difference across R_2?

24. A 15Ω resistor is connected in series to a 120-V generator and two 10Ω resistors connected in parallel.

a. Draw the circuit diagram.
b. What is the total resistance of the load?
c. What is the magnitude of the circuit current?
d. What is the current in one of the 10 Ω resistors?
e. What is the potential difference across the 15 Ω resistor?

Answers

1a) 1.2 Ω	1b) 7 Ω	1c) 14 Ω
2a) 13 V	2b) 12 V	6) 3.0 Ω
7) 2.0 A	8) C	9) 240 Ω
10) 190 Ω	11) 2.4 Ω	12) 4.0 A
13) 12 Ω	14a) 350. Ω	14b) 0.143 A
14c) 17.9 V	15b) 60. Ω	15c) 0.20 A
16a) 0.10 A	16b) 400 Ω	16c) 40. Ω
17a) 680 Ω	17b) 0.018 A	17c) 2.2 V
18) 7.2 Ω	19) 4.0 Ω	20a) 36 Ω

20b) $I_{62\Omega} = 0.19$ A; $I_{88\Omega} = 0.14$ A

21) I = 23.6 A so fuse will melt 22a) 17 Ω

22b) $I_{35\Omega}$ = 1.0 A; $I_{55\Omega}$ = 0.64 A; $I_{85\Omega}$ = 0.41 A

23b) 6.0 Ω 23c) 0.75 A 23d) 2.3 V

24b) 20.0 Ω 24c) 6.0 A 24d) 3.0 A

24e) 90. V

How to read Resistor Color Codes

First the code

Black	Brown	Red	Orange	Yellow	Green	Blue	Violet	Gray	White
0	1	2	3	4	5	6	7	8	

How to read the code

First, find the tolerance band. It will typically be Gold (5%) and sometimes Silver (10%).

Starting from the other end, identify the first band - write down the number associated with that color.

Now 'read' and record the next color.

Now read the third or 'multiplier' band and write down that number of zeros.

If the 'multiplier' band is Gold, move the decimal point one to the left. If the 'multiplier' band is Silver, move the decimal point two places to the left. If the resistor has one more band past the tolerance band, it is a quality band.

Read the number as the '% Failure rate per 1000 hours of operation at full power load'. (To get better failure rates, resistors are typically specified to have twice the needed wattage dissipation that the circuit produces) 1% resistors have three bands to read digits to the left of the multiplier. They have a different temperature coefficient to provide the 1% tolerance.

6.2 Thevenin Theorem

Thevenin's theorem is a fundamental principle in electrical circuit theory that allows us to simplify complex circuits into a simpler equivalent circuit. It is named after French engineer Léon Charles Thévenin.

According to Thevenin's theorem, any linear circuit can be replaced by an equivalent circuit consisting of a single voltage source in series with a single resistance. This equivalent circuit is known as the Thevenin equivalent circuit, and it has the same voltage-current characteristics as the original circuit between its two terminals. Thevenin's theorem is a powerful tool in circuit analysis, as it allows us to analyze complex circuits by replacing them with simpler equivalent circuits. This can greatly simplify calculations and make circuit analysis more efficient.

Steps of solution

1-Remove the load resistance.
2-Make all voltage sources short circuit and all current sources open circuit,
3-Calculate Thevenin equivalent resistance ($R_{th.}$) viewing from removed load resistance.
4-Calculate voltage terminals between the removed resistance terminals.
5- Calculate the current from the circuit shown

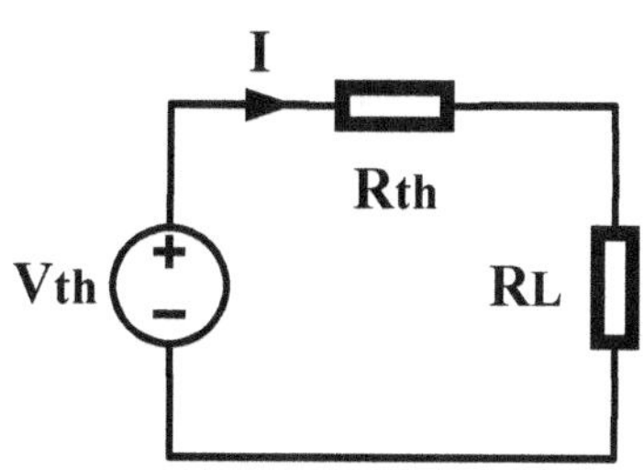

$$I = \frac{V_{th}}{R_{th} + R_L}$$

6.3 Kirchhoff's Laws

Kirchhoff's laws are more comprehensive than Ohm's and are used for solving electrical networks, which the latter may not solve readily.

Kirchhoff's laws, two in number, are particularly useful in determining the equivalent resistance of a complicated network of conductors and for calculating the currents flowing in the various conductors.

1. Kirchhoff` s Point Law or Current Law (KCL)

In any electrical network, the algebraic sum of the currents meeting at a point (or junction) is Zero. The total current *entering* a junction equals the total current *leaving* that junction. Consider the case of a network shown in the figure:

$$I_1+(-I_2)+(I_3)+(+I_4)+(-I_5) = 0$$

$$I_1+I_4-I_2-I_3-I_5 = 0$$

Or

$$I_1+I_4 = I_2+I_3+I_5$$

Or

Incoming currents =Outgoing currents

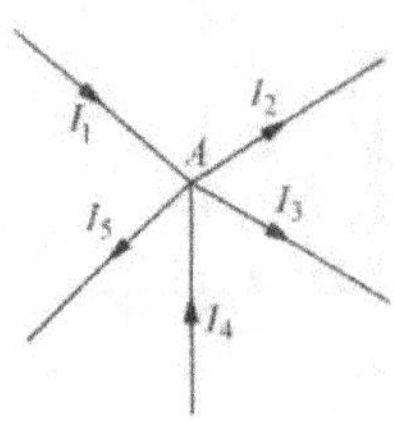

2. *Kirchhoff's Mesh Law or Voltage Law (KVL)*

In any electrical network, the algebraic sum of the products of currents and resistances in each of the conductors in any closed path (or mesh) in a network plus the algebraic sum of the e.m.f. 's. in that path is zero.

That is, $\sum IR + \sum e.m.f = 0$ round a mesh

It should be noted that algebraic sum is the sum that considers the polarity of the voltage drops. If we start from a particular junction and go around the mesh till we return to the starting point, then we must be at the same potential with which we started.

Hence, it means that all the sources of emf met on the way must necessarily be equal to the voltage drops in the resistances, every voltage being given its proper sign, plus or minus.

Determination of Voltage Sign

In applying Kirchhoff's laws to specific problems, particular attention should be paid to the algebraic signs of voltage drops and e.m.fs.

(a) Sign of Battery EMF.

A *rise* in voltage should be given a + ve sign and *a fall* in voltage a -ve sign. That is, if we go from the -ve terminal of a battery to its +ve terminal, there is a *rise* in potential. Hence this voltage should be given a + ve sign.

And on the other hand, if we go from the +ve terminal to the -ve terminal, there is *a fall* in potential. Hence this voltage should be preceded by a -ve sign.

The sign of the battery, e.m.f, is independent of the direction of the current through that branch.

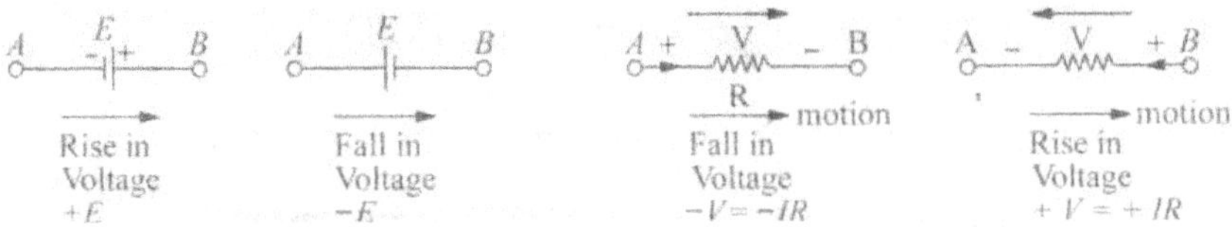

(b) Sign of *IR* Drop

Now, take the case of a resistor (Fig. 2.4). If we go through a resistor in the *same* direction as the current, then there is a fall in potential because the current flows from a higher to a lower potential. Hence, this voltage fall should be taken -ve. However, if we go in a direction opposite to that of the current, there is a voltage rise. Hence, this voltage rise should be given a positive sign. Consider the closed path *ABCDA* in Fig.

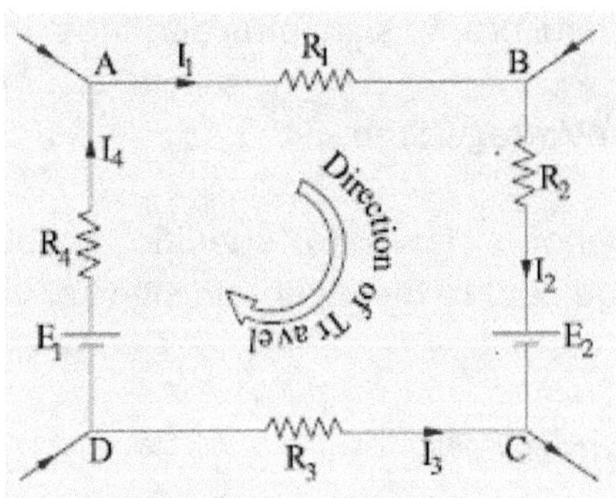

As we travel around the mesh in the clockwise direction, different voltage drops will have the following signs:

I_1R_1 is - ve (fall in potential)
I_2R_2 is - ve (fall in potential)
I_3R_3 is + ve (rise in potential)
I_4R_4 is - ve (fall in potential)
E_2 is - ve (fall in potential)
E_1 is + ve (rise in potential)

Using Kirchhoff's voltage law, we get

$-I_1R_1 - I_2R_2 - I_3R_3 - I_4 R_4 - E_2 + E_1 = 0$

Or $I_1R_1 + I_2R_2 - I_3R_3 + I_4R_4 = E_1 - E_2$

Assumed Direction of Current:

In applying Kirchhoff's laws to electrical networks, the current flow direction may be either clockwise or anticlockwise. If the assumed direction of the current is not the actual direction, then on solving the question, the current will be found to have a minus sign.

If the answer is positive, the assumed direction is the same as the actual one. However, the important point is that once a particular direction has been assumed, the same should be used throughout the solution of the question.

Kirchhoff's laws apply both to d.c. and a.c. voltages and currents. However, in the case of alternating currents and voltages, any e.m.f. of self-inductance or that existing across a capacitor should also be considered.

Resistance in Series:

If three conductors with resistances R1, R2 and R3 are joined end on end, as shown in Fig below, they are connected in series. It can be proved that the equivalent resistance between points A & D equals the sum of the three individual resistances.

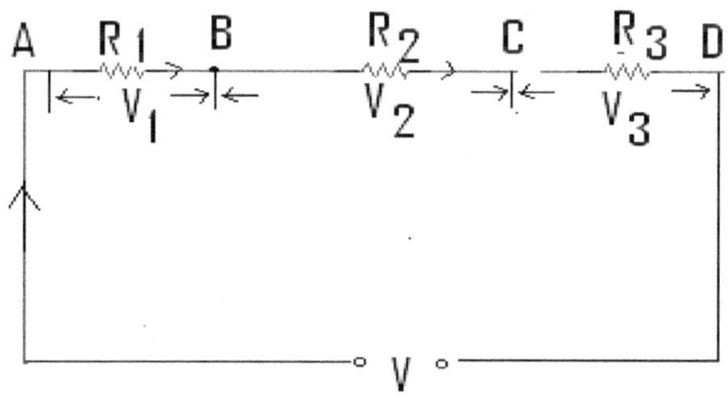

For a series circuit, the current is the same through all three conductors, but the voltage drop across each is different due to its different values of resistances and is given by Ohm's Law, and the sum of the three voltage drops is equal to the voltage supplied across the three conductors.

$\therefore V = V_1 + V_2 + V_3 = IR_1 + IR_2 + IR_3$

But $\quad V = IR$

where R is the equivalent resistance of the series combination.

$$IR = IR_1 + IR_2 + IR_3$$
$$\text{or } R = R_1 + R_2 + R_3$$

The main characteristics of a series circuit are

1. the same current flows through all parts of the circuit.
2. Different resistors have their voltage drops.
3. Voltage drops are addictive.
4. Applied voltage equals the sum of different voltage drops.
5. Resistances are additive.
6. Powers are additive

Voltage Divider Rule

In a series circuit, the same current flows through each of the given resistors, and the voltage drop varies directly with its resistance.

Consider a circuit in which a 24- V battery is connected across a series combination of three resistors of 4Ω each. Determine the voltage drops across each resistor.

Total resistance $R = R_1 + R_2 + R_3 = 12\ \Omega$

According to Voltage Divider Rule, voltages divide by the ratio of their resistances, and hence the various voltage drops are

$$V_1 = V\frac{R_1}{R} = 24\ X\ \frac{2}{12} = 4V$$

$$V_2 = V\frac{R_2}{R} = 24\ X\ \frac{4}{12} = 8V$$

$$V_3 = V\frac{R_3}{R} = 24\ X\ \frac{6}{12} = 12V$$

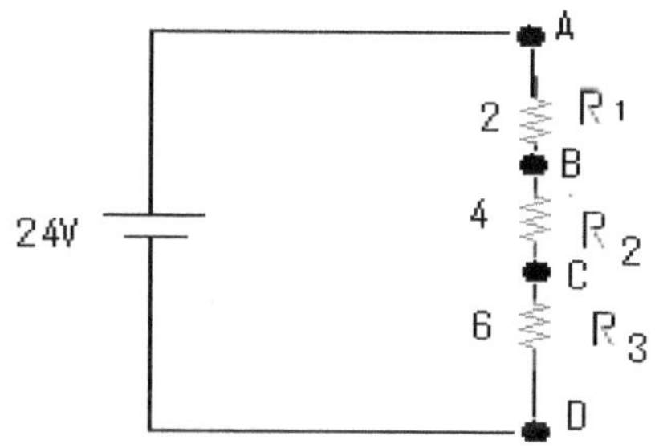

Resistances in Parallel

Three resistances, as joined in Fig, are said to be connected in parallel. In this case

i. The potential difference across all resistances is the same
ii. Current in each resistor is different and is given by Ohm's Law
iii. The total current is the sum of the three separate currents.

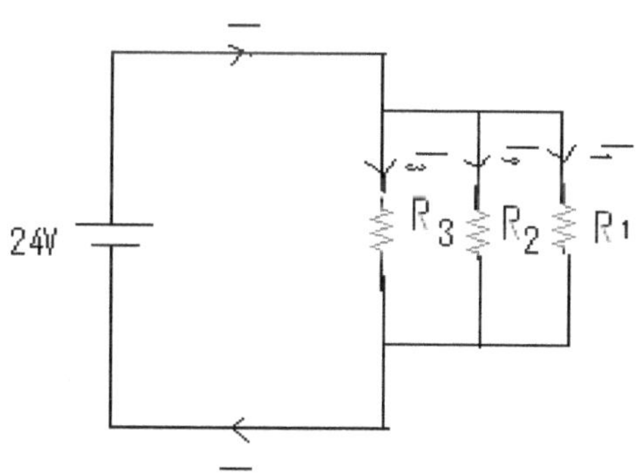

$$I = I_1 + I_2 + I_3 = \frac{V}{R_1} + \frac{V}{R_2} + \frac{V}{R_3}$$

$I = \frac{V}{R}$ Where V is the applied voltage.

R = equivalent resistance of the parallel combination.

$$\frac{V}{R} = \frac{V}{R_1} + \frac{V}{R_2} + \frac{V}{R_3}$$

$$\frac{1}{\quad} \frac{1}{R_1} + \frac{1}{R_2} + \frac{1}{R_3}$$

$$G = G_1 + G_2 + G_3$$

The main characteristics of a parallel circuit are:

1. Same voltage acts across all parts of the circuit
2. Different resistors have their current.
3. Branch currents are additive.
4. Conductances are additive.
5. Powers are additive

Division of Current in Parallel Circuits

Two resistances are joined in parallel across a voltage *V. The current in each* branch, given by Ohm's law, is

$I_1 = \frac{VI}{R_1}$ and $I_2 = \frac{V}{R2}$

$\frac{I_1}{I_2} = \frac{R_1}{R_2}$

As $\frac{I}{R_1} = G_1 \; and \; \frac{I}{R_2} = G_2$

$\frac{I_1}{I_2} = \frac{G_1}{G_2}$

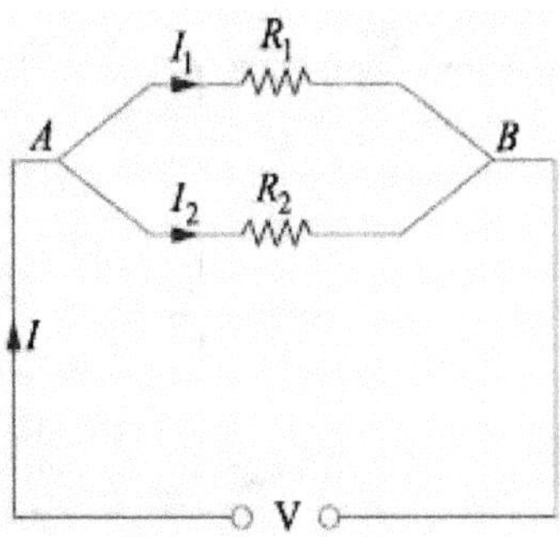

Hence, the division of current in the branches of a parallel circuit is directly proportional to the conductance of the branches or inversely proportional to their resistances. The branch currents are also expressed in terms of the total circuit current.

Now $\quad I_1 + I_2 = I; \; \therefore \; I_2 = I - I_1 \; \therefore \; \frac{I_1}{I - I_1} = \frac{R_2}{R_{12}}$

or $\quad I_1R_1 = R_2(I - I_1)$

$$I_1 = I\frac{R_1}{R_1 - R_2} = I\frac{G_1}{G_1 + G_2}$$

and $\quad I_2 = I\frac{R_1}{R_1 - R_2} = I\frac{G_1}{G_1 + G_2}$

This Current Divider Rule has direct application in solving electric circuits by Norton's theorem.

Take the case of three resistors in parallel connected across a voltage V

The total current is

$$I = I_1 + I_2 + i_3$$

Let the equivalent resistance be *R*. Then

$$V = Ir$$

Also $\quad V = I_1R,\ IR = I_1R$

Or $\quad \frac{I}{I_1} = \frac{R_1}{R}\ or\ I_1 = \frac{IR}{R_1}$

$$\frac{1}{R} = \frac{1}{R_1} + \frac{1}{R_2} + \frac{1}{R_3}$$

$$R = \frac{R_1R_2R_3}{R_1R_3 + R_2R_3 + R_1R_2}$$

From .... (i) above,

$$I_1 = I\left[\frac{R_2R_3}{R_1R_3 + R_2R_3 + R_1R_2}\right]$$

$$= I.\frac{G_1}{G_1 + G_2 + G_3}$$

$$I_2 = I\left[\frac{R_1R_3}{R_1R_3 + R_2R_3 + R_1R_2}\right]$$

$$= I.\frac{G_2}{G_1 + G_2 + G_3}$$

$$I_3 = I\left[\frac{R_1R_2}{R_1R_3+R_2R_3+R_1R_2}\right]$$

$$= I.\frac{G_3}{G_1 + G_2 + G_3}$$

1. A 10 Ω, a 15 Ω, and a 5 Ω resistor are connected in series across a 90 V battery. Find the equivalent resistance (R_{eq}) of the circuit. What is the current (I) through each resistor? *Ans: 30 Ω, 3 A*

	V (V)	I (A)	R (Ω)	P (W)
R1				
R2				
R3				
Total				

2. A 20 Ω and a 30 Ω resistors are connected in series across a 110 V potential difference. What is the total resistance of the circuit? What is the current through each resistor? What is the voltage drop across each resistor? *Ans: 50 Ω; 2.2 A; 44 V and 66 V*

	V (V)	I (A)	R (Ω)	P (W)
R1				
R2				
Total				

3. A 3 Ω, a 5 Ω, and a 4 Ω resistors are connected in series across a 12 V battery. What is the total resistance of the circuit? What is the

current through each resistor? What is the voltage drop across each resistor? *Ans: 12 Ω; 1 A; 3 V, 5 V, and 4 V*

	V (V)	I (A)	R (Ω)	P (W)
R1				
R2				
R3				
Total				

4. A 60 Ω, a 90 Ω, and an unknown resistor are connected in series across a 130 V battery. An ammeter in the circuit reads 0.67 A. What is the total resistance of the circuit? What is the resistance of the unknown resistor? What is the power dissipated in each resistor? *Ans: 195 Ω; 45 Ω; 26.7 W, 40 W, and 20 W*

	V (V)	I (A)	R (Ω)	P (W)
R1				
R2				
R3				
Total				

Parallel Circuits

For each problem, draw the circuit diagram using appropriate symbols and meters.

1. Three 15 Ω resistors are connected in parallel across a 3 V battery. Find the total resistance of the circuit. What is the total current? What is the current through each resistor? *Ans: 5 Ω; 0.6 A; 0.2 A in each*

	V (V)	I (A)	R (Ω)	P (W)
R1				
R2				
R3				
Total				

2. A 120 Ω, a 60 Ω, and a 40 Ω resistors are connected in parallel and placed across a 120 V potential difference. What is the total resistance of the circuit? What is the total current? What is the current through each resistor? *Ans: 20 Ω; 6 A; 1 A. 2 A, and 3 A*

	V (V)	I (A)	R (Ω)	P (W)
R1				
R2				
R3				
Total				

3. A 40 Ω, an 8 Ω, and a 10 Ω resistors are connected in parallel across a 120 V battery. What is the total resistance of the circuit? What is the circuit current? What is the current through each resistor? What is the power dissipated? *Ans: 4 Ω; 30 A; 3 A, 15 A, 12 A; 360 W, 1800 W, 1440 W*

	V (V)	I (A)	R (Ω)	P (W)
R1				
R2				
R3				
Total				

4. A 12 Ω, a 15 Ω, and an unknown resistance are connected in parallel. The total current of the combination is 12 A. The current across the 15 Ω the branch is 4 A. What is the total voltage? What is the total resistance? What is the unknown resistance? *Ans: 60 V; 5 Ω; 20 Ω*

	V (V)	I (A)	R (Ω)	P (W)
R1				
R2				
R3				
Total				

6.4 Star - Delta (Y- Δ) Transformation

The methods of series, parallel, and series-parallel combinations of elements do not always lead to the simplification of networks. Such networks are handled by Star Delta transformation.

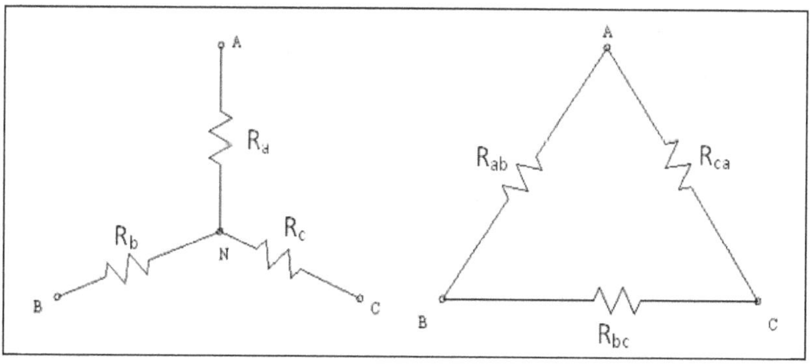

The figure shows three resistances, R_a, R_b, and R_c, connected in star to three nodes, A, B, and C. A common point N & figure b shows three resistances connected in the delta between the same three nodes A, B, C. If these two networks are to be equivalent, then the resistance between any pair of nodes of the delta-connected

network of a) must be the same as that between the same pair of nodes of the star–a connected network of fig b).
Star resistances in terms of delta

Equating resistance between node pair AB.

$$R_a + R_b = R_{ab} \,//\, (R_{bc} + R_{ca}) = \frac{R_{ab}\,R_{bc}+R_{ab}\,R_{ca}}{R_{ab}+R_{bc}+R_{ca}} \qquad (1)$$

Similarly, for node pair BC.

$$R_b + R_c = R_{bc} \,//\, (R_{ca} + R_{ab}) = \frac{R_{bc}\,R_{ca}+R_{bc}\,R_{ab}}{R_{ab}+R_{bc}+R_{ca}} \qquad (2)$$

For Node pair CA

$$R_c + R_a = R_{ca} \,//\, (R_{ab} + R_{bc}) = \frac{R_{ca}\,R_{ab}+R_{ca}\,R_{bc}}{R_{ab}+R_{bc}+R_{ca}} \qquad (3)$$

Subtracting equation 2 from 3 gives

$$R_a - R_b = \frac{R_{ca}\,R_{ab}+R_{bc}\,R_{ab}}{R_{ab}+R_{bc}+R_{ca}} \qquad (4)$$

Adding 1 and 4 gives

$$R_a = \frac{R_{ab}R_{ca}}{R_{ab}+R_{bc}+R_{ca}} \qquad (5)$$

Similarly

$$R_b = \frac{R_{bc}\,R_{ab}}{R_{ab}+R_{bc}+R_{ca}} \qquad (6)$$

$$R_c = \frac{R_{ca}R_{bc}}{R_{ab}+R_{bc}+R_{ca}} \qquad (7)$$

Thus, the equivalent star resistance connected to a node is equal to the product of the two delta resistances connected to the same node decided by the sum of delta resistance. Delta resistances in terms of star resistances:

Dividing (5) by (6) gives

$$\frac{R_a}{R_b} = \frac{R_{ca}}{R_{bc}} \qquad \therefore R_{ca} = \frac{R_a R_{bc}}{R_b}$$

Dividing (5) by (6) gives

$$\frac{R_a}{R_c} = \frac{R_{ab}}{R_{bc}}$$

$$\therefore R_{ab} = \frac{R_a R_{bc}}{R_c}$$

Substituting for R_{ab} & R_{ca} in equation (5) simplifies and gives

$$R_a = \frac{\frac{R_a R_{bc}}{R_c} \frac{R_a R_{bc}}{R_b}}{\frac{R_a R_{bc}}{R_c} + \frac{R_a R_{bc}}{R_b} + R_{bc}}$$

$$R_a = \frac{R_a\left(\frac{R_{bc}^{\ 2}}{R_b R_c}\right)}{R_a\left(\frac{R_{bc}}{R_c} + \frac{R_{bc}}{R_b} + \frac{R_{bc}}{R_a}\right)}$$

$$R_a\left(\frac{R_{bc}}{R_c} + \frac{R_{bc}}{R_b} + \frac{R_{bc}}{R_a}\right) = R_a\left(\frac{R_{bc}^{\ 2}}{R_b R_c}\right)$$

$$\frac{R_{bc}(R_b R_c + R_a R_c + R_a R_b)}{R_a R_b R_c} = \frac{R_{bc}^{\ 2}}{R_a R_b R_c}$$

$$R_{bc} = \frac{(R_b R_c + R_a R_c + R_a R_b)}{R_a}$$

$$= R_b + R_c + \frac{R_c R_a}{R_b}$$

Similarly

$$R_{ca} = R_c + R_a + \frac{R_c R_a}{R_b}$$

$$R_{ab} = R_a + R_b + \frac{R_a R_b}{R_c}$$

Thus, the equivalent Delta resistance between two nodes is the sum of two-star resistances connected to those nodes plus the product of the same two-star resistances divided by the third-star resistance.

$$R_1 = \frac{R_A R_B}{R_A + R_B + R_C} \qquad R_A = R_1 + R_2 + \frac{R_1 R_2}{R_2}$$

$$R_2 = \frac{R_A R_C}{R_A + R_B + R_C} \qquad R_B = R_1 + R_3 + \frac{R_1 R_3}{R_2}$$

$$R_1 = \frac{R_B R_C}{R_A + R_B + R_C} \qquad R_C = R_2 + R_3 + \frac{R_2 R_3}{R_1}$$

If all are similar equal to R

$$R_1 = \frac{R^2}{3R} = \frac{R}{3} \qquad R_A = 3R$$

6.5 Mesh Current and Node Voltage Analysis

The simple series & parallel circuits can be solved by using Ohm's law & Kirchhoff's law. If the circuits are complex, conducting several sources & a large numbers of elements, they may be simplified using star-delta transformation. There are also other effective solving complex electric circuits.

Mesh current or loop current analysis & node voltage analysis are very effective methods of solving complex electric circuits. We have various network theorems, which are effective alternate methods for solving complex electrical circuits.

1) Mesh current or loop current analysis

2) Node voltage analysis

6.5.1 Mesh Current Analysis

This method, particularly applied to complete networks, employs a loop or mesh currents system instead of branch currents, as in Kirchhoff's law. Here, the currents in different meshes are assigned another path so they do not split at a junction into branch currents.

If `b` is the number of branches & j is the number of junctions in each network, then the number of independent equations to be solved reduces from `b` by Kirchhoff's law b-(j-1) for loop current method.

The above circuit consists of two batteries, E_1, and E_2, connected with five resistors.
Let the mesh currents for three meshes be I_1, I_2, & I_3

Mesh-1 --- $E_1 - I_1R_1 - R_4(I_1 - I_2) = 0$
$I_1[R_1 + R_4] - I_2R_4 - E_1$--------2

Mesh-2---- $-I_2R_2 - R_5[I_2 - I_3] - R_4[I_2 - I_1] = 0$
$I_1R_4 - I_2(R_2+R_4+R_5) + I_3R_5 = 0$---2

Mesh-3 --- $I_3R_3 - E_2 - R_5[I_3 - I_2] = 0$
$I_2R_5 - I_3[R_3 + R_5] - E_2 = 0$-----3

The above three equations can be solved to find three mesh and branch currents.

1. Determine the current supplied by each battery in the circuit shown.

Mesh-1--- $20 - 5I_1 - 3(I_1 - I_2) - 5 = 0$
$8I_1 - 3I_2 = 15$ ---1

Mesh-2 --- $-4I_2 + 5 - 2(I_2 - I_3) + 5 - 3(I_2 - I_1) = 0$
$3I_1 - 9I_2 + 2I_3 = -15$ ----2

Mesh-3 --- $-8I_3 - 30 - 5 - 2(I_3 - I_2) = 0$

$2I_2 - 10\ I_3 = 35$ --- 3

$24I_1 - 9I_2 = 45$ ---1+3

$24I_2 - 72I_2 + 16I_3 - 120$ ---2+8

$63I_2 - 16I_3 = 165$ ------4

$32I_2 - 160I_3 = 35+16$ ------3+16

$630I_2 - 160I_3 = 1650$ -------4X10

$-598I_2 = -10990$

$I_3 = -\ 3.135$

$I_1 = 2.558$

$I_2 = 1.82$

Discharge Current of B_1 = 765/299 A

Discharge Current of $B_2 = I_1 - I_2$ = 220/299

Discharge Current of $B_3 = I_2 - I_3$ = 2965/598 A

Discharge Current of $B_4 = I_2$ = 545/299 A

Discharge Current of B_5 = 1875/598

6.5.2 Node Voltage Analysis

The nodal analysis applies Kirchhoff's Current Law (KCL) to solve for the voltages at each node in an equation. A **node voltage** is a potential difference between the given node and a designed reference node (ground). Since one node is defined as ground, a circuit with N nodes will require N-1 equations to solve completely.

A node is a point in a network when two or more elements meet.

➢ Node voltage analysis is one of the most effective methods of solving an electrical network.

➢ This method's number of equations to be solved is one less than the number of equations required in mesh current analysis.

➢ For the application of this method, every junction in the network where three or more branches meet is regarded as a mode.

➢ One of these modes is regarded as the reference node, datum node, or zero–potential node. Hence, the number of simultaneous equations will be solved because (n-1), where `n` is the number of independent nodes.

➢ A three-node network ill here two unknown voltages & hence, two equations have to be solved, as the voltage of the third node is assumed to be at zero potential.

Consider a general electrical circuit shown below where E_1 and E_2 are the e.m.f's of the sources.

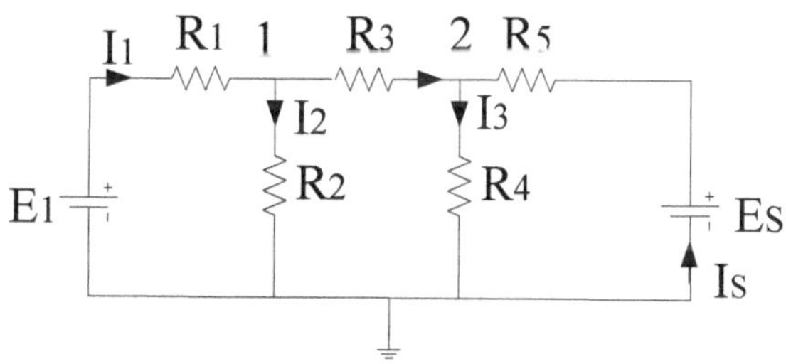

In this circuit, there are three nodes 1.2&3. One of these modes says three, is taken as the reference node & is assumed to be zero potential. The other two nodes, i.e., 1&2, are assigned node voltages V_1 & V_2, respectively.

Currents in the various branches are assumed arbitrarily in any direction. Applying KCL to node 1& 2, we get
At node 1,

$$I_1 = I_2 + I_3$$

$$\frac{E_1-V_1}{R_1} = \frac{V_1}{R_2} + \frac{V_1-V_2}{R_3}$$

$$\left(\frac{1}{R_1}+\frac{1}{R_2}+\frac{1}{R_3}\right)V_1 - \frac{1}{R_3}V_2 = \frac{E_1}{R_1} \ldots\ldots\ldots\ldots.1$$

At node ---2,

$$I_3{}^2 + I_5 = I_4$$

$$\frac{V_1-V_2}{R_3} + \frac{E_2-V_2}{R_5} + \frac{V_2}{R_4}$$

$$-\frac{1}{R_3}V_1 + \left(\frac{1}{R_3}+\frac{1}{R_4}+\frac{1}{R_5}\right)V_2 = \frac{I_2}{R_5} \ldots\ldots\ldots\ldots.2$$

Equations 1 & and 2 are written as

$G_{11}V_1+G_{12}V_2 = I_1$

$G_{21}V_1+G_{12}V_2 = I_2$

Where G_{11} = sum of all the conductance connected to node 1, which is always positive

$G_{12} = -\frac{1}{R_3}$ = Mutual conductance between node 1 and node 2 is always negative.

$I_1 = \frac{E_1}{R_1}$ = sum of all source currents to node 1. The source current is positive if it is flowing towards the negative. If it is flowing away from the node

$G_{21} = -\frac{1}{R_3}$ = mutual conductance between node 2 and node 1. It is always negative.

$G_{22} = \frac{1}{R_3}+\frac{1}{R_4}+\frac{1}{R_5}$ = sum of all the conductance connected to node 2. It is always positive.

$I_2 = \frac{E_2}{R_2}$ = sum of all source current to node 2

Procedure for solving an electrical circuit using node voltage analysis:

1) All the nodes of the network are identified. One of them is taken as a reference node at zero potential, usually a node to which a maximum number of branches are connected taken as a reference node.

2) The remaining nodes are assigned with node voltages V_1, V_2, V_3 ..... etc.

3) The node voltage equations are written using the general node equations.

4) The node voltage equations are solved.

5) Once the node voltages are known, the current in all the network branches can be found.

Example: Consider the circuit shown below and write down the node voltage equations.

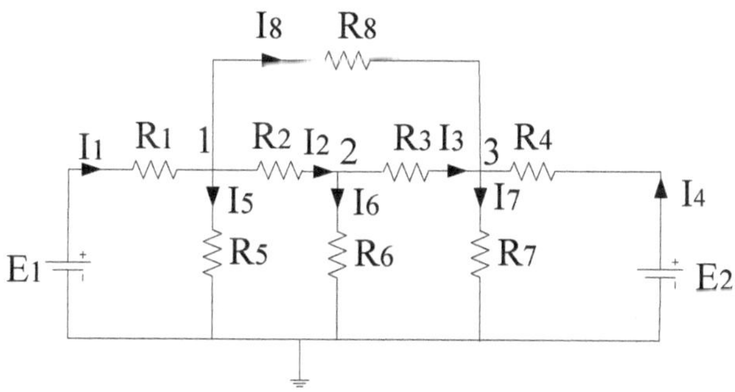

Network node

$$V_1\left(\frac{1}{R_1}+\frac{1}{R_2}+\frac{1}{R_5}+\frac{1}{R_8}\right)-\frac{V_2}{R_2}-\frac{V_3}{R_8}=\frac{E_1}{R_1}$$

$$-\frac{V_1}{R_2}+V_3+\left[\frac{1}{R_2}+\frac{1}{R_3}+\frac{1}{R_8}\right]-\frac{V_3}{R_3}=0$$

$$-\frac{V_1}{R_8}-\frac{V_2}{R_3}+V_3\left[\frac{1}{R_3}+\frac{1}{R_4}+\frac{1}{R_8}\right]=\frac{E_2}{R_4}$$

Write the Node equation by inspection method for the network shown below.

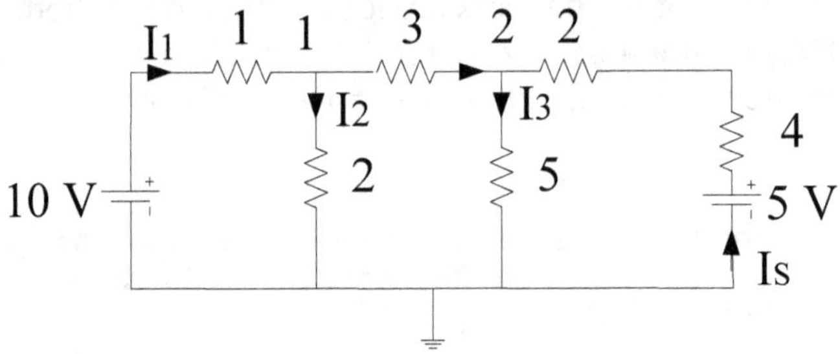

The general equations are

$$G_{11}V_1 + G_{12}V_2 = I_1$$

$$G_{21}V_1 = G_{12}V_2 = I_2$$

$$G_{11} = \left(1 + \frac{1}{2} + \frac{1}{3}\right)\text{Mho}$$

= self-conductance at node 1

$$G_{22} = \left(\frac{1}{3} + \frac{1}{5} + \frac{1}{3}\right)\text{mho}$$

G_{12} = -(1/3) G_{21} = -(1/3) sum of mutual conductance between 1 & 2 & 2 & 1

I_1 = Source current at node 1=10/1=10A

I_2 = Source current at node 2=2/5 +5/6=1.23

Node equations are

$$1.83V_1 - 0.33V_2 = 10$$

$$-0.33V_1 + 0.7V_2 = 1.23$$

Write the node voltage equations and determine the currents in each branch for the network shown

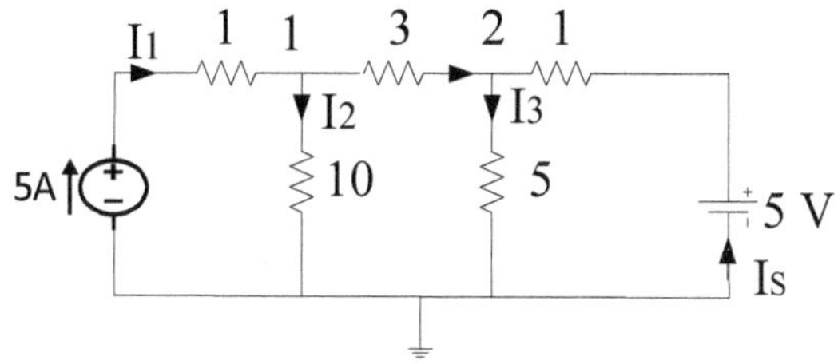

$$V_1\left(\frac{1}{3}+\frac{1}{10}\right) - V_2\left(\frac{1}{3}\right) = 5$$

$$-V_1\left(\frac{1}{3}\right)+V_2\left(1+\frac{1}{5}+\frac{1}{3}\right) = 10$$

Solving V_1 & V_2 we get V_1 = 19.85 Volts V_2= 10.9 Volts

$$I_{10\Omega} = \frac{V_1}{10} = 1.985\ A \quad I_{3\Omega} = \frac{V_1 - V_2}{3} = 2.98\ A$$

$$I_{5\Omega} = \frac{V_1}{5} = 2.18\ A \quad I_{1\Omega} = \frac{V_1 - 10}{1} = 0.9A$$

Example: Given the circuit below, find the voltages at all nodes.

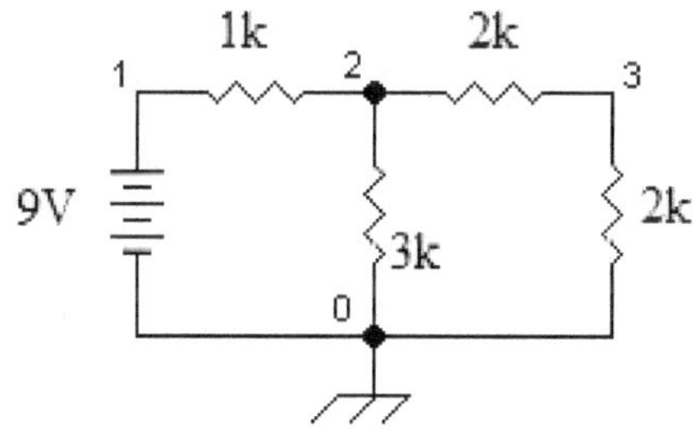

An example circuit for nodal analysis example.

- node 0: $V_0 = 0V$ (defined as a ground node)
- node 1: $V_1 = 9V$ (free node voltage)
- node 2: $\frac{V_1 - V_2}{1k} = \frac{V_2 - V_0}{3k} + \frac{V_2 - V_3}{2k}$

- node 3: $\dfrac{V_2 - V_3}{2k} = \dfrac{V_3 - V_0}{2k}$

which results in the following system of linear equations:

$$\begin{cases} +11V_2 & -3V_3 & = & 54 \\ +1V_2 & -2V_3 & = & 0 \end{cases}$$

Therefore, the solution is:

$$\begin{cases} V_0 = 0.00V \\ V_1 = 9.00V \\ V_2 = 5.68V \\ V_3 = 2.84V \end{cases}$$

Mesh Current Analysis

Mesh analysis applies Kirchoff's Voltage Law (KVL) to solve mesh currents. A *mesh current* is defined as the current in a mesh: a loop not containing any other loops. For *M* meshes, there will be *M* equations. If all sources are voltage sources, all *M* equations will be KVL.

If the circuit has *N* current sources, there will be *N* KCLs and *M-N* KVLs.

Mesh analysis is often easier as it requires fewer unknowns but can only be used on planar circuits.

Example

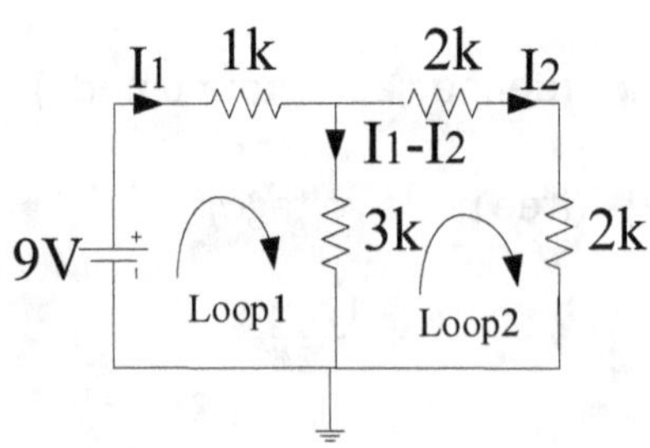

Circuit diagram for use with the Mesh Current example problem. The circuit has two loops indicated on the diagram. Using KVL, we get the following:

Loop 1:

$$0 = 9 - 1000I_1 - 3000(I_1 - I_2)$$

Loop 2:

$$0 = 3000(I_1 - I_2) - 2000I_2 - 2000I_2$$

Simplifying, we get the simultaneous equations:

$$0 = 9 - 4000I_1 + 3000I_2$$

$$0 = 0 + 3000I_1 - 7000I_2$$

Solving to get:

$$I_1 = 3.32mA$$

$$I_2 = 1.42mA$$

Alternating current power

Alternating Current (AC) is used in large areas and all different life facilities to generate easily. AC generators are electrical machines that operate on the principle of electromagnetic induction to generate electric power. AC is the current whose value and direction change continuously over a period on a sinusoidal wave, as shown in Figure 6.3.

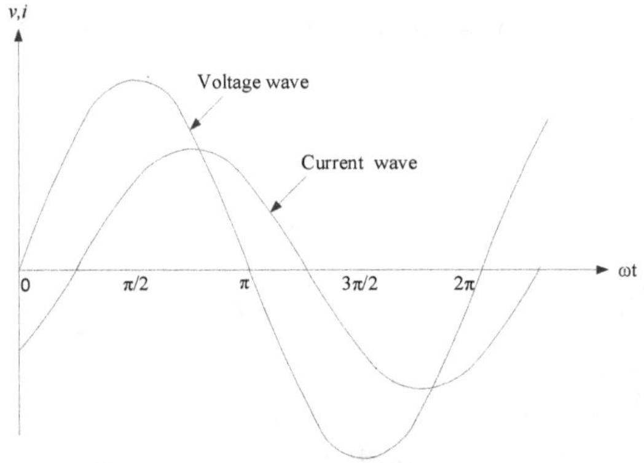

Figure 6.3 Voltage and current sine wave.

The advantages of alternating current are as follows

I. Changes in value and direction and the polarity are not fixed.
II. Generates mechanical methods by cutting magnetic fields such as generators.
III. Widely used.
IV. Low cost of production, especially when generated with large power.
V. It can be changed to constant current using electrical components.
VI. It can be converted from low to high voltage, and vice versa, using electric transformers.
VII. It can transfer to long distances using high-voltage towers.

Sinusoidal Wave Cycle and Frequency

The complete cycle of the sinusoidal, or the sine wave, is called *the period*, and the number of oscillations per second is called *the frequency*, and it is denoted by the letter *F*. From Figure 6.4, the portion between the two points (A-B) is called *the full wave* and the part bound between (A-C) is called half the positive wave and (C - B) is called half the negative wave. The relationship between the time of the wave *T* and the frequency *F* for one cycle is

$$f = \frac{1}{T}$$

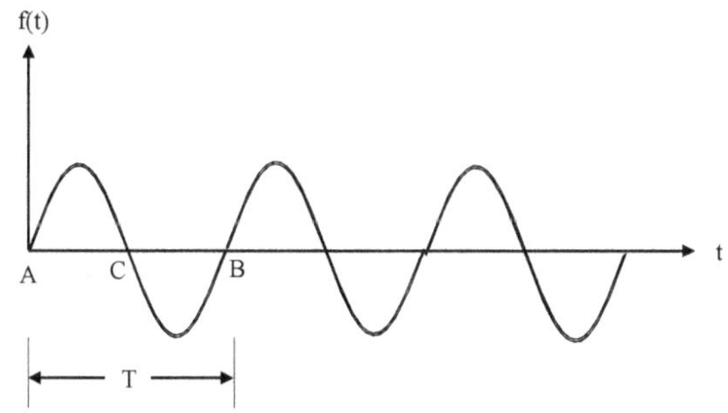

Figure 6.4 Sinusoidal waveform.

Example: What is the frequency if the one cycle is for a sine wave of 10 msec?

Solution

$$f = 1/T = 1/(10 \times 10-3) = 1000/10 = 100\,Hz$$

Example: Find the time of the one cycle if the frequency is 50 Hz.

Solution

$$T = 1/f = 1/50 = 0.02\,sec$$

Electric Power Generation

Figure 6.5 shows the model of a generator that generates alternating currents.

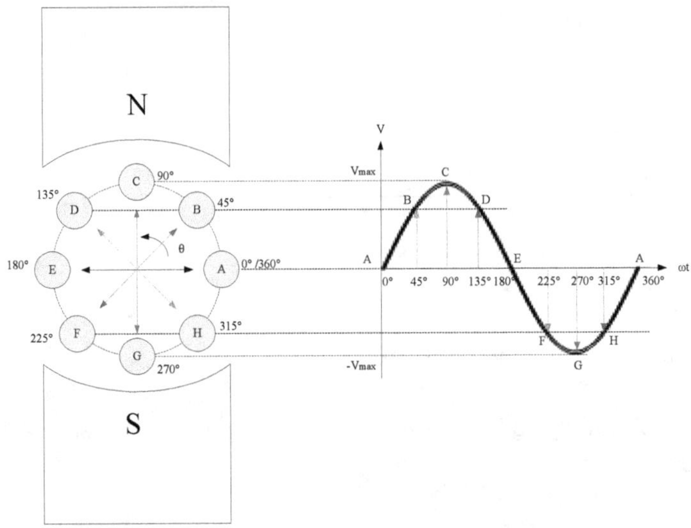

Figure 6.5 The model of a generator that produces alternating current.

The above generator consists of a coil with copper wire in a rectangular frame that connects its ends to isolated copper conductors installed on a rotational axis moving at a constant speed within two magnetic poles, north, and south.

Because of the conductor rotation and a side intersection with the lines of the magnetic forces in which an electric motive force is produced, as in Figure 6.6, the value depends on the magnetic flux density (B), the length of the conductor within the magnetic field (L), the velocity of the conductor (V). The angle formed by the direction of rotation with the direction of the lines of the magnetic field and the following expression for any moment is:

The instantaneous value of electric motive force.

$e = B \, . \, L \, . \, V \, . \, \sin \theta$

As the coil revolves (n) of the cycles per minute, the frequency calculates by the following equation.

$$F = \frac{n}{60}$$

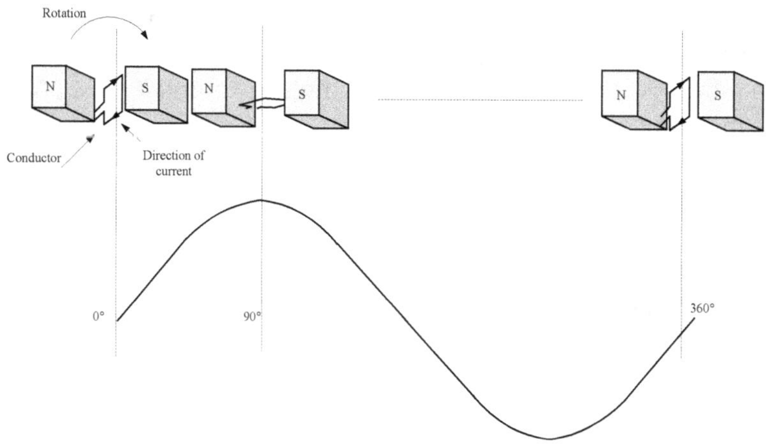

Figure 6.6 The process of generating (emf) in the conductor.

Terms and Concepts

Each cycle generates one period of the emf within two poles of the magnet for any moment, called *the instantaneous value* of the electric motive force as in the following equation.

$$E = 2\pi . \phi . f . N . \sin\theta$$

Where: N is the number of coil turns.

When the direction of the movement of the coil is vertical on the lines of the magnetic field, i.e., the angle is equal to 90°, i.e. ($\theta = 90°$), so (sin 90° = 1).On this basis, the maximum value of the voltage is

$$E_{max} = 2\pi\ \phi\ f\ N$$

Therefore, the expression of the real value of the voltage is

$$E = E_{max} \sin\theta$$

The actual value of the electric voltage (V_{eff}) is the value indicated by the measuring instruments and is equal to

$$V_{eff} = 0.707\ V_{max}$$

Example: A coil containing 100 turns rotates at a speed of 1500 rpm within the two magnetic poles, calculate (i) Frequency, (ii) Time (iii) Immediate electric motive force if you know that the angle between the movement of the conductor and the lines of the field is 30°, and the maximum value of voltage 3.14 V.

Solution

i) $f = n / 60$

$= 1500/60 = 25$ Hz

ii) $T = 1/f$

$= 1/25 = 0.04$ sec.

iii) $e = Emax\ sin\ \theta$

$= 3.14 \times \sin 30°$

$= 3.14 \times 0.5 = 1.57$ V

Example: A coil containing 50 turns rotates continuously inside a magnetic field. Calculate

(a) The speed at which the frequency of the electric motive force 25Hz must rotate the coil.
(b) Several poles.
(c) The maximum value of electric motive force if the current value is 15 V and the amount of angle 30°.
(d) The actual value of electric motive force.

Solution

a) $F = n/60$

$N = 60 \times F = 60 \times 35 = 1500$ rpm

b) $p = (120F) / n = (120 \times 25) / 1500 = 2$ poles

c) $e = E_{max} . \sin\theta$

$$E_{max} = \frac{e}{\sin\theta} = \frac{e}{\sin 30°} = \frac{15}{0.5} = 30V$$

d) E_{eff} = 0.707 E_{max}

Circuit elements	Voltage (V)	Current(A)	Power(W)
Resistor R (Ohms Ω)	$V = RI$	$I = \frac{V}{R}$	$P = i^2R$
Inductor L (Henry H)	$V = L\frac{di}{dt}$	$I = \frac{1}{L}\int vdt + io$	$P = Li\frac{di}{dt}$
Capacitor C (Farad F)	$I = \frac{1}{C}\int idt + v_0$ where v_0 is the initial voltage across the capacitor	$I = C\frac{dv}{dt}$	$P = CV\frac{dv}{dt}$

= 0.707 × 30 = 12.21 V

V-I Relation of circuit elements

Example: The current function for a pure resistor of 5 Ω is a repeating saw tooth, as shown below. Find v(t), P(t).

$V(t) = R\,i(t) = 5\,X\,10 = 50\,V$

$0‹t‹2ms$

$\frac{i}{t} = \frac{10}{2X10^{-3}} = 5X10^3$ $\qquad i = 5X10^3t$

$V = 5 \times 5 \times 10^3t = 25 \times 10^3$

$P = 125 \times 10^6t^2$

Example: A pure inductance L = 0.02H has an applied voltage $V(t) = 150\ sin\ 1000t$ volts. Determine the current i(t) and draw their wave forms.

$V(t) = 150\, sin\, 1000t \qquad L = 0.02H$

$$I(t) = \frac{1}{L}\int V dt = \frac{1}{0.02}\int 150 \sin 1000t dt$$

$$\frac{150}{0.02}\left(\frac{-\cos 1000t}{1000}\right)$$

$$\therefore -7.5 \cos 1000t\, Amps$$

$$= V(t)i(t)$$

$$= (-150)\,(7.5)\frac{1}{2}\sin 2000t$$

$$= -562.5\, sin\, 2000t$$

AC Current Values

The alternating current changes in value and direction on a sine wave. Some important values of the AC can be calculated as follows.

1 - The maximum value of the alternating current

It is the highest value of the current direction and arises in the case where the direction of the movement of the terminals of the coil is vertical on the lines of the magnetic field and symbolizes it (I_{max}) and shall be calculated from the following formula:

$$I_{max} = \sqrt{2}\ \ I_{eff}$$

$$I_{max} = 1.414\ \ I_{eff}$$

2. Average value of alternating current (mean value):

The ac value of the sinusoidal wave can be calculated according to the following equation and symbolized by the symbol (I_{av}):

$$I_{av} = 0.637\, I_{max}$$

3. Actual ac value:

It is the reading that is indicated by the measuring instruments that are read on these devices, symbolized by the symbol (I_{eff}) from the following equation:

$$I_{eff} = 0.707\, I_{max}$$

4- The Instantaneous value of alternating current:

Which is the value we get at any moment of the current, symbolized by the letter (i) and calculated from the following equation:

$$I = I_{max}\ \sin\theta$$

Example: What is the maximum value voltage of ac source has a voltage of 220 V?

Solution

$$V_{eff} = 0.707\ V_{max}$$

$$V_{max} = V_{eff} / 0.707$$

$$V_{max} = 220 / 0.707 = 311V$$

Example: An ammeter is connected to a circuit through which an alternating current passes. If the reading recorded by this device is 2 A, find the maximum value of the current.

Solution

$$I_{eff} = 0.707\ I_{max}$$

$$I_{max} = I_{eff} / 0.707$$

$$I_{max} = 2 / 0.707$$

$$I_{max} = 2.82\ A$$

Example: A wire rotates at a speed of 300 rpm in a uniform magnetic field. Suppose the frequency 50 Hz. Calculate the number of poles.

Solution

$$n = \frac{60\,f}{p}$$

$$P = \frac{60\,f}{n} = \frac{60 \times 50}{3000}$$

$P = 3000 / 3000 =$ 1 pole pair or two poles

AC Circuits

AC Circuit Containing Pure Resistance

Figure 6.7a shows a pure resistance connected to an ac source, where the current wave is in phase with the voltage wave, as shown in Figure 6.5b. Both waves start from zero and reach the greatest value at a 90° angle, then fall to zero at 180°, and so the waves are completed simultaneously, as in Figure 6.5 b is the representative of the two waves with directional lines.

(a)

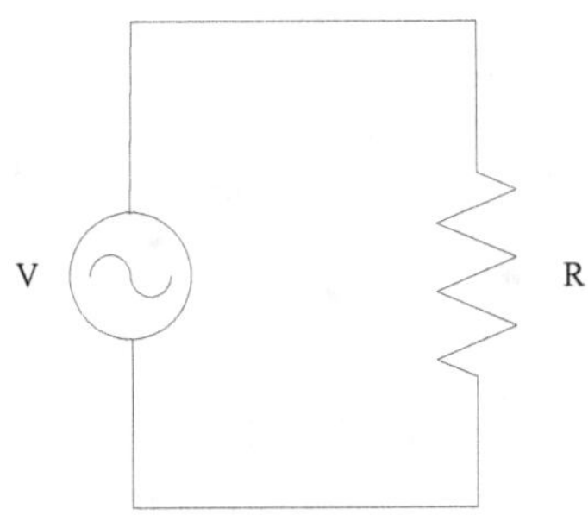

(b)

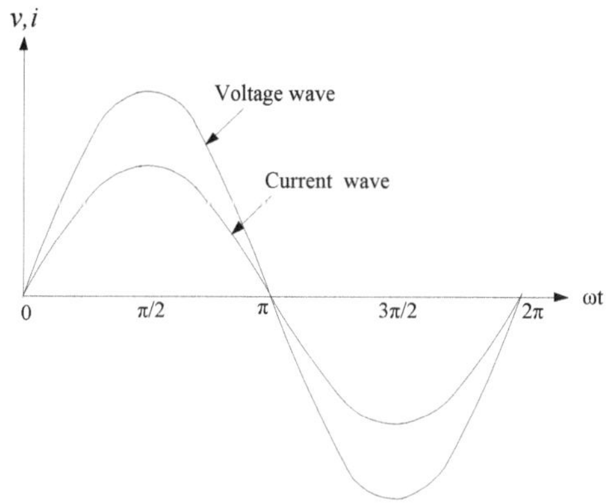

Figure 6.7 AC circuit containing only pure resistance.
a. Circuit diagram. b. Voltage and current waveforms.

The resistance in the circuit shall be calculated as follows

$$R = \frac{V}{I}$$

or of the following formula

$$R = \frac{\rho L}{A}$$

Where:
ρ = resistivity and its unit $\Omega.mm^2/m$
L = Conductor length in meters
A = cross Section area in mm^2

AC Circuit with Inductive Reactance

Figure 6.8 represents the inductive connection in the circuit of the alternating current, as the current wave is delayed from the voltage wave at an angle of 90° because of the phenomenon of self-induction. As the back emf at zero, the current wave passes until the voltage wave reaches the upper value at an angle of 90°. The current wave starts to appear because the back emf equals zero. Figure 6.7 represents the circuit voltage and current waves containing inductive load connected to the ac source.

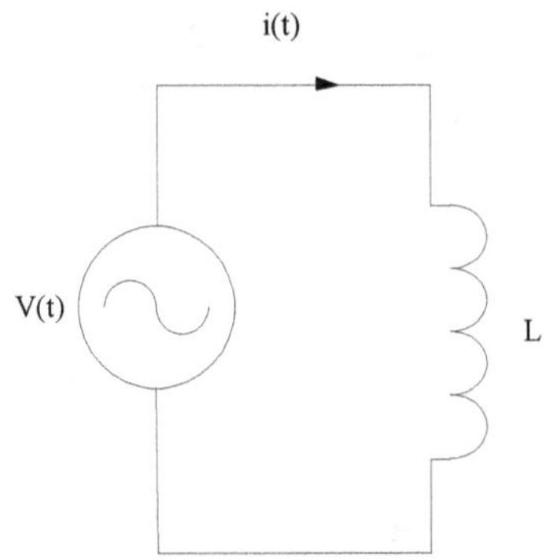

Figure 6.8 AC circuit containing inductive reactance.

The inductive reactance is represented by the symbol "X_L," the unit is measured by the Ω. It is calculated from the following formula:

$$XL = 2\pi . f . L$$

Where

f = frequency.
L = Self-inductive factor.

$$\omega = 2\pi . f$$

ω = Angler Speed (rad/sec).

$$XL = \omega . L$$

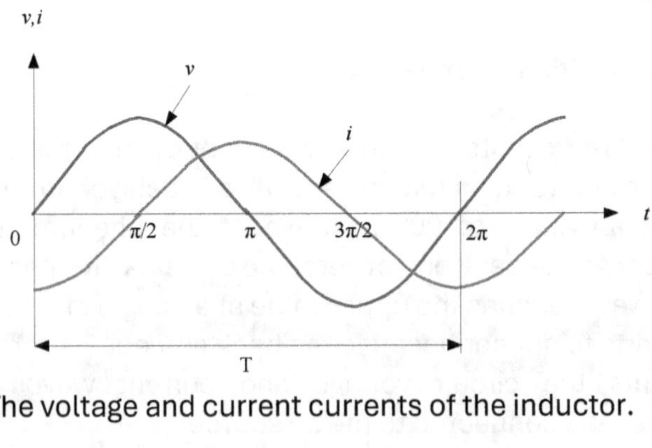

Figure 6.9 The voltage and current currents of the inductor.

From the above, scientifically, each coil has a natural resistivity ($R = \rho.L/A$) and magnetic resistance ($X_L = 2\pi.f.L$), and these two resistors are always connected to the series.

AC Circuit with Capacitive Reactance

Figure 6.10 represents a capacitor in the circuit of the alternating current, as the current wave, in this case, ahead of the voltage wave at an angle of 90°, for the passage of current in the capacitor to charge, and then shows the voltage on the ends gradually, as shown in Figure 6.11a. Figure 6.11b represents the voltage and current waves with directional lines.

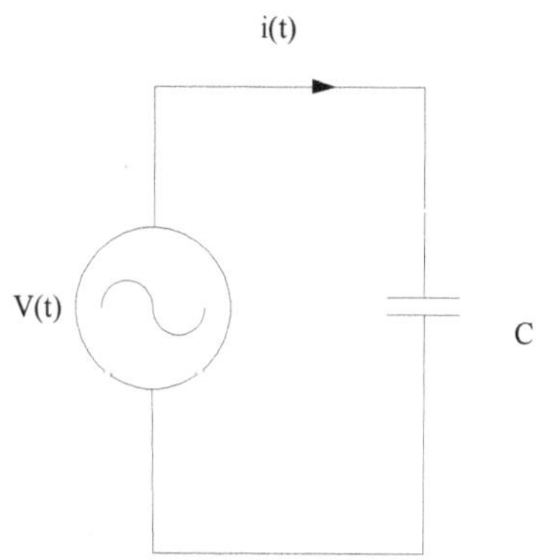

Figure 6.10 Capacitor in the circuit of the alternating current.

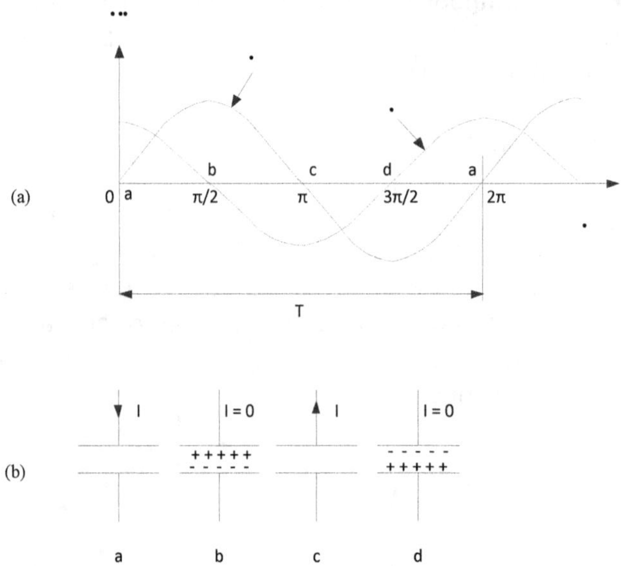

Figure 6.11 Voltage and current waveforms and the direction of the charges and current follow.

The capacitive reactance (Xc) is a unit of measurement (Ω), and its value depends on frequency (f) and capacitor (C) and is calculated from the following formula.

$$XC = \frac{1}{2\pi.F.C}$$

Where: C = Capacitor capacity and calculated by F (Farad)

$$\omega = 2\,\pi.f$$

$$XC = \frac{1}{\omega C}$$

ω = angular speed in (rad/sec.)

Series Impedance Connection to the Ac Circuit

1 R-L Series Circuit

Figure 6.12 represents a series of resistance and inductor connections with an ac voltage source. From Figure 6.10 b, note that the voltage (V_R) on the resistance (R) is in phase with the angle of the current in the circuit, and the voltage across the inductor (V_L) ahead of the current at an angle of 90°, so the total voltage of the circuit (V) is the sum of both voltages.

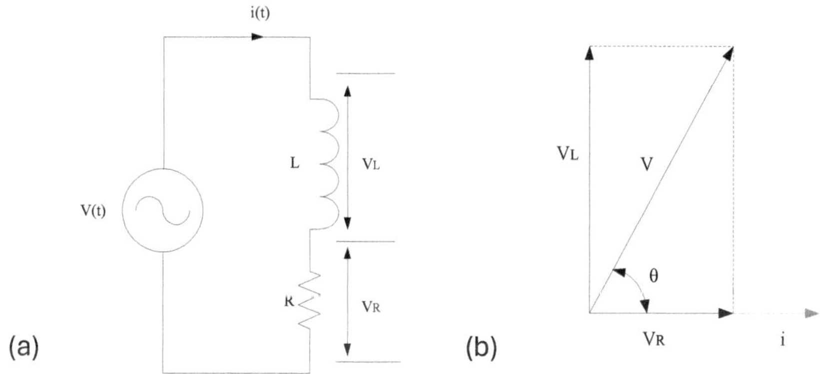

Figure 6.12 Circuit (a) and phasor diagram (b) of R-L series circuit.

From the right-angled triangle in Figure 6.10 b, according to Pythagoras' theory, the total voltage is equal to

$$V = \sqrt{V_R{}^2 + V_L{}^2}$$

Thus, the power factor can be calculated from this triangle.

$$\cos\theta = \frac{VR}{VL}$$

where cos θ = power factor of the circuit.

The power factor can also be calculated from the Impedance triangle Figure 6.13.

$$\cos\theta = \frac{R}{Z}$$

From Figure 6.12a, the current is equal in all parameters of the circuit, so the total impedance of the circuit is equal to:

$$Z = \frac{V}{I}$$

where:

Z = total impedance of the circuit and calculated by Ω.

V = the total voltage of the circuit and calculated by volts.

I = total circuit current and calculated in amperes.
and

$R = V_R / I$

$X = V_L / I$

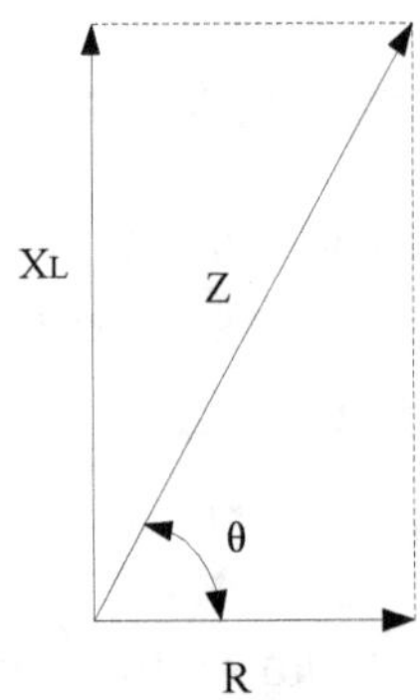

Figure 6.13 The impedance triangle.

Example: A self-inductance of 10 mH and its resistance 5 Ω connect to an AC source of 200 V and frequency of 60 Hz, calculate the current in the circuit and voltage on both ends of resistance and reactance.

Solution

$$X_L = 2\pi.f.L$$
$$X_L = 2\times 3.14 \times 60 \times 0.01$$

$$X_L = 3.77\ \Omega$$

$$Z = \sqrt{R^2 + X_L^2} = \sqrt{5^2 + 3.77^2} = 6.262\ \Omega$$

$$I = V / Z = 200 / 6.262 = 31.93\ A$$

$$V_R = I \times R = 31.93 * 5 = 159.7\ V$$

$$V_L = I \times X_L = 33.9 * 3.77 = 120.41\ V$$

R-C Series Circuit

Figure 6.14a; represents a series connection of resistance and capacitor with an ac voltage source. From Figure 6.14b, the voltage on the resistance is aligned with the current with the circuit's current, and the voltage on the capacitive reactance is 90° below the current. Therefore, the total voltage of the circuit is the sum of both voltages.

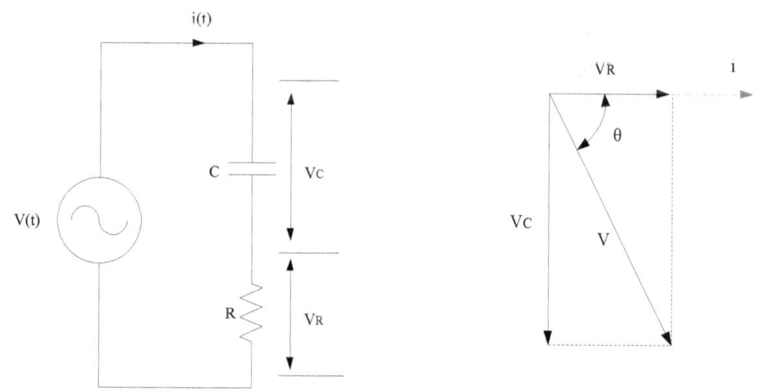

Figure 6.14 Resistors and capacitor in series. Left: Circuit diagram and Voltage and current directions (right).

Figure 6.14b represents the voltage triangle and can be found in the power factor formula.

$$\cos\theta = \frac{VR}{V}$$

As for the impedance triangle, as in Figure 6.15, the power factor formula can be calculated.

$$\cos\theta = \frac{R}{Z}$$

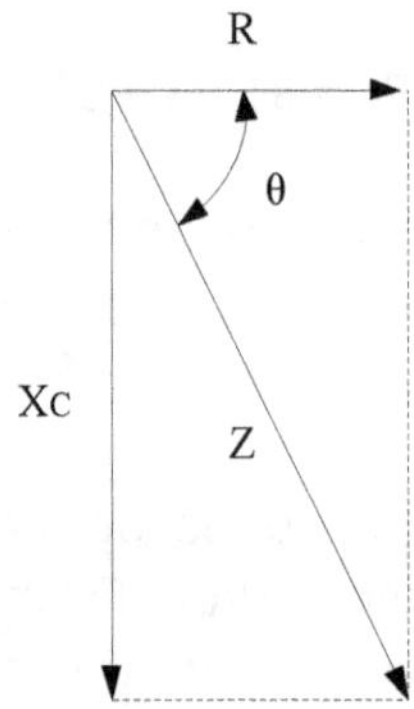

Figure 6.15 RC impedance triangle.

Example: A resistance of 30 Ω and a capacitor of 80 μF are connected in series to an AC source of voltage 200 V, 50 Hz. Calculate the total current of the circuit and the voltage on the capacitor.

Solution

$$X_C = \frac{1}{2\pi f C} = \frac{1}{2\times3.14\times50\times80}$$

$$= 40\ \Omega$$

$$Z = \sqrt{R^2 + {X_C}^2}$$

$$= \sqrt{30^2 + 40^2} = 50\ \Omega$$

$$I = \frac{V}{Z} = \frac{200}{50} = 4\ A$$

$$V_C = I * X_C = 4 \times 40$$

$$= 160\ V$$

R-L-C series circuit

Figure 6.16a; represents a series connection of resistance, inductor, and capacitor with an ac voltage source. From the phasor diagram shown in Figure 6.16b, we note that the voltage on the resistance is in phase with the current. The phase angle between the voltage on the inductance and the current is 90°, while the angle between the voltage on the capacitance and current is 90°, as shown in Figure 6.16b. The voltage on the inductive reactance, the voltage on the capacitance reactance on one straightness, and at an angle of 180° is reversed In the direction.

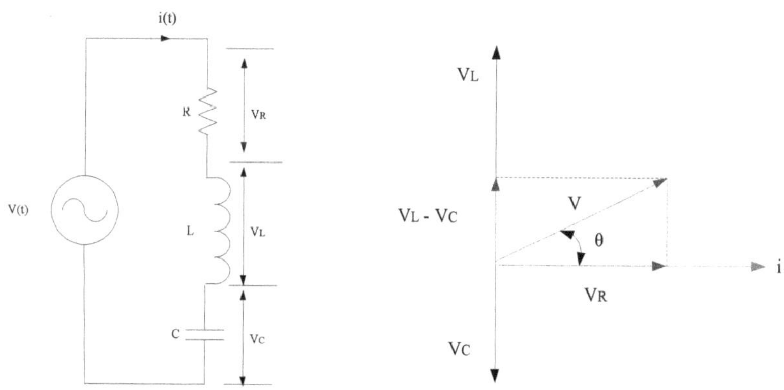

Figure 6.16 Circuit diagram (left) and voltage vectors (right) for the (R-C-L) circuit.

From Figure 6.16b, we note

$$V_{LC} = V_L - V_C.$$

$$V = \sqrt{V_R^2 + V_{LC}^2}$$

$$V = \sqrt{V_R^2 + (V_L - V_C)^2}$$

$$I.Z = \sqrt{I^2.R^2 + (I.X_L - I.X_C)^2}$$

$$I.Z = I\sqrt{R^2 + (X_L - X_C)^2}$$

$$Z = \sqrt{R^2 + (X_L - X_C)^2} \quad \text{if } X_L > X_C$$

$$Z = \sqrt{R^2 + (X_C - X_L)^2} \quad \text{if } X_L < X_C$$

$$\cos\theta = \frac{R}{Z}$$

Example: An RLC circuit containing a resistance of 20 Ω, inductance of 0.16 H, and capacitor of 91 μF has connected an alternating current source of 200 V and its frequency of 50 Hz. Calculate the total impedance and the total current in the circuit.

Solution

$$X_L = 2\pi.f.L$$

$$X_L = 2 \times 3.14 \times 50 \times 0.16 = 50\Omega$$

$$X_C = 1/(2\pi f\, C)$$

$$X_C = 1/(2 \times 3.14 \times 50 \times 91 \times 10^{-6})$$

$$X_C = 10^6/(91 \times 314) = 35\ \Omega$$

$$Z = \sqrt{R^2 + (X_L - X_C)^2}$$

$$Z = \sqrt{20^2 + (50 - 35)^2} = \sqrt{20^2 + 15^2} = \sqrt{400^2 + 225^2}$$

$$Z = 25\ \Omega$$

$$I = 200/25 = 8\ A$$

Parallel Connection

Parallel R-L Circuit

Figure 6.17a; represents a parallel connection of resistance and inductor with an ac voltage source. In Figure 6.17b, we note that the voltage (V) of the circuit is equally on terminals of both the resistance and inductance. The drawing of the directional lines shows the horizontal line representing the circuit's voltage. With the voltage, the current passes through the resistance in phase with voltage while the current following in the reactance is later the voltage at (90°).

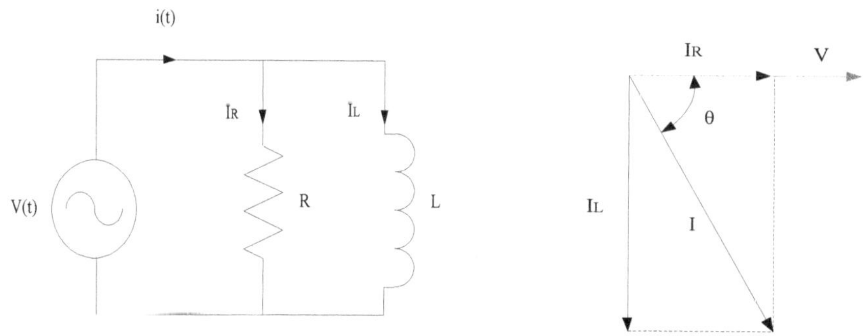

(a) (b)

Figure 6.17 The R-L parallel circuit. Left: Circuit diagram and right: Phasor diagram.

From Figure 6.17b and according to the Pythagoras theory

$$I^2_T = I^2_R + I^2_L$$

$$I_T = \sqrt{I_R^{\ 2} + I_L^{\ 2}}$$

$$\sin\theta = I_L / I_T$$

$$I_L = I_T \sin\theta$$

$$\cos\theta = I_R / I_T$$

$$I_R = I_T . \cos\theta$$

where I_R = Resistive current.
I_L = Inactive current.
I_T = Total Current.

Example: A circuit with 50 Ω of resistance, parallel to the self-induction coefficient 89 mH, connected to the source of the alternating voltage 200V and its frequency 50Hz.Calculating the total current of the circuit.

Solution

$$I_R = V/R$$

$$I_R = 200/50 = 4\ A$$

$$X_L = 2\pi f L$$

$$X_L = 2\times\pi\times 50 \times 0.089$$

$$X_L = 28\ \Omega$$

$$I_L = 200/28 = 7.12\ A$$

$$I_T = \sqrt{{I_R}^2 + {I_L}^2}$$

$$I = \sqrt{4^2 + 7.12^2} = 8.18\ A$$

Parallel R-C circuit

The electric circuit in Figure 6.18 represents an R-C circuit connected in parallel to the voltage supply. The voltage of the circuit (V) is the same at terminals of both resistance and capacitance. The current in the resistance is in phase with the voltage and current in the capacitance ahead of the voltage by an angle of 90°, and the sum of both currents is the total current (I) as in Figure 6.18b.

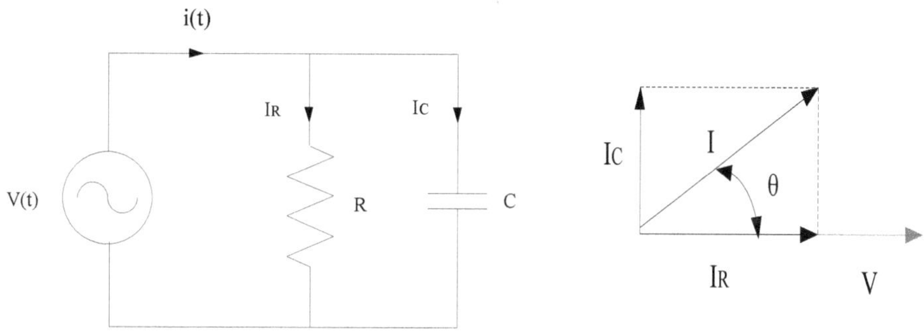

(a) (b).

Figure 6.18 The voltage and current vectors of the parallel circuit R-C.

From Figure 6.18b.

$$I^2_T = I^2_R + I^2_C$$

$$I_T = \sqrt{I_R{}^2 + I_C{}^2}$$

where IT = The total current.
I_R = The current in the resistance.
I_C = The current in the capacitance.

Example: A parallel R-C circuit containing 25 Ω of resistance, and a capacitor of 95.5 μF, connected to a voltage source of 200 V and a frequency of 50. Calculate the total current of the circuit.

Solution

$$X_C = \frac{1}{2\pi.F.C}$$

$$= 1/\left(2\times 3.14\times 50 \times 95.5 \times 10^{-6}\right)$$

$= 10^6 / 29987$

$= 33.34\ \Omega$

$I_C = V/X_C$

$= 200/33.34 = 6\ A$

$I_R = V/R$

$= 200/25 = 8\ A$

$I_T = \sqrt{I_R^2 + I_C^2}$

$= \sqrt{8^2 + 6^2}$

$= \sqrt{64 + 36}$

$= 10\ A$

Example: The following Figure 6.19, the power system has two loads attached to a 6 Ω line with a supply current of values of $5\angle 0°A$, if the load's data are given as:

Load 1: P=100 W at 0.8 pf leading.
Load 2: P=1000 W at 0.7 pf lagging. Determine

a) The value of the circuit's total watts, VAR and VA.
b) Value of the supply voltage E.
c) The power factor for the circuit.
d) Type of element and their impedance in each box.

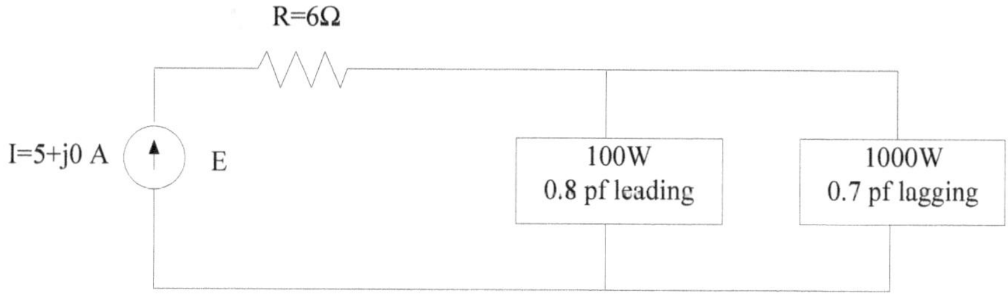

Figure 6.19 Circuit.

Solution

Given Load 1: P_1=100 W, $\cos\phi_1$=0.8 pf leading.

Reactive power is

$$Q_1=P_1 \tan \phi_1=100*\tan (\cos^{-1}0.8) =75 \text{ VAR.}$$

The total complex power for load 1 is

$$S_1=P_1-jQ_1=100-j75 \text{ VA}$$

For Load 2: P_2=1000 W, $\cos\phi_2$=0.7 pf lagging.

Reactive power is

$$Q_2=P_2 \tan \phi_2=100*\tan (\cos^{-1}0.7) =1020.2 \text{ VAR}$$

The total complex power for load 2 is

$$S_2=P_2-jQ_2=1000+j1020.2 \text{ VA}$$

The P and Q of a 6-ohm line is

$$P_r=I^2*R=52*6=150 \text{ W}$$

The reactive power of the resistor is zero, i.e., Q_r=0 V.A.R.

a) The total watts is:

$$P=P_1+P_2+P_3=150+100+1000=1250\ W$$

The total VAR is

$$Q=Q_2-Q_1=1020.2-75=945.2\ VAR$$

The total VA is

$$S=P+jQ=1250+j945.2\ VA=1567.13\angle 37.1°\ VA$$

b) The supply voltage E

$$E=\frac{S}{I^*}=\frac{1567.13\angle 37.1°}{5\angle 0°}=313.43\angle 37.1°\ V$$

c) The power factor for the circuit

$$pf=\cos(\tan^{-1}\frac{945.2}{1250})=0.798\ \text{lagging.}$$

d) The voltage across loads is

$$V=E-I*R=313.43\angle 37.1°-5\angle 0°*6=290.06\angle 40.67°\ V$$

The current through load 1 is

$$I_1=(\frac{S_1}{V})^*=(\frac{100-j75}{290.06\angle 40.67°})^*=0.431\angle 77.54°\ A$$

The impedance of load 1 is

$$Z_1=\frac{V}{I_1}=\frac{290.06\angle 40.67°}{0.431\angle 77.54°}=538.4-j403.8\ \Omega$$

Resistor and capacitor as load as power factor is leading.

Current through load two is

$I_2 = I - I_1 = 5\angle 0° - 0.431\angle 77.54° = 4.925\angle - 4.9°\ A$

The impedance of load 2 is

$$Z_2 = \frac{V}{I_2} = \frac{290.06\angle 40.67°}{4.925\angle - 4.9°} = 41.23 + j42.06\ \Omega$$

The resistor and inductor as load as power factor is lagging.

Example: Three loads are connected in parallel across 660V(rms),60 Hz line as shown in Figure 6.20

Load 1: absorbs 18 kW and 10kVAR with a lagging power factor.
Load 2: absorbs 6 kVA at 0.96 leading power factor.
Load 3: absorbs 22.4kW at unity power factor. Determine

a) The value of the total watts, VAR and VA for the circuit.
b) The power factor for the circuit.

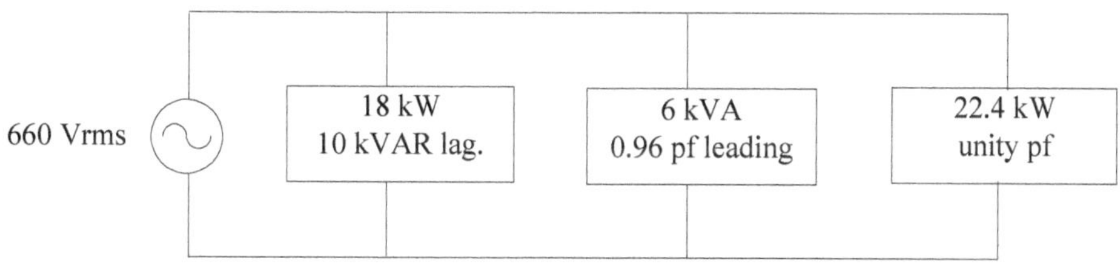

Figure 6.20 Circuit.

Solution

Given:
Load 1: absorbs 18 kW and 10kVAR with a lagging power factor.
Load 2: absorbs 6 kVA at 0.96 leading power factor.
Load 3: absorbs 22.4kW at unity power factor.

For load 2, the reactive power is

$$Q_2 = S_2 \sin \phi_2 = 6 * \sin (\cos^{-1} 0.96) = 5.76 \text{ kVAR}.$$

The total complex power for load 2 is

$$S_2=P_2+jQ_2=6+j5.76 \text{ kVA}$$

For Load 3: P_3=22.4 kW, $\cos\phi_3$=1.

Reactive power is Q_3=0 VAR (unity power factor).

The total complex power for load 3 is

$$S_3=P_3-jQ_3=22.4+j0 \text{ kVA}$$

a) The total watts is:

$$P=P_1+P_2+P_3=18+5.76+22.4=46.16 \text{ kW}$$

The total VAR is

$$Q=Q_1+Q_2+Q_3=10+(-1.68)=8.32 \text{ kVAR}$$

The total VA is

$$S=P+jQ=46.16+j8.32 \text{ kVA}=46.9\angle 10.21^\circ\, kVA$$

b) The power factor for the circuit

$$pf=\cos\left(\frac{46.16}{46.9}\right)=0.798 \text{ lagging.}$$

Example: Three loads are connected in parallel across a 500V(rms),60 Hz line, as shown in Figure 6.21.

Load 1: absorbs 20 kW and 10kVAR with a lagging power factor.
Load 2: absorbs 6 kVA at 0.6 leading power factor.
Load 3: absorbs 10kW at unity power factor. Determine
a) The value of the total watts, VAR and VA for the circuit.
b) The power factor for the circuit.

c) The value of capacitance C connected in parallel across the loads that will raise the power factor to unity.

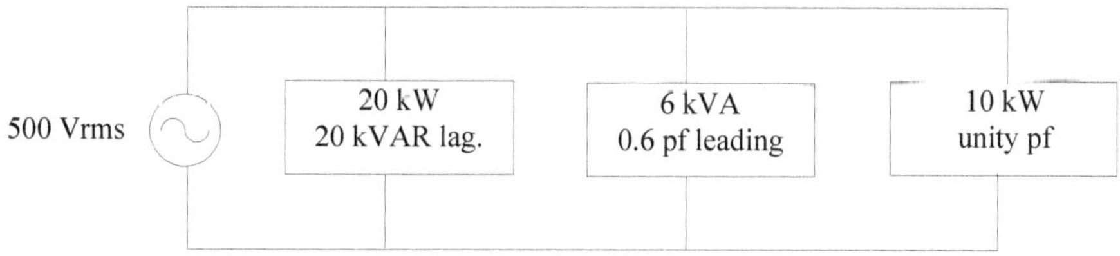

Figure 6.21 Circuit.

Solution

Given:
Load 1: absorbs 20 kW and 20kVAR with a lagging power factor.
Load 2: absorbs 6 kVA at 0.6 leading power factor.
Load 3: absorbs 10 kW at unity power factor.

For load 2, the reactive power is

$$Q_2=S_2 \sin \phi_2=6*\sin(\cos^{-1}0.6)=4.8 \text{ kVAR}.$$

The total complex power for load 2 is

$$S_2=P_2+jQ_2=6+j4.8 \text{ kVA}$$

For Load 3: $P_3=10$ kW, $\cos\phi_3=1$.

Reactive power is $Q_3=0$ VAR (unity power factor).

The total complex power for load 3 is

$$S_3=P_3-jQ_3=10+j0 \text{ kVA}$$

a) The total watts is:

$$P=P_1+P_2+P_3=20+6+10=36 \text{ kW}$$

The total VAR is

$$Q=Q_1+Q_2+Q_3=20+(-4.8)=15.2\ kVAR$$

The total VA is

$$S = P + jQ = 36 + j15.2\ kVA = 39\angle 22.89°\ kVA$$

b) The power factor for the circuit

$$pf = \cos(22.89°) = 0.921\ lagging.$$

c)

$$pf = \frac{P}{S}$$

With the required unity power factor S =P=36 kVA, Q=0, Q_c=15.2 kVAR.

$$X_C = \frac{V}{Q_c} = \frac{500}{15.2 * 10^3} = 0.033\ \Omega$$

$$C = \frac{1}{2\pi f X_c}$$

$$C = \frac{1}{2\pi * 60 * 0.03} = 88.42\ mF$$

Multiple Choice Questions

1. A current of 3.0 μA flows through a resistor in 1.5 minutes. How much charge flows through the resistor at this time?

   A. 4.5 x 10-3 C
   B. 2.7 x 10-4 C
   C. 4.5 x 10-6 C
   D. 2.7 x 10-1 C

Ans: B

2. A potential difference of 6V is applied across XY in the circuit shown.

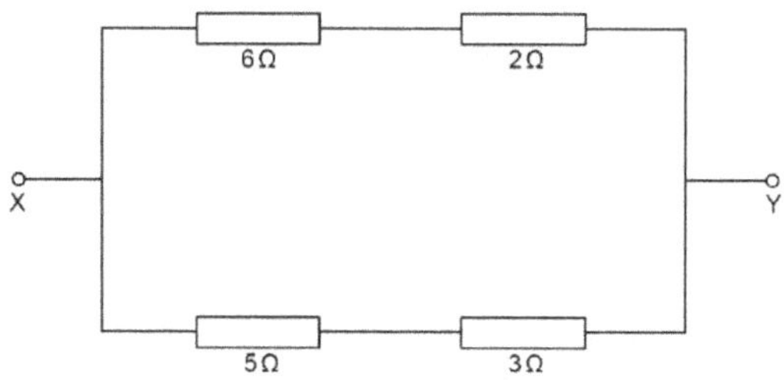

What is the current in the 5 Ω resistor?

A $\frac{15}{4}$A

B $\frac{3}{4}$A

C $\frac{3}{5}$A

D $\frac{3}{8}$A

Ans: B

3. As shown, A potential divider is used to give outputs of 2 V and 3 V from a 5 V source.

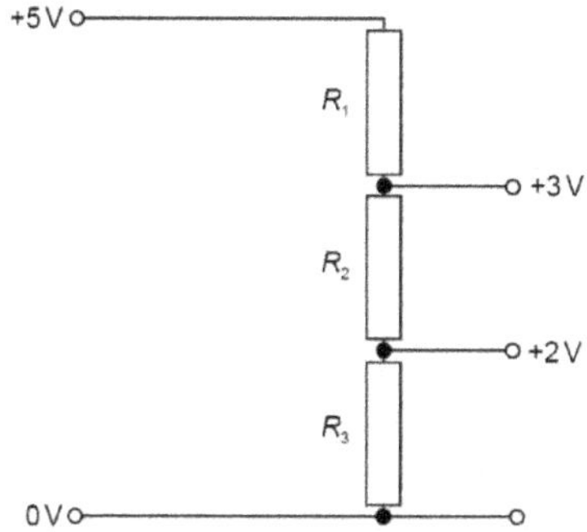

Which combination of resistors R_1, R_2, R_3, gives the correct voltages?

	R_1 / kΩ	R_2 / kΩ	R_3 / kΩ
A	2	1	2
B	1	1	2
C	3	2	2
D	3	2	3

Ans: A

4. The circuit's resistors P, Q, and R have equal resistance.

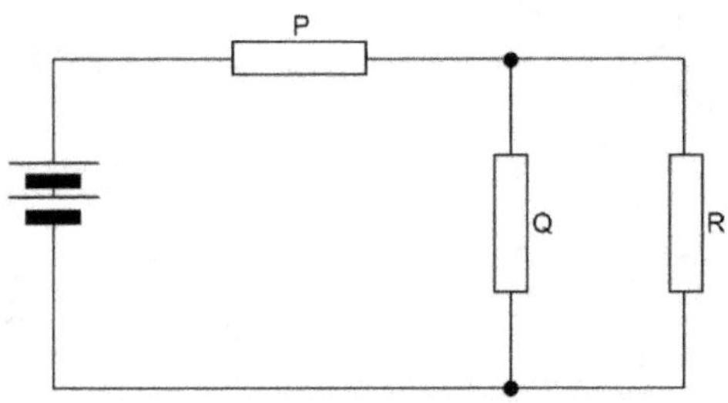

5. The battery, of negligible internal resistance, supplies a total power of 12W. What is the power dissipated in resistor R?

   A 4 W
   B 3 W
   C 2 W
   D 5 W

Ans: C

6. Two conductors of conductance 0.24S and 0.36S, are connected in parallel to a 4.5 V battery of negligible internal resistance, as shown below.

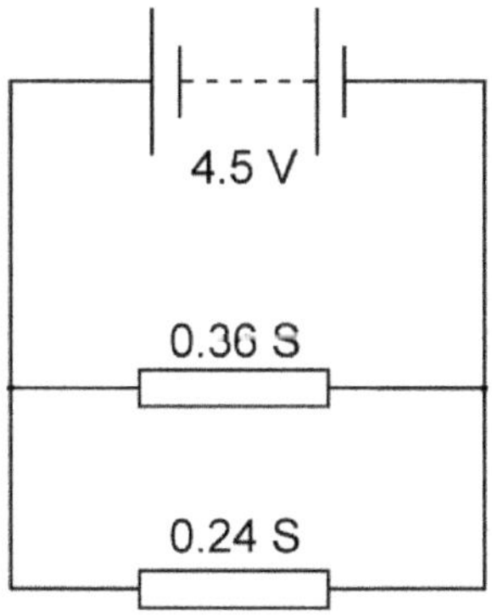

What is the total power dissipated in the two conductors?

A 12 W

B 2.9 W

C 7.3 W

D 4.9 W

Ans: A

7. An unknown resistor R and a 220 Ω resistor are connected to a 6.0 V battery of negligible internal resistance, as shown in the diagram below.

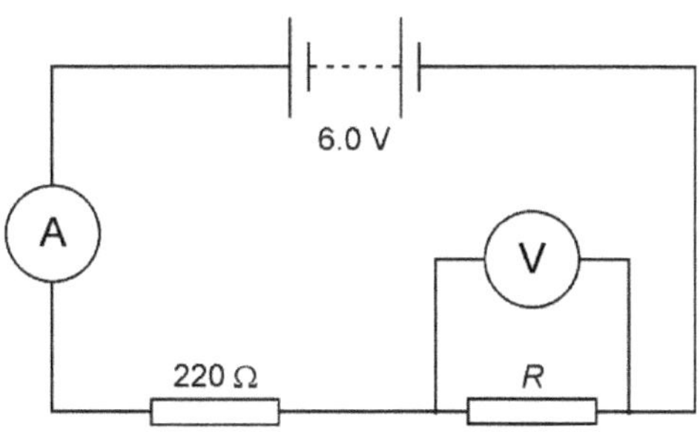

The reading on the voltmeter is 4.1 V. What is the reading on the ammeter?

A 19 mA

B 15 mA

C 8.6 mA

D 30 mA

Ans: C

8. A battery is connected across resistor R via a switch S.

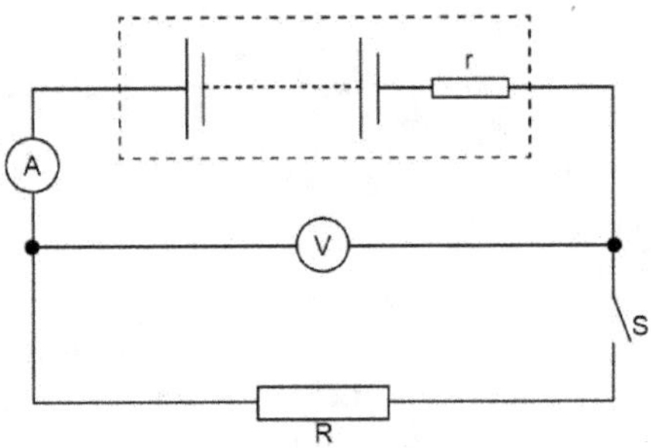

The switch S is open. The voltmeter reading is 9.0 V, and the ammeter reading is zero. With S closed, the voltmeter reading is 6.0 V, and the ammeter reading is 2.0 A. What is the internal resistance of the battery?

A 3.0 Ω

B 1.5 Ω

C 6.0 Ω

D 4.5 Ω

Ans: B

9. A 14 V d.c. supply is used to charge a 12 V car battery of internal resistance 0.80 Ω at a rate of 2.5 A for 6 hours.

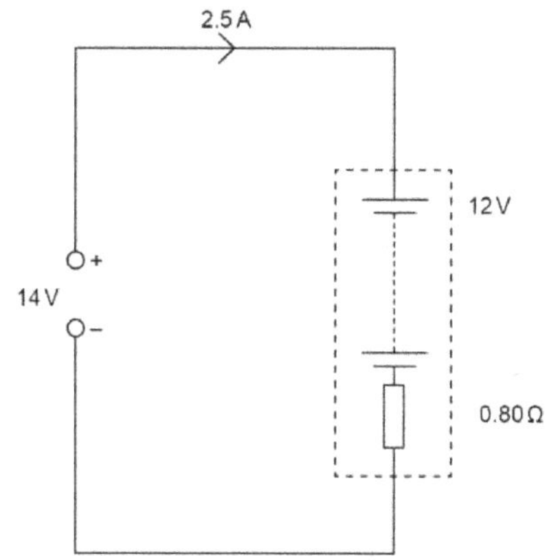

How much electrical energy is provided by the charging supply?

A 110 kJ

B 13 kJ

C 760 kJ

D 650 kJ

Ans: C

10. The y-input terminals of an oscilloscope are connected to a voltage supply of peak value 5.0 V and frequency 50 Hz. The time base is 10 ms per division, and the y-axis is 5.0 V per division. Assuming that these diagrams are to scale, which trace will be obtained?

A

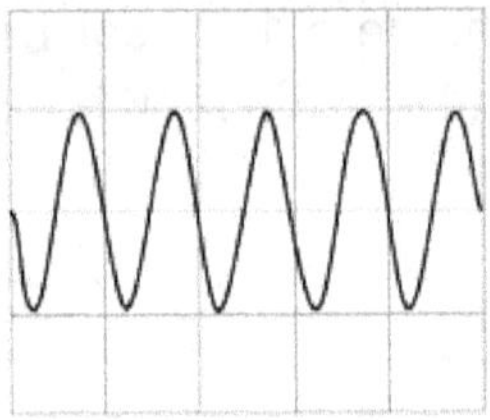

B

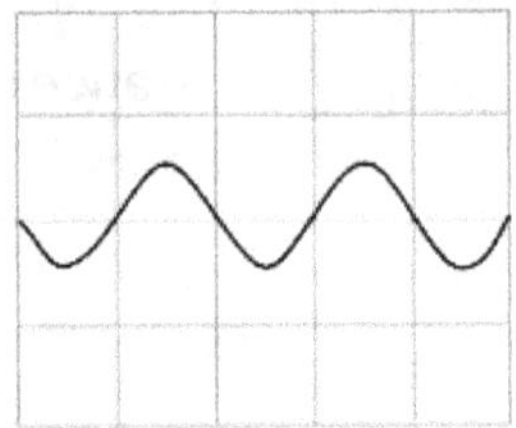

C

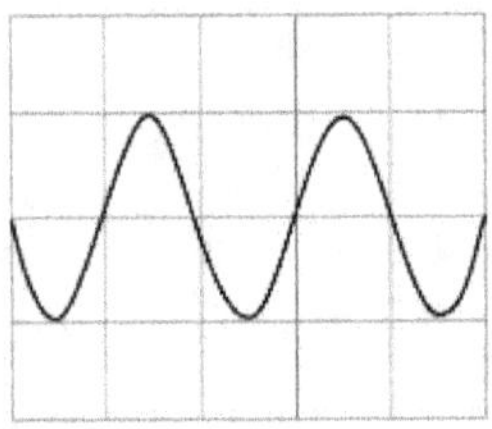

D

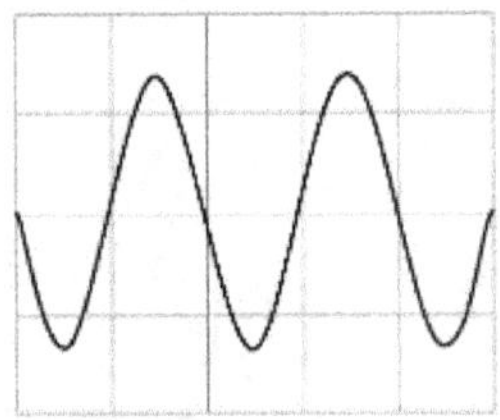

Ans: A

Solved Examples

1. For the circuit shown below, calculate
a. The total resistance
b. The total current
c. The total power dissipated in each resistor

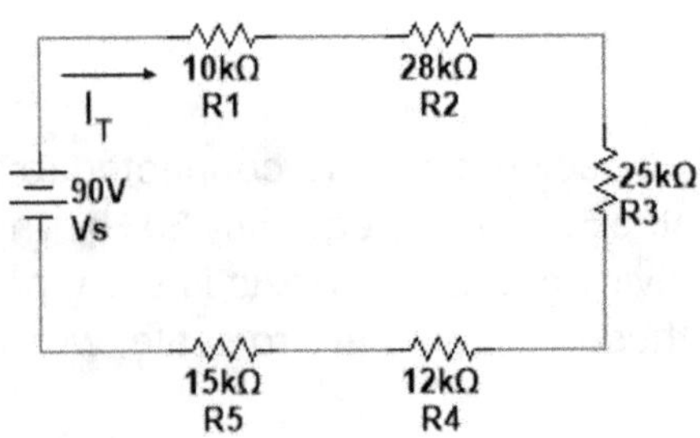

Solution:

a. Rt=R1+R2+R3+R4+R5=10+28+25+12+15=90 kΩ

b. $It = \frac{Vs}{Rt} = \frac{90}{90*10^3} = 1\ mA =$
$the\ current\ in\ each\ resistor\ (series\ connection)$

c. $P_{R1} = I_t^2 R_1 = (1 * 10^{-3})^2 * 10 * 10^3 = 0.01\ W$

$$P_{R2} = I_t^2 R_2 = (1 * 10^{-3})^2 * 28 * 10^3 = 0.028\ W$$

$$P_{R3} = I_t^2 R_3 = (1 * 10^{-3})^2 * 25 * 10^3 = 0.025\ W$$

$$P_{R4} = I_t^2 R_4 = (1 * 10^{-3})^2 * 12 * 10^3 = 0.012\ W$$

$$P_{R5} = I_t^2 R_5 = (1 * 10^{-3})^2 * 15 * 10^3 = 0.015\ W$$

$$\begin{aligned} Total\ Power\ &= P_{R1} + P_{R2} + P_{R3} + P_{R4} + P_{R5} \\ &= 0.01 + 0.028 + 0.025 + 0.012 + 0.015 \\ &= 0.09\ Watt = 90\ mW \end{aligned}$$

Or Ptotal =Vs It=90*1 mA=90 mW

2. For the circuit shown below, calculate

a. The total resistance
b. The total current
c. The power dissipated in each resistor

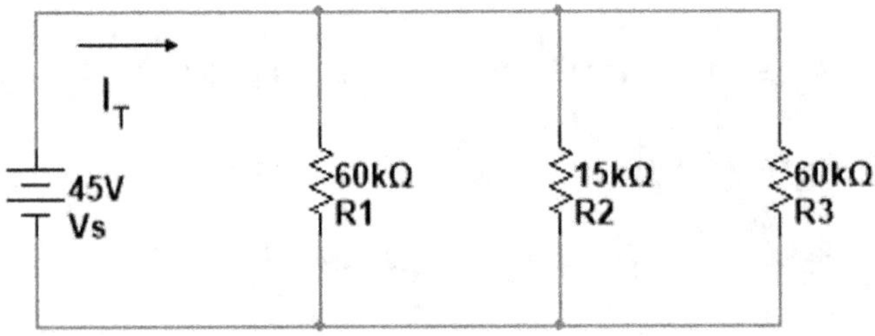

Solution:

a. $$\frac{1}{R_t} = \frac{1}{R_1} + \frac{1}{R_2} + \frac{1}{R_3} = \frac{1}{60\ k} + \frac{1}{15\ k} + \frac{1}{60\ k}$$

$$R_t = 10\ k\Omega$$

b.

$$I_t = \frac{V_s}{R_t} = \frac{45}{10 * 10^3} = 4.5\ mA$$

c. $$I_1 = \frac{V_s}{R_1} = \frac{45}{60\ k} = 0.75\ mA = I_3$$

$$I_2 = \frac{V_s}{R_2} = \frac{45}{15\ k} = 3\ mA$$

$$P_{R1} = I_1^2 R_1 = (0.75 * 10^{-3})^2 * 60 * 10^3 = 0.03375\ W$$

$$P_{R2} = I_2^2 R_2 = (3 * 10^{-3})^2 * 15 * 10^3 = 0.135\ W$$

$$P_{R3} = I_3^2 R_3 = (0.75 * 10^{-3})^2 * 60 * 10^3 = 0.03375\ W$$

$$Total\ Power = P_{R1} + P_{R2} + P_{R3} = 0.2025W$$

3. For the circuit shown below, calculate the following:

a. The total resistance

b. The total current

c. The total power dissipated in all resistors

d. The current in the 40kΩ?resistor ($R_?$)

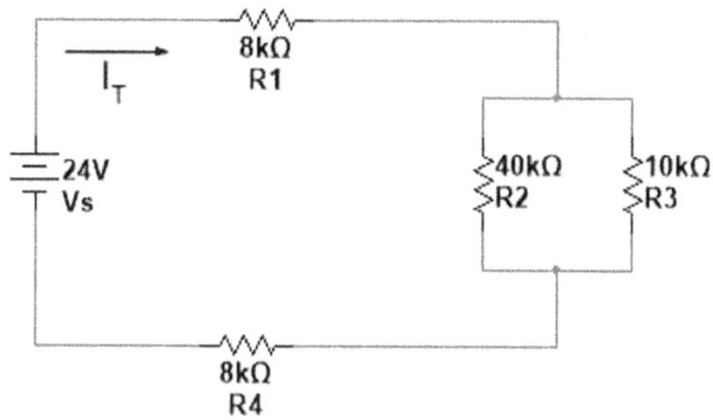

Solution:

a. $R_t = R_1 + R_2 \, // \, R_3 + R_4 = 8 + \frac{40*10}{40+10} + 8 = 24 \; k\Omega$

b.

$$I_t = \frac{V_S}{R_t} = \frac{24}{24*10^3} = 1 \; mA$$

$$I_1 = I_t = 1 \; mA = I_4$$

$$I_2 = I_t \frac{R_3}{R_? + R_3} = 1 \; mA \frac{10}{40 + 10} = 0.2 \; mA$$

$$I_3 = I_t \frac{R_2}{R_2 + R_3} = 1 \; mA \frac{40}{40 + 10} = 0.8 \; mA$$

$$P_{R1} = I_1^2 R_1 = (1 * 10^{-3})^2 * 8 * 10^3 = 8 \; mW$$

$$P_{R2} = I_2^2 R_2 = (0.2 * 10^{-3})^2 * 40 * 10^3 = 1.6 \; mW$$

$$P_{R3} = I_3^2 R_3 = (0.8 * 10^{-3})^2 * 10 * 10^3 = 6.4 \; mW$$

$$P_{R4} = I_4^2 R_4 = (1 * 10^{-3})^2 * 8 * 10^3 = 8 \; mW$$

$$Total \; Power = P_{R1} + P_{R2} + P_{R3} + P_{R4} = 24 \; mW = Vs \; It$$

4. For the circuit shown below, calculate V_{AB}.

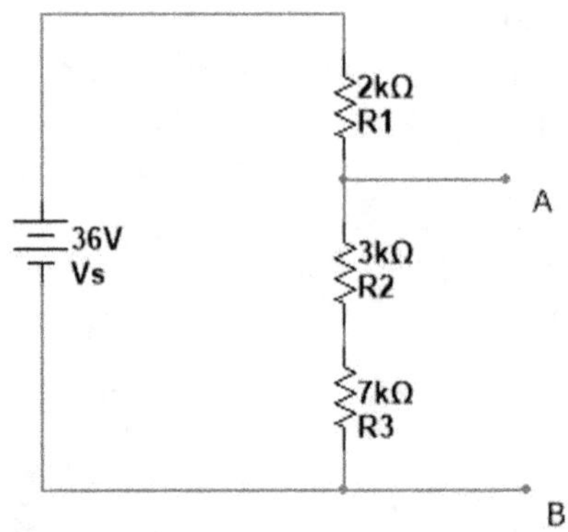

Solution:

$$V_{AB} = V_s \frac{R_2 + R_3}{R_1 + R_2 + R_3}$$

$$= 36 \frac{(3+7)k\Omega}{2+3+7}$$

$$= 30V$$

5. For the circuit shown below, calculate the following:

a. The total resistance

b. The total current

c. The power dissipated in the 20k resistor

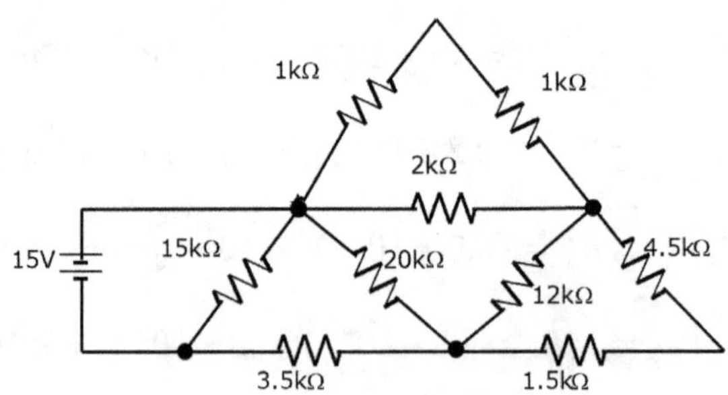

6. For the circuit shown below, if the current through R_3 is 1.5 mA and the voltage across R_1 is 10V, calculate:

a. R_2

b. R_3

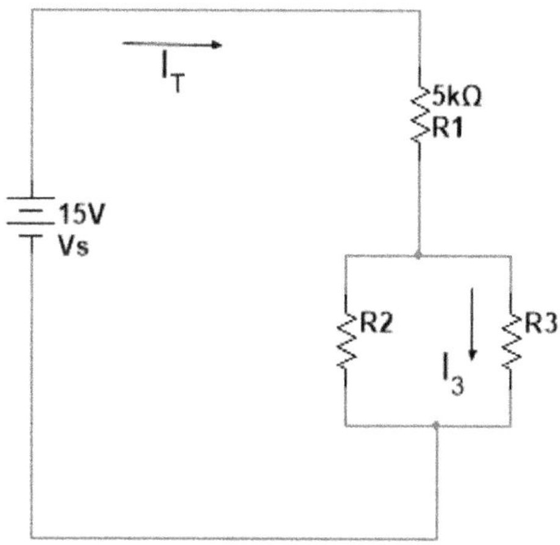

Solution:

Since Vs=15V and the voltage across R1 =10 V, the voltage across parallel resistors is 5 V, according to KVL.

So,

$$V_{R2} = V_{R3} = 5\ V$$

And

$$I_t = \frac{V_{R1}}{R1} = \frac{10}{5\ k} = 2\ mA$$

$$I_2 = I_t - I_3 = 2mA - 1.5\ mA = 0.5\ mA$$

a. $R_2 = \frac{V_{R2}}{I_2} = \frac{5}{0.5\ mA} = 10\ k\Omega$

b. $R_3 = \frac{V_{R3}}{I_3} = \frac{5}{1.5\ mA} = 3.333\ k\Omega$

7. Calculate the total resistance between points A & B for the circuit shown below.

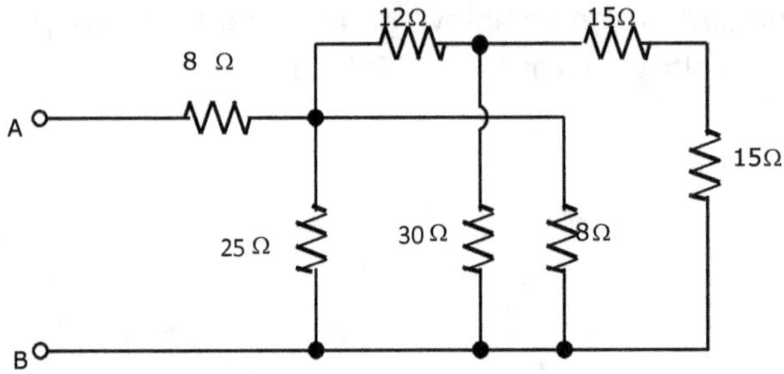

8. Determine the value of the current through R_2 in the circuit shown. Given I_T is 12 mA, I_1 is 3 mA, and I_3 is 2 mA.

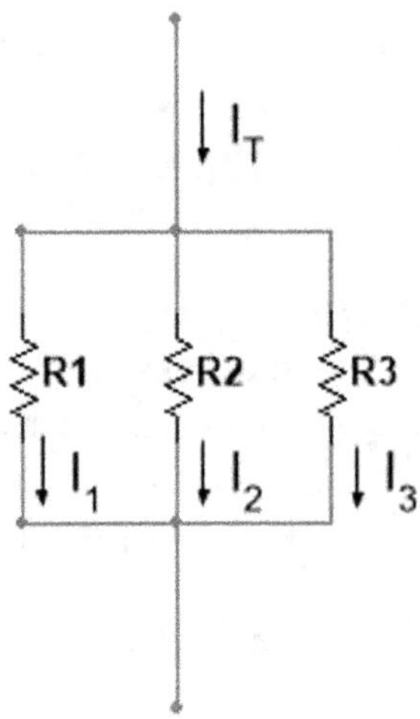

Solution:

Applying KCL:

$$I_T = I_1 + I_2 + I_3$$

$$12\ mA = 3\ mA + I_2 + 2\ mA$$

$$I_2 = 12 - 3 - 2 = 7\ mA$$

9. For the circuit shown below, the total resistance is 2 kΩ Given that R3 is double R2, and the parallel equivalent of R2 & R3 is double R1, then calculate:

a. R_1, R_2, R_3

b. The current through R_2

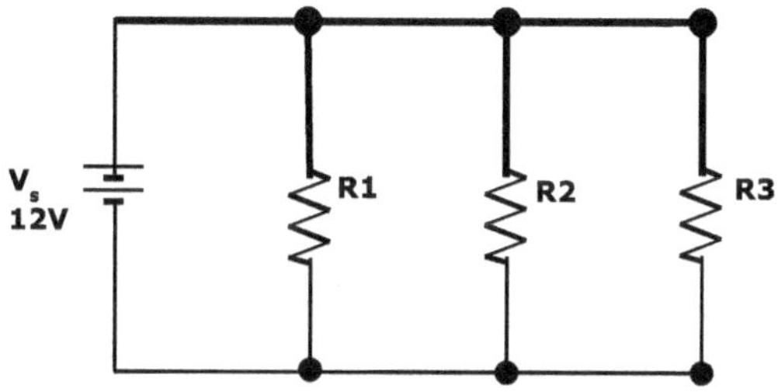

Solution:

a.

$$Rt=2\ k\Omega$$

$$R3=2R2$$

$$Rp1=R2*R3/(R2+R3)=2\ R1$$

$$R1=(1/2)\ Rp1$$

$$Rt=R1//(R2//R3)=R1//Rp1=R1//2R1$$

$$2=\frac{R1*2R1}{R1+2R1}$$

$$R1=3k\Omega$$

$$Rp1=R2*R3/(R2+R3)=2\ R1=2*3=6\ k\Omega$$

$$R2*R3/(R2+R3)=6\ k\Omega$$

$$R2*2R2/(R2+2R2)=6\ k\Omega$$

$$R2=9\ k\Omega$$

R3=2R2=2*9=18 kΩ

b.

I2=Vs/R2=12/9 k= 1.333

10. Find the indicated values for the following circuit.

   a. Determine the voltage across each circuit element.
   b. Determine the total circuit current and the current through each branch.
   c. Find the values of R3 and R4 for extra credit.

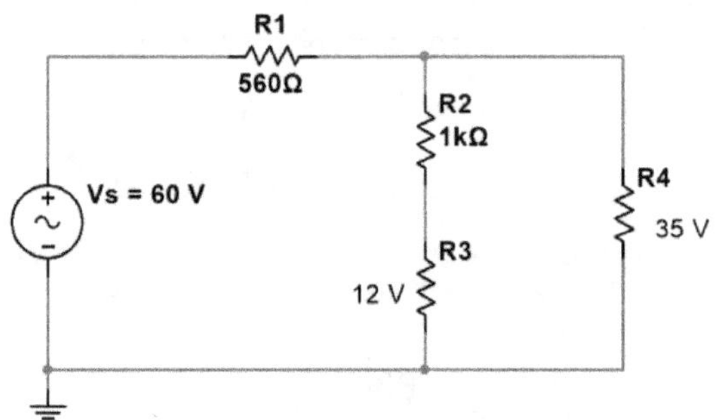

Solution:

a. The voltage across each circuit element.

$$V_{R1} = V_s - V_{R4} = 60 - 35 = 25\ V$$

$$V_{R2} = V_{R4} - V_{R3} = 35 - 12 = 23\ V$$

b. The total circuit current and the current through each branch

$$I_T = \frac{V_{R1}}{R_1} = \frac{25}{560} = 44.643\ mA$$

$$I_{R2} = \frac{V_{R2}}{R_2} = \frac{23}{1\ k\Omega} = 23\ mA$$

$$I_{R4} = I_T - I_{R2} = 44.643\ mA - 23\ mA = 21.643\ mA$$

c. The values of R3 and R4

$$I_{R2} = I_{R3} = 23\ mA$$

$$R_3 = \frac{V_{R3}}{I_{R3}} = \frac{12}{23\ mA} = 521.74\ \Omega$$

$$R_4 = \frac{V_{R4}}{I_{R4}} = \frac{35}{21.643\ mA} = 1.617\ k\Omega$$

11. For the given circuit, determine the following parameters.

a. The total capacitance
b. The voltage across each capacitor
c. The current through each capacitor.

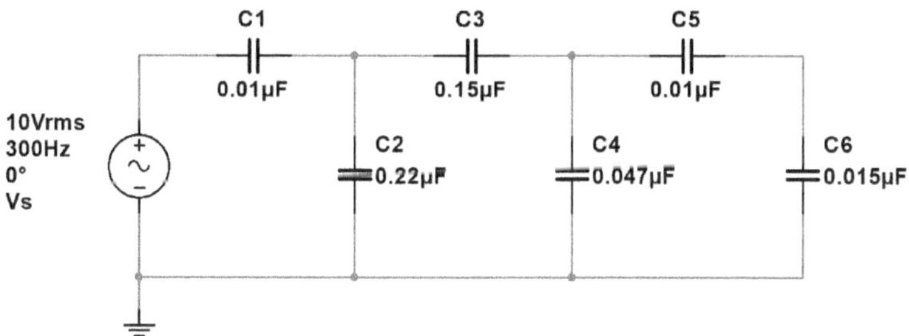

Solution:

a. The total capacitance

$$C_a = \frac{C_5 C_6}{C_5 + C_6} = \frac{0.01 * 0.015}{0.01 + 0.015} = 6 * 10^{-3}\ \mu F$$

$$C_b = C_a + C_4 = 6 * 10^{-3}\ \mu F + 0.047\ \mu F = 0.053\ \mu F$$

$$C_c = \frac{C_3 C_b}{C_3 + C_b} = \frac{0.15\mu F * 0.053\ \mu F}{0.15\mu F + 0.053\ \mu F} = 0.03916\ \mu F$$

$$C_d = C_2 + C_c = 0.22\ \mu F + 0.03916\ \mu F = 0.25916\ \mu F$$

$$C_T = \frac{C_1 C_d}{C_1 + C_d} = \frac{0.01\mu F * 0.25916\ \mu F}{0.01\mu F + 0.25916\ \mu F} = 0.0096284\ \mu F$$

b. The voltage across each capacitor

$$V_{c1} = V_s \frac{C_d}{C_1 + C_d} = 10 * \frac{0.25916\ \mu F}{0.01\mu F + 0.25916\ \mu F} = 9.628\ V$$

$$V_{cd} = V_s - V_{c1} = 10 - 9.628 = 0.3715\ V$$

$$V_{c3} = V_{cd} \frac{C_b}{C_3 + C_b} = 0.3715\ V * \frac{0.053\ \mu F}{0.15\mu F + 0.053\ \mu F} = 9.699\ mV$$

$$V_{c4} = V_{cd} - V_{c3} = 0.3715\ V - 9.699\ mV = 0.3618\ V$$

$$V_{c5} = V_{c4} \frac{C_6}{C_5 + C_6} = 0.3618\ V * \frac{0.015\ \mu F}{0.01\mu F + 0.015\ \mu F} = 0.217\ V$$

$$V_{c6} = V_{c4} - V_{c5} = 0.3618\ V - 0.217\ V = 0.14472\ V$$

c. The current through each capacitor.

$$I_c = \frac{V_c}{X_c} = \frac{V_c}{\frac{1}{j2\pi f C}} = j2\pi f C V_c$$

$$I_{c1} = j2\pi f C_1 V_{c1} = j2\pi * 300 * 0.01\mu F * 9.628\ V = 0.181\ mA$$

$$I_{c2} = j2\pi f C_2 V_{c2} = j2\pi * 300 * 0.22\mu F * 0.3715\ V = 0.154\ mA$$

$$I_{c3} = j2\pi f C_3 V_{c3} = j2\pi * 300 * 0.15\mu F * 9.699\ mV = 0.00274\ mA$$

$$I_{c4} = j2\pi f C_4 V_{c4} = j2\pi * 300 * 0.047\mu F * 0.3618\ V = 0.03205\ mA$$

$$I_{c5} = j2\pi f C_5 V_{c5} = j2\pi * 300 * 0.01\mu F * 0.217\ V$$
$$= 0.00409\ mA$$

$$I_{c6} = j2\pi f C_6 V_{c6} = j2\pi * 300 * 0.015\mu F * 0.14472\ V$$
$$= 0.00409\ mA$$

12. For the circuit shown below, calculate:
    a. The voltage across each inductor.
    b. The current through each inductor.
    c. The inductance value of L_2

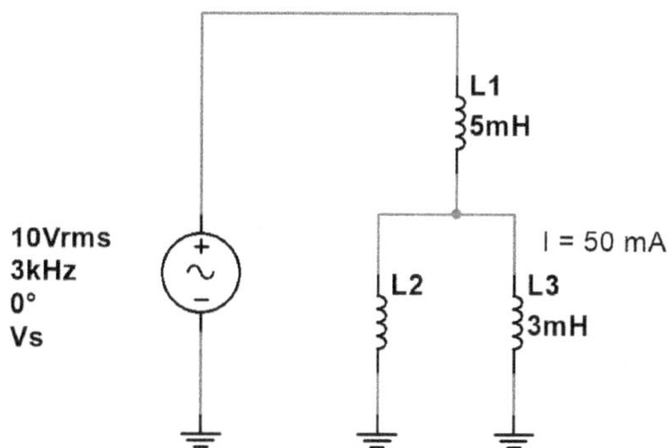

Solution:

a. The voltage across each inductor.

$$V_{L3} = I_{L3}\, X_{L3} = 50\ mA * 2\pi * 3000 * 3mA = 2.827 = V_{L2}$$

$$V_{L1} = V_s - V_{L3} = 10 - 2.827 = 7.1725\ V$$

b. The current through each inductor.

$$I_{L1} = \frac{V_{L1}}{X_{L1}} = \frac{7.1725}{2\pi * 3000 * 5mH} = 76.102\ mA$$

$$I_{L2} = I_{L1} - I_{L3} = 76.102\ mA - 50\ mA = 26.102\ mA$$

c. The inductance value of L_2

$$X_{L2} = \frac{V_{L2}}{I_{L2}} = \frac{2.827}{26.102\ mA} = 108.306\ \Omega$$

$$L_2 = \frac{X_{L2}}{2\pi f} = \frac{108.306\ \Omega}{2\pi * 3000} = 5.745\ mH$$

13. For the circuit shown below, perform the following tasks.

a. Find the voltage across the load (secondary voltage).
b. Find the current in the secondary.

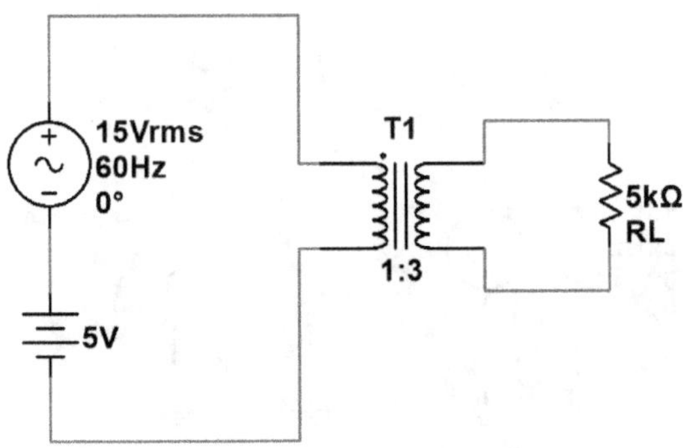

Solution:

a. The voltage across the load (secondary voltage).

$$V_p = \sqrt{2} * 15 = 21.21V$$

$$V_{p-p} = 2\sqrt{2} * 15 = 42.42V$$

The peak values:

$$V_{pri+} = 5 + (+21.21) = 26.21\ V$$

$$V_{pri-} = 5 + (-21.21) = -16.21\ V$$

$$V_{sec} = \frac{3}{1} V_{pri}$$

$$V_{sec+} = \frac{3}{1} * 26.21\,V = 78.63\,V$$

$$V_{sec-} = \frac{3}{1} * (-16.21\,V) = -48.63\,V$$

b. The current in the secondary.

The peak values:

$$I_L = \frac{V_{sec}}{R_L}$$

$$I_{L+} = \frac{78.63\,V}{5000} = 15.726\,mA$$

$$I_{L-} = \frac{-48.63V}{5000} = -9.720\,mA$$

14. For the circuit below, calculate the following:

a. Determine the value to which the variable resistor must be set to make the total current 9 mA.
b. What are the resulting impedance magnitude and phase angle?

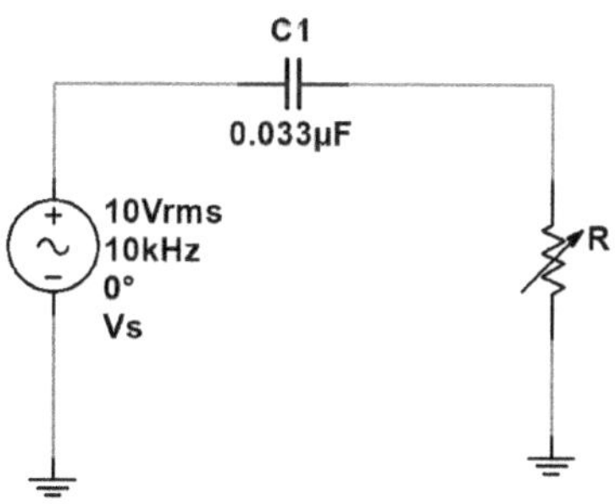

a.

$$X_C = \frac{1}{2\pi f C} = \frac{1}{2\pi * 10000 * 0.033\mu F} = 482.287\,\Omega$$

$$|Z| = \frac{V}{I} = \frac{10}{9\,mA} = 1111.111\Omega$$

$$|Z| = \sqrt{R^2 + X_c^2}$$

$$1111.111 = \sqrt{R^2 + (482.287)^2}$$

$$R = \sqrt{(1111.111)^2 - (482.287)^2}$$

$$R = 1000\ \Omega$$

b.

$$Z = R - jX_c = 1000 - j482.287\ \Omega$$

$$|Z| = \sqrt{R^2 + X_c^2} \angle -\tan^{-1}\frac{X_c}{R} = 1111.111\angle - \tan^{-1}\frac{482.287}{1000}$$
$$= 1111.111\angle - 25.747°\ \Omega$$

15. For the circuit below, calculate the following:

a) The total impedance in rectangular coordinates.
b) The voltage across each circuit element.
c) The total circuit current.

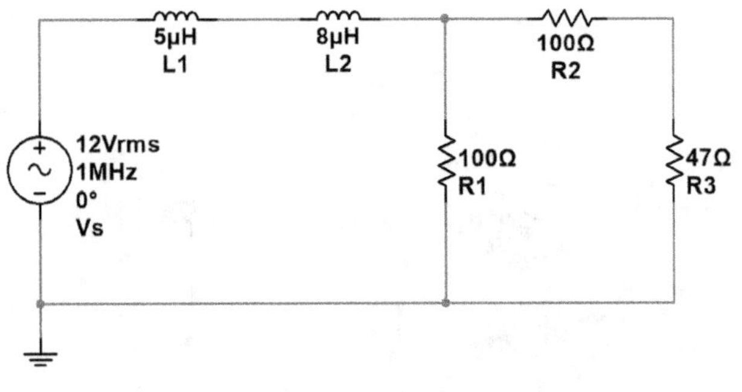

a.

$$R_T = R_1//(R_2 + R_3) = \frac{100 * (100 + 47)}{100 + 100 + 47} = 59.514\ \Omega$$

$$L_T = L_1 + L_2 = 5 + 8 = 13\ \mu H$$

$$X_T = 2\pi f L_T = 2\pi * 1 * 10^6 * 13 * 10^{-6} = 81.681\ \Omega$$

$$Z_T = R_T + jX_T = 59.514 + j81.681\ \ \Omega$$

b.

$$V_{L1} = V_s \frac{jX_{L1}}{Z_T} = 12 * \frac{j2\pi * 1 * 10^6 * 5 * 10^{-6}}{59.514 + j81.681} = 3.73\angle 36°\ V$$

$$V_{L2} = V_s \frac{jX_{L2}}{Z_T} = 12 * \frac{j2\pi * 1 * 10^6 * 8 * 10^{-6}}{59.514 + j81.681} = 5.968\angle 36°\ V$$

$$V_{R_T} = V_s \frac{R_T}{Z_T} = 12 * \frac{59.514}{59.514 + j81.681} = 7.066\angle - 53.922°\ V$$
$$= V_{R_1}$$

$$V_{R_2} = V_{R_T} \frac{R_2}{R_2 + R_3} = 7.066\angle - 53.922° \ * \frac{100}{100 + 47}$$
$$= 4.807\angle - 53.922°\ V$$

$$V_{R_3} = V_{R_T} \frac{R_3}{R_2 + R_3} = 7.066\angle - 53.922° \ * \frac{47}{100 + 47}$$
$$= 2.259\angle - 53.922°\ V$$

c.

$$I_T = \frac{V_s}{Z_T} = \frac{12}{59.514 + j81.681} = 118.737\angle - 53.922°\ \ mA$$

Chapter 7

Circuit Analysis and Network

Circuit analysis is the process of using mathematical tools and techniques to understand and predict the behavior of electrical circuits. It involves analyzing the flow of current and voltage through the circuit components and determining their effects on the overall circuit performance.

Circuit analysis is used to optimize the performance of electrical circuits, ensuring that they operate efficiently and safely. It is also used to troubleshoot and diagnose circuit problems and to design new circuits for specific applications. The tools and techniques used in circuit analysis are essential for designing and operating a wide range of electronic devices, from simple household appliances to complex industrial systems.

7.1 Definition of the Laplace Transform

There are two types of Laplace transforms.

1) The two-sided (or bilateral) Laplace transform.

The two-sided Laplace transform allows time functions to be non-zero for negative time.

$$\mathcal{L}[x(t)] = X(s) = \int_{-\infty}^{\infty} x(t)e^{-st}\, dt$$

where s is the complex frequency given by

$$s = \sigma + j\omega$$

where σ (in nepers/s) and ω (in radians/s) are the real and imaginary parts of s, respectively. The advantage of bilateral Laplace transform is that it can handle causal and non-causal signals over $-\infty$ to ∞.

2) The one-sided (or unilateral) form

A one-sided function is zero for negative time; that is, - t < 0, where - 0 denotes a time just before 0.

$$\mathcal{L}[x(t)] = X(s) = \int_{0^-}^{\infty} x(t)e^{-st}\,dt$$

The one-sided Laplace transform is more commonly used than the two-sided Laplace transform because here, we encounter only positive-time signals in practice, defined as positive-time signals. The t-domain signal x(t) is changed into the s-domain function X(s).

The **Laplace transform** of a signal x(t) is the integration of the product of x(t) and e^{-st} over the interval from 0 to +∞.

In the inverse Laplace transform, the s-domain X(s) is changed back to t-domain x(t) and is given as

$$\mathcal{L}^{-1}[X(s)] = x(t) = \frac{1}{2\pi j}\int_{\sigma_1 - j\infty}^{\sigma_1 + j\infty} X(s)e^{st}ds$$

where the integration is performed along a straight line $(\sigma_1 + j\omega, -\infty < \omega < \infty)$.

The Laplace transform pair x(t) and X(s) are represented as

$$x(t) \Leftrightarrow X(s)$$

because the transform goes both ways, there is a one-to-one correspondence between x(t) and X(s).

A signal x(t) is Laplace transformable , must converge. The integral converges when

$$\left|\int_{0^-}^{\infty} x(t)e^{-(\sigma + j\omega)t}dt\right| < \infty$$

This can be simplified as

$$\left|\int_{0^-}^{\infty} x(t)e^{-(\sigma+j\omega)t}dt\right| \leq \int_{0^-}^{\infty} \left|x(t)e^{-(\sigma+j\omega)t}dt\right|$$
$$\leq \int_{0^-}^{\infty} |x(t)|\left|e^{-(\sigma+j\omega)t}dt\right| < \infty$$

Since $|e^{j\omega t}| = 1$ for any value of t,

$$\int_{0^-}^{\infty} |x(t)|e^{-\sigma t}dt < \infty$$

Example: Find the Laplace transform of the following functions and establish the region of convergence (ROC) for each case.

(a) e^{-5t} u(t),
(b) δ(t).

Solution

(a) For the exponential function, the Laplace transform is

$$L[e^{-5t}u(t)] = \int_{0^-}^{\infty} e^{-5t}e^{-st}dt = -\frac{1}{s+5}e^{-(s+5)t}\Big|_{0^-}^{\infty} = \frac{1}{s+5}$$

Notice that $\lim_{t\to\infty} e^{-(s+5)t} = 0$ only if Re(s+5) > 0 or Re(s) > -5. Alternatively, the ROC is obtained from

$$|e^{-(s+5)t}| = |e^{-(\sigma+5)t}||e^{-j\omega t}| = |e^{-(\sigma+5)t}| < \infty$$

which is valid when σ + 5> 0 or σ > -5. The result in this equation is so important that it deserves boxing.

In general

$$L[e^{-at}u(t)] = \frac{1}{s+a}$$

A special case of this is when a=0. $e^{-at}u(t)$ becomes u(t) and we have

$$L[u(t)] = \frac{1}{s}$$

(b) For the unit impulse function,

$$L[\delta(t)] = \int_{0^-}^{\infty} \delta(t)e^{-st}dt = e^{-0} = 1 \quad \text{for all s}$$

The sifting property has been applied.In this case, the transform does not depend on s. The convergence region is the entire s plane.

Example: Find the Laplace transform of x(t) = sin ωt u(t).

Solution

The Laplace transform of the cosine function is

$$X(s) = L[\sin \omega t] = \int_0^{\infty} (\sin \omega t)e^{-st}dt$$

$$= \int_0^{\infty} \left(\frac{e^{j\omega t} - e^{-j\omega t}}{2j}\right) e^{-st}dt$$

$$= \frac{1}{2j}\int_0^{\infty} \left(e^{-(s-j\omega)t} - e^{-(s+j\omega)t}\right)dt$$

$$= \frac{1}{2j}\left(\frac{1}{s-j\omega} - \frac{1}{s+j\omega}\right) = \frac{\omega}{s^2+\omega^2}$$

7.2 Properties of the Laplace Transform

The Laplace transform properties are summarized below, which are useful for obtaining the transform of many signals. Some properties are linearity, Scaling, Time shifting, Frequency shifting, Time Differentiation, Time Convolution, Time Integration, Frequency Differentiation, Time Periodicity, Modulation, and Initial and Final Values.

7.2.1 Linearity

Let the Laplace transform of two signals $x_1(t)$ and $x_2(t)$ be $X_1(s)$ and $X_2(s)$, respectively. Then the linearity property is defined as the Laplace transform of the sum of two functions of time is equal to the sum of the transforms of each function and is given as

$$\mathcal{L}[a_1 x_1(t) + a_2 x_2(t)] = a_1 X_1(s) + a_2 X_2(s)$$

where a_1 and a_2 are constants.

The linearity property is explained using the following example.

$$\mathcal{L}[a_1 x_1(t) + a_2 x_2(t)] = \int_0^{\infty} [a_1 x_1(t) + a_2 x_2(t)]\, e^{-at}\, dt$$

$$= a_1 \int_0^{\infty} x_1(t)\, e^{-at}\, dt + a_2 \int_0^{\infty} x_2(t)\, e^{-at}\, dt$$

$$= a_1 X_1(s) + a_2 X_2(s)$$

Example:

$x(t) = 3e^{4t} + 5e^{-6t}$.

By using linearity property,

$$X(s) = L[3e^{4t} + 5e^{-6t}] = L[3e^{4t}] + L[5e^{-6t}]$$

$$X(s) = \frac{3}{s-4} + \frac{5}{s+6}$$

7.2.2 Scaling

Let X(s) be the Laplace transform of the x(t) signal. Let *a* be any positive real constant, then Laplace transform of x(*a*t) is given as

$$\mathcal{L}[x(at)] = \int_0^{\infty} x(at)\, e^{-st}\, dt$$

Let λ=*a*t, dλ = *a* dt, we obtain

$$\mathcal{L}[x(at)] = \int_0^{\infty} x(\lambda)\, e^{-\lambda(\frac{s}{a})} \frac{d\lambda}{a} = \frac{1}{a} \int_0^{\infty} x(\lambda)\, e^{-\lambda(\frac{s}{a})}\, d\lambda$$

Thus, the scaling property is obtained as

$$\mathcal{L}[x(at)] = \frac{1}{a}\, X\left(\frac{s}{a}\right), \quad a > 0$$

The unilateral Laplace transform is valid only for causal signals, not non-causal signals (when *a* is negative, i.e., *a* < 0).

Example:

Consider $$L[r(t)] = \frac{1}{s^2}$$

By applying the scaling property,

$$L[r(2t)] = \frac{1}{2}\frac{1}{(s/2)^2} = \frac{2}{s^2}$$

7.2.3 Time Shifting

We often see a signal delay in control systems due to transmission delays or other effects. Let X(s) be the Laplace transform of signal x(t). Then time shifting is given as

$$\mathcal{L}[x(t-a)u(t-a)] = e^{-as}\, X(s)$$

If a function is delayed in time by *a*, then the Laplace transform of the function is multiplied by e^{-as}.

The example below describes the time-shifting property.

$$\mathcal{L}[x(t-a)u(t-a)] = \int_0^{\infty} x(t-a)u(t-a)\, e^{-st} dt, \quad a \geq 0$$

If u(t-a) = 0 for t < a and u(t-a) = 1 for t > a.

Then,

$$\mathcal{L}[x(t-a)u(t-a)] = \int_0^{a} 0\, e^{-st} dt + \int_a^{\infty} x(t-a)(1)e^{-st} dt$$

By substituting λ= t-a, $d\lambda$ = dt and t = λ + a. As

$$t \to a, \lambda \to 0 \text{ and as } t \to \infty, \lambda \to \infty.$$

Thus,

$$\mathcal{L}[x(t-a)u(t-a)] = \int_0^\infty x(\lambda)\, e^{-s(\lambda+a)} d\lambda$$

$$= e^{-as} \int_0^\infty x(\lambda)\, e^{-s\lambda} d\lambda$$

$$= e^{-as} X(s)$$

Example:

Consider g(t) = u(t-4) – u(t-6).

We know that $\mathcal{L}[u(t)] = \frac{1}{s}$

Using the linearity property and the time-shifting property,

$$G(s) = L[u(t-4) - u(t-6)] = \frac{1}{s} e^{-4s} - \frac{1}{s} e^{-6s} = \frac{1}{s} (e^{-4s} - e^{-6s})$$

7.2.4 Frequency Shifting

Let X(s) be the Laplace transform of the x(t) signal. Then the frequency shifting is given as

$$\mathcal{L}[e^{-at} x(t)u(t)]$$
$$= \int_0^\infty e^{-at} x(t) e^{-st} dt = \int_0^\infty x(t) e^{-(s+a)t}\, dt$$
$$= X(s+a)$$

$$\mathcal{L}[e^{-at} x(t)u(t)] = X(s+a)$$

The Laplace transform turns multiplication by e^{-at} in the time domain into s+*a* in the transform X(s).

Example:

Consider $\cos \omega\, t u(t) \Leftrightarrow \frac{s}{s^2+\omega^2}$

By applying the frequency property and replacing every s by s+a, the Laplace transform of the cosine function is given by

$$L[e^{-at}\cos\omega\, t u(t)] = \frac{s+a}{(s+a)^2+\omega^2}$$

7.2.5 Time Differentiation

Let X(s) be the Laplace transform of the x(t) signal. The time differentiation property is important for solving differential equations. The derivative of x(t) and its Laplace transform is given as

$$\mathcal{L}\left[\frac{dx}{dt}u(t)\right] = \int_{0^-}^{\infty}\frac{dx}{dt}e^{-st}\,dt$$

By derivative by parts, $U = e^{-st}, dU = -se^{-st}\,dt$ and $dV = \left(\frac{dx}{dt}\right)dt = dx,\ V = x(t)$, then

$$\mathcal{L}\left[\frac{dx}{dt}u(t)\right] = x(t)e^{-st}|_{0^-}^{\infty} - \int_{0^-}^{\infty} x(t)[-se^{-st}]dt$$

$$= 0 - x(0^-) + s\int_{0^-}^{\infty} x(t)e^{-st}dt$$

$$= sX(s) - x(0^-)$$

$$\mathcal{L}[x'(t)] = sX(s) - x(0^-)$$

By differentiating the above equation, we get

$$\mathcal{L}\left[\frac{d^2x}{dt^2}\right] = s\mathcal{L}[x'(t)] - x'(0^-) = s[sX(s) - x(0^-)] - x'(0^-)$$

$$= s^2X(s) - sx(0^-) - x'(0^-)$$

$$\mathcal{L}[x''(t)] = s^2X(s) - sx(0^-) - x'(0^-)$$

In the same way, for the nth derivative, we get

$$\mathcal{L}\left[\frac{d^nx}{dt^n}\right] = s^nX(s) - s^{n-1}x(0^-) - s^{n-2}x'(0^{-1}) - L - s^0x^{(n-1)}(0^-)$$

Example:

Consider x(t) = cos ωt u(t)

Then at t=0 , we have x(0) = 1.

By applying time differentiation, the Laplace transform of x(t) is

$$L\left[\frac{d}{dt}\cos\omega\, tu(t)\right] = s\left(\frac{s}{s^2+\omega^2}\right) - \cos(0^-)u(0^-)$$
$$= \frac{s^2}{s^2+\omega^2}$$

7.2.6 Time Convolution

Let X(s) and H(s) be the Laplace transform of the signal x(t) and h(t).

Then the convolution property is given as

$$\mathcal{L}[x(t) * h(t)] = X(s)H(s)$$

In the time domain, the convolution property,

$$x(t) * h(t) = \int_0^\infty x(t-\tau)h(\tau)d\tau$$

Considering the Laplace transform on both sides

$$\mathcal{L}[x(t) * h(t)] = \int_0^\infty \left[\int_0^\infty h(\tau)x(t-\tau)d\tau\right] e^{-st}dt$$
$$\mathcal{L}[x(t) * h(t)] = \int_0^\infty h(\tau)\left[\int_0^\infty x(t-\tau)e^{-st}dt\right] d\tau$$

Let λ=t-λ so that t = τ+ λ and dt=dλ

$$\mathcal{L}[x(t) * h(t)] = \int_0^{\infty} h(\tau) \left[\int_0^{\infty} x(\lambda) e^{-s(t+\lambda)} d\lambda \right] d\tau$$

$$= \int_0^{\infty} h(\tau) e^{-st} d\tau \int_0^{\infty} x(\lambda) e^{-s\lambda} d\lambda = H(s)X(s)$$

The Convolution property of a Laplace transform turns into a product of transforms.

Example:

i) Consider g(t) = u(t-3) – u(t-4).

By applying the Laplace transform on both sides.

$$G(s) = \frac{1}{s}(e^{-3s} - e^{-4s})$$

Considering the convolution of g(t)*g(t), which is $G^2(s)$,

$$G^2(s) = \frac{1}{s^2}(e^{-3s} - e^{-4s})^2 = \frac{1}{s^2}(e^{-6s} - 2e^{-7s} + e^{-8s})$$

By applying the inverse Laplace transform, we get

x(t) = g(t)*g(t) = (t-6)u(t-6) – 2(t-7)u(t-7) + (t-8)u(t-8)

= r(t-6) -2r(t-7) + r(t-8)

7.2.7 Time Integration

Let X(s) be the Laplace transform of the x(t) signal. Then Laplace transform of the integral is given as

$$\mathcal{L}\left[\int_0^{t} x(t)dt\right] = \frac{1}{s} X(s)$$

The example below describes the Time integration property.

From the convolution property, we get

$$\mathcal{L}\left[\int_0^t x(\lambda)d\lambda\right] = \mathcal{L}\left[\int_0^t x(\lambda)u(t-\lambda)d\lambda\right] = \mathcal{L}[u(t) * x(t)] = \frac{1}{s} X(s)$$

$$\mathcal{L}\left[\int_0^t x(t)dt\right] = \frac{1}{s} X(s)$$

Examples:

i) Consider x(t) = u(t).

We know that X(s) = 1/s.

By applying the time integration property,

$$L\left[\int_0^t u(t)dt\right] = L[t] = \frac{1}{s}\frac{1}{s}$$

ii) Consider r(t) = tu(t)

We know that $L[t] = \frac{1}{s^2}$

By applying the time integration property,

$$L\left[\int_0^t tdt\right] = L\left[\frac{t^2}{2}\right] = \frac{1}{s}\frac{1}{s^2}$$

Similarly, $L[t^2] = \frac{2}{s^3}$

$$L\left[\int_0^t t^2dt\right] = L[t^2] = \frac{2}{s^3}$$

or $L[t^3] = \frac{6}{s^4} = \frac{3!}{s^4}$

Continuing this process leads to

$$L[t^n] = \frac{n!}{s^{n+1}}$$

7.2.8 Frequency Differentiation

Let X(s) be the Laplace transform of the x(t) signal. Then Laplace transform of the Frequency differentiation is given as

$$\mathcal{L}[tx(t)] = -\frac{dX(s)}{ds}$$

The example below describes the Frequency differentiation property; X (s) is given as

$$X(s) = \int_{0^-}^{\infty} x(t)e^{-st}dt$$

Differentiating on both sides, we get

$$\frac{dX(s)}{ds} = \int_{0^-}^{\infty} x(t)(-te^{-st})dt = \int_{0^-}^{\infty} (-tx(t))e^{-st}dt = \mathcal{L}[-tx(t)]$$

The Frequency differentiation property is given as

$$\mathcal{L}[tx(t)] = -\frac{dX(s)}{ds}$$

Similarly, the general form of frequency differentiation property is given as

$$\mathcal{L}[t^n x(t)] = (-1)^n \frac{d^n X(s)}{ds^n}$$

Example:

$$L(t\cos\omega t) = -\frac{d}{ds}\left[\frac{s}{s^2+\omega^2}\right] = \frac{s^2-\omega^2}{(s^2+\omega^2)^2}$$

$$L(t\sin\omega t) = -\frac{d}{ds}\left[\frac{\omega}{s^2+\omega^2}\right] = \frac{2\omega s}{(s^2+\omega^2)^2}$$

7.2.9 Time Periodicity

Consider a periodic signal x(t), and it can be expressed as a sum of time-shifted functions as

$$x(t) = x_1(t) + x_2(t) + x_3(t) + L$$

$$= x_1(t) + x_1(t-T)u(t-T) + x_1(t-2T)u(t-2T) + L$$

where 0 < t < T.

By applying the Laplace transform and the time-shifting property, we get

$$X(s) = X_1(s) + X_1(s)e^{-Ts} + X_1(s)e^{-2Ts} + X_1(s)e^{-3Ts} + L$$

$$= X_1(s)[1 + e^{-Ts} + e^{-2Ts} + e^{-3Ts} + L]$$

But

$$1 + a + a^2 + a^3 + L = \frac{1}{1-a}$$

if |a| < 1.

$$X(s) = \frac{X_1(s)}{1 - e^{-Ts}}$$

Example:

Consider $x_1(t) = \begin{cases} \sin \pi t, 0 < t < 1 \\ 0, 1 < t < 2 \end{cases}$

T=2 and $X_1(s) = \frac{\pi}{s^2+\pi^2}$

The Laplace transform of the periodic signal is

$$X(s) = \frac{X_1(s)}{1 - e^{-Ts}} = \frac{\pi}{(1 - e^{-2s})(s^2 + \pi^2)}$$

7.2.10 Modulation

Let X(s) be the Laplace transform of the signal x(t), then for any real number ω,

$$\mathcal{L}[x(t)cos\omega t] = \frac{1}{2}[X(s+j\omega)+X(s-j\omega)]$$

$$\mathcal{L}[x(t)sin\omega t] = \frac{j}{2}[X(s+j\omega)-X(s-j\omega)]$$

The Laplace transform of x(t)cosωt and x(t)sinωt are

$$x(t)cos\omega t = \frac{1}{2}x(t)[e^{-j\omega t}+e^{j\omega t}]$$

$$x(t)sin\omega t = \frac{1}{2j}x(t)[e^{-j\omega t}-e^{j\omega t}]$$

$$\mathcal{L}\left[e^{\pm j\omega t}x(t)\right] = X(smj\omega)$$

Example:

We know that

$$u(t) \Leftrightarrow \frac{1}{s}$$

$$u(t)\,cos(\,\omega_o t) \;\Leftrightarrow \frac{1}{2}\left[\frac{1}{s+j\omega_o}+\frac{1}{s-j\omega_o}\right] = \frac{s}{s^2+\omega_o^2}$$

$$u(t)\,sin(\,\omega_o t) \;\Leftrightarrow \frac{1}{2j}\left[\frac{1}{s+j\omega_o}-\frac{1}{s-j\omega_o}\right] = \frac{\omega_o}{s^2+\omega_o^2}$$

7.2.11 Initial and Final Values

The initial value x(0) and a final value x(∞)relate frequency domain expressions to the time domain as the time approaches zero and infinity, respectively. These properties can be derived by using the differentiation property as

$$sX(s)-x(0^-) = \mathcal{L}\left[\frac{dx}{dt}\right] = \int_{0^-}^{\infty}\frac{dx}{dt}e^{-st}dt$$

By taking limits, we get

$$\lim_{s\to\infty}[sX(s) - x(0)] = 0$$

Since x(0) is not dependent on s, it is given as

$$x(0) = \lim_{s\to\infty} sX(s)$$

This is called the initial-value Theorem. x(0) can be found using X(s), without using the inverse transform of x(t).

If we consider $s \to 0$, then

$$sX(s) - x(0^-) = \int_{0^-}^{\infty} \frac{dx}{dt} e^{-0t}\, dt = \int_{0^-}^{\infty} dx \;=\; x(\infty) - x(0^-)$$

$$x(\infty) = \lim_{s\to 0} sX(s)$$

This is known as the final-value Theorem. In the final-value Theorem, all X(s) poles must be located in the left half of the s-plane.

Example:

$$x(t) = 4 + e^{-2t}\cos 3t\, u(t) \Leftrightarrow X(s) = \frac{4}{s} + \frac{s+2}{(s+2)^2 + 3^2}$$

By applying the initial-value Theorem,

$$x(0) = \lim_{s\to\infty} sX(s) = \lim_{s\to\infty} 4 + \frac{s^2+2s}{s^2+4s+13}$$

$$= \lim_{s\to\infty} 4 + \frac{1+2/s}{1+4/s+13/s^2} = 4 + \frac{1+0}{1+0+0} = 5$$

x(0) = 5

Example:

i) $x(t) = (5 + e^{-2t})u(t) \Leftrightarrow X(s) = \frac{5}{s} + \frac{1}{s+2}$

Using the final-value Theorem,

$$x(\infty) = \lim_{s\to 0} sX(s) = \lim_{s\to 0}\left(5 + \frac{s}{s+2}\right) = 5 + 0 = 5$$

ii) $x(t) = \cos 2\,tu(t) \Leftrightarrow X(s) = \frac{s}{s^2+4}$

$$x(\infty) = \lim_{s\to 0} sX(s) = \lim_{s\to 0}\frac{s^2}{s^2+2} = 0$$

T

he final-value Theorem cannot be used to find the value of $x(t) = \cos 2t$ because $X(s)$ has poles at $s = \pm j\sqrt{2}$, which are not in the left half of the s plane.

Table 7.1 Table of General Properties of Laplace Transform

F(s) = $\int_0^\infty e^{-st} f(t)dt$

S No.	Name	Laplace Transform	Inverse Laplace Transform
01.	Definition	L{f(t)} = f(s)	L^{-1}\{f(s)\} = f(t)
02.	Linearity	$af_1(t) + b\,f_2(t)$	$a\,f_1(s)+bf_2(s)$
03.	Change of scale	f(at)	$\frac{1}{a}f(\frac{s}{a})$
04.	First shifting Th.	e^{at} f(t)	f(s-a)
05.	second shifting	u(t-a) = $\begin{cases} f(t-a), t > a \\ 0, t < a \end{cases}$	e^{-as} f(s)
06.	Derivative		
	(multiply by s)	$f'(t)$	$sf(s) - f(0)$
07.	Second derivative		
	(multiply by s^2)	$f''(t)$	$s^2f(s) - s\,f(0) - f'(0)$
08.	n the derivative		
	(multiply by s^n)	$f^{(n)}(t)$	$s^nf(s) - s^{n-1}f(0) - s^{n-2}f'(0)\ldots - f^{n-1}(0)$
09.	Integral(division by s)	$\int_0^t f(u)du$	$\frac{f(s)}{s}$
10.	Multiple integrals	$\int_0^t \cdots \int_0^t f(u)du^n \approx \int_0^t \frac{(t-u)n-1}{(n-1)!} f(u)du$	$\frac{f(s)}{s^n}$
	(division by s^n)		
11.	Multiply by t	- t f(t)	$f'(s)$
12.	Multiply by t^2	t^2 f(t)	$f''(s)$
13.	Multiply by t^n	$(-1)^n t^n f(t)$	$f^n(s)$
14.	Division by t	$\frac{f(t)}{t}$	$\int_1^\infty f(u)du$
15.	Convolution	f(t) * g(t) = $\int_0^t f(u)g(t-u)du$	
		= $\int_0^t f(t-u)g(u)du$	f(s)*g(s)=L(f * g)
		= L^{-1}(f(s) g(s)}	
16.	f-periodic with period p	f(t) = f(t + p)	$\frac{1}{1-e^{-sp}}\int_0^p e^{-su} f(u)du$

Table 7.2 Table of some Important Laplace Transforms

S No.	Laplace Transform	Inverse Laplace Transform
01.	L{f(t)} = f(s)	L^{-1} {f(s)} = f(t)
02.	1	$\frac{1}{s}$
03.	t	$\frac{1}{s^2}$
04.	t^2	$\frac{2!}{s^3}$
05.	t^n n = 0,1,2,3,...	$\frac{n!}{s^{n+1}}$
06.	e^{at}	$\frac{1}{s-a}$
07.	e^{-at}	$\frac{1}{s+a}$
08.	$\frac{t^{(n-1)}\, e^{at}}{(n-1)!}$	$\frac{1}{(s-a)^n}$ n = 1,2,3,.....
09.	$\frac{t^{(k-1)}\, e^{at}}{\Gamma(k)}$	$\frac{1}{(s-a)^k}$ k = 1,2,3,.....
10.	$sin\ at$	$\frac{a}{s^2+a^2}$
11.	$cos\ at$	$\frac{s}{s^2+a^2}$
12.	e^{bt} sin at	$\frac{a}{(s-b)^2+a^2}$
13.	e^{bt} cos at	$\frac{s-b}{(s-b)^2+a^2}$
14.	sinh at	$\frac{a}{s^2-a^2}$
15.	cosh at	$\frac{s}{s^2-a^2}$
16.	e^{bt} sinh at	$\frac{a}{(s-b)^2-a^2}$
17.	e^{bt} cosh at	$\frac{s-b}{(s-b)^2-a^2}$
18.	u(t - a)	$\frac{e^{-as}}{s}$
19.	f(t – a). u(t – a)	e^{-as} f(s)
20.	$\frac{e^{bt}-e^{at}}{b-a}$	$\frac{1}{(s-a)(s-b)}$ $a \neq b$
21.	$\frac{be^{bt}-ae^{at}}{b-a}$	$\frac{s}{(s-a)(s-ab)}, a \neq b$
22.	$\frac{sin\ at-at\ cos\ at}{2a^3}$	$\frac{1}{(s^2+a^2)^2}$
23.	$\frac{t\ sin\ at}{2a}$	$\frac{s}{(s^2+a^2)^2}$

S No.	Laplace Transform	Inverse Laplace Transform
24.	$\frac{\sin at + at\cos at}{2a}$	$\frac{s^2}{(s^2+a^2)^2}$
25.	$\cos at - \frac{1}{2}at\sin at$	$\frac{s^3}{(s^2+a^2)^2}$
26.	t cos at	$\frac{s^2-a^2}{(s^2+a^2)^2}$
27.	$\frac{at\ \cos hat - \sinh at}{2a^3}$	$\frac{1}{(s^2-a^2)^2}$
28.	$\frac{t\sinh at}{2a}$	$\frac{s}{(s^2-a^2)^2}$
29.	$(\sinh at\ +\ at\cosh at)/\ 2a$	$\frac{s^2}{(s^2-a^2)^2}$
30.	$(\cosh at\ +\ 1/2at\sinh at)$	$\frac{s^3}{(s^2-a^2)^2}$
31.	$t\cosh at$	$\frac{s^2+a^2}{(s^2-a^2)^2}$
32.	$\frac{t^2\sin at}{2a}$	$\frac{3s^2-a^2}{(s^2-a^2)^3}$
33.	$\frac{t^2\cos at}{2}$	$\frac{s^3-3a^2s}{(s^2+a^2)^3}$
34.	$\frac{t^3\cos at}{6}$	$\frac{s^4-6a^2s^2+a^4}{(s^2+a^2)^4}$
35.	$\frac{t^3\sin at}{24a}$	$\frac{s^3-a^2s}{(s^2+a^2)^4}$
36.	$\frac{e^{bt}-e^{at}}{t}$	$Log\left\lvert\frac{s-b}{s-a}\right\rvert$
37.	$\frac{\sin \square t}{t}$	$\tan^{-1} \square \frac{1}{s}$
38.	Γ' (1) – log t	$\frac{Logs}{s}$

Example: Solve $y'' - 3y' + 2y = 2e^{-t}, y(0) = 2, y'(0) = -1$

Solution. Taking the Laplace transform of the given differential equation,

$$[s^2Y - sy(0) - y'(0)] - 3[sY - y(0)] + 2Y = \frac{2}{s = 1}$$

Then using initial conditions $y(0) = 2,\ y'(0) = -1$ and solving the algebraic, the equation for Y, we find using partial fractions

$$Y = \frac{2s^2-5s-5}{(s+1)(s-1)(s-2)} = \frac{1/3}{(s+1)} + \frac{4}{(s-1)} + \frac{-7/3}{(s-2)}$$

Taking the inverse Laplace transform, we obtain the required solution.

$$y = \frac{1}{3}e^{-t} + 4e^{t} - \frac{7}{3}e^{2t}$$

7.3 Fourier Series

The Fourier series is named after the man who discovered it in his research on heat flow involving partial differential equations. The series has many applications for Physical problems. For example, in the heat conduction (and diffusion)equation

$$\frac{\partial U}{\partial t} = \frac{1}{k}\frac{\partial^2 U}{\partial x^2}$$

Where k constant and $U(x,t)$ is the temperature at place x at time t.

Definition 7.3.1

Given a function $f(x)$ defined in the interval $-L \leq x \leq L$, evaluate the coefficients called Fourier coefficients, given by

$$a_k = \frac{1}{L}\int_{-L}^{L} f(x)\cos\frac{k\pi x}{L}dx\ ,\ b_k = \frac{1}{L}\int_{-L}^{L} f(x)\sin\frac{k\pi x}{L}dx \qquad (7.3.1a)$$

Using those coefficients in (4.3) and assuming the series converges to $f(x)$, the required Fourier series is given by

$$f(x) = \frac{a_0}{2} + \sum_{k=1}^{\infty}\left(a_k\cos\frac{k\pi x}{L} + b_k\sin\frac{k\pi x}{L}\right) \qquad (7.3.1b)$$

7.3.2 Application to Heat Conduction Equation

Example: A metal bar 100cm long has ended $x = 0$ and $x = 100$ kept at $0^0\ C$. Initially, half of the bar was at 60^0While the other half is at $40^0\ C$· Assuming a heat conduction coefficient of 0.16 units and that the bar's surface is insulated, find the temperature everywhere in the bar at time t.

Mathematical formulation

The heat conduction equation is

$$\frac{\partial U}{\partial t} = 0.16\frac{\partial^2 U}{\partial x^2}$$

(7.3.2a)

Where $U(x,t)$ is the temperature at place x at time t. The boundary conditions are

$$U(0,t) = 0, U(100,t) = 0\ , U(x,0) = \begin{cases} 60, & 0 < x < 50 \\ 40, & 50 < x < 100 \end{cases}$$

(7.3.2b)

Solution

Assuming $U = XT$ in (7.3.2a)and from the previous discussion in section 4.3 on partial differential, we get

$$XT' = 0.16X''T \qquad \text{or}\ \frac{T'}{0.16T} = \frac{X''}{X} \tag{7.3.2c}$$

Setting these equal to a constant, which, as our previous experience indicated, was negative, and which we denote by $-\psi^2$, we have

$$\frac{T'}{0.16T} = \frac{X''}{X} = -\psi^2$$

or $\quad T' + \psi^2 T = 0\ \ ,\ X'' + \psi^2 X = 0$

(7.3.2d)

The solution to (7.3.2d) is

$$U(x,t) = e^{-0.16\psi^2 t}(A\cos\psi\, x + B\sin\psi\, x) \tag{7.3.2e}$$

The first two conditions in (7.3.1b) show that $A = 0, \psi = n\pi/100$. To satisfy the last condition of (7.3.1b), we use superposition of the solutions to obtain

$$U(x,t) = b_1 e^{-16(10)^{-6}\pi^2 t}\sin\frac{\pi x}{100} + b_2 e^{-64(10)^{-6}\pi^2 t}\sin\frac{2\pi x}{100} + \cdots$$

(7.3.2f)

For $t = 0$,

$$b_1 \, sin\frac{\pi x}{100} + b_2 \, sin\frac{2\pi x}{100} = U(x,0) \qquad (7.3.1g)$$

Thus, we have

$$b_n = \frac{2}{100}\int_0^{100} U(x,0)\, sin\frac{n\pi x}{100}\, dx$$

$$= \frac{2}{100}\int_0^{50}(60)\, sin\frac{n\pi x}{100}\, dx\; b_n + \frac{2}{100}\int_{50}^{100}(40)\, sin\frac{n\pi x}{100}\, dx$$

$$= \frac{120}{n\pi}\left(1 - cos\frac{n\pi}{2}\right) + \frac{80}{n\pi}\left(cos\frac{n\pi}{2} - cos\, n\,\pi\right)$$

Thus $b_1 = \frac{200}{\pi}$, $b_2 = \frac{40}{\pi}$... and equation becomes

$$U(x,t) = \frac{200}{\pi} e^{-16(10)^{-6}\pi^2 t}\, sin\frac{\pi x}{100} + \frac{40}{\pi} e^{-64(10)^{-6}\pi^2 t}\, sin\frac{2\pi x}{100} + \cdots$$

Which is the required solution.

Example

Solve using Laplace Transforms:

$$y'' - 3y' + 2y = 12\, e^{4t}, y(0) = 1, y'(0) = 0$$

Solution:

The Laplace transformations of the given equations yield:

$$(s^2 Y(s) - sY_0 - Y_1) - 3(sY(s) - Y_0) + 2Y(s) = \frac{12}{s-4}$$

Putting the initial conditions into the equation: $Y_0 = 1$ and $Y_1 = 0$

$$Y(s)[s^2 - 3s + 2] - s + 3 = \frac{12}{s-4}$$

$$Y(s) = \frac{s^2 - 7s + 24}{(s-1)(s-2)(s-4)}$$

Now:

$$\frac{s^2 - 7s + 24}{(s-1)(s-2)(s-4)} \equiv \frac{A}{(s-1)} + \frac{B}{(s-2)} + \frac{C}{(s-4)}$$

Solving the partial fractions, we obtain: $A = 6, B = -7, C = 2$

$$Y(s) = \left\{\frac{6}{s-1} - \frac{7}{s-2} + \frac{2}{s-4}\right\}$$

Thus:

$$y(t) = L^{-1}\{Y(s)\}$$

$$= L^{-1}\left\{\frac{6}{s-1} - \frac{7}{s-2} + \frac{2}{s-4}\right\}$$

$$= 6e^t - 7e^{2t} + 2e^{4t}$$

7.4 Applications of Laplace Transform

(1) System analysis

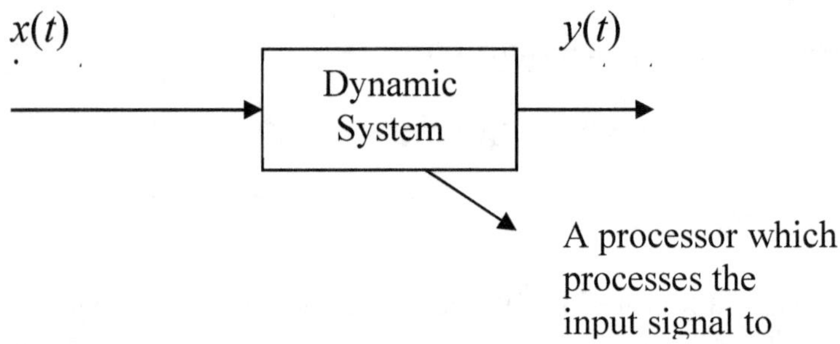

static system: *y(t) = ax(t)* => easy (simple processing)

Dynamic system:

$$\frac{dy^{(n)}(t)}{dt^n} + a_1\frac{dy^{(n-1)}(t)}{dt^{n-1}} + \ldots + a_n y(t) = b_0\frac{dx^{(m)}(t)}{dt^m} + \ldots + b_m x(t)$$

Can we easily determine y(t) for given u(t)?

Easier solution method

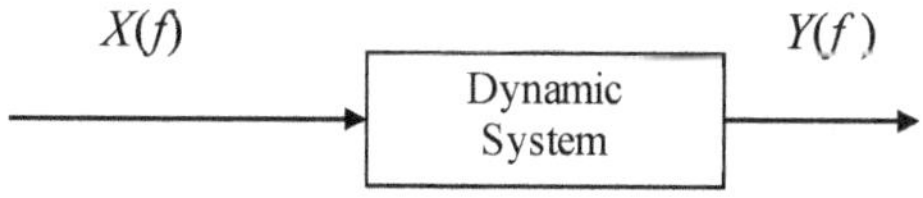

$H(f)$

$$Y(f) = H(f)X(f)$$
$$y(t) = F^{-1}(Y(f))$$

→ Algebraic equation, not differential

(1) we have a systematic way to obtain *H(f)* based on the differential equation
(2) we can obtain *X(f)*

➔ Fourier transforms an easier way.

Example:

Fourier transform of the input signal:

$$X(f) = \int_{-\infty}^{\infty} x(t)e^{-j2\pi ft}dt$$

$$|e^{-j2\pi ft}| = 1$$

if $x(t)$ does not go to zero when $t \to \infty$ and $t \to -\infty$

⇨ *X(f)* typically does not exist! (the existence of *X(f)* if not guaranteed).

⇨ A very strong condition can not be satisfied by many signals!

Solution:

Why should we care about *t*<0 for system analysis? We do not care!

$x(t)$ does not go to zero (when $t \to \infty$), but $x(t)e^{-\sigma t}$ May! (It will be much easier.)

⇨ use "single-sided" Fourier transform of $x(t)e^{-\sigma t}$, instead of a "double-sided" Fourier transform of *x*(*t*).

$$\Rightarrow \int_0^\infty (x(t)e^{-\sigma t})e^{-j\omega t}dt = \int_0^\infty x(t)e^{-(\sigma+j\omega)t}dt$$

Very useful, let's use a new name for it: Laplace transform.

Definition: Laplace transform of *x*(*t*)

$$L[x(t)] = X(s) = \int_0^\infty x(t)e^{-st}dt \quad (s = \sigma + j\omega)$$

What is a Laplace transform of *x*(*t*)?

A time function? No, *t* has been eliminated by the integral concerning *t*!

- A function of *s* (*s* is a complex variable)

7.5 System Analysis using Laplace Transform

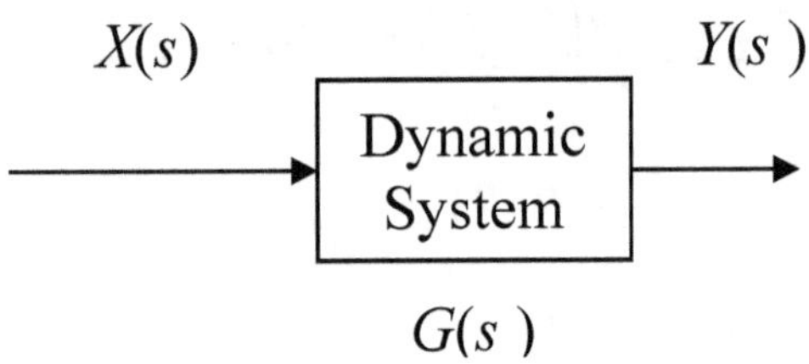

$$Y(s) = G(s)X(s)$$

$$y(t) = L^{-1}(Y(s))$$

(6) How is the inverse transform defined?

$$x(t) = \frac{1}{2\pi j}\int_{\sigma-j\infty}^{\sigma+j\infty} X(s)e^{st}ds$$

Example:

Evaluating Laplace Transforms using the definition.

(1) $x(t)=1$ and step function $x(t)=u(t)$

$$L[x(t)=u(t)] = \int_0^{\infty} x(t)e^{-st}dt = \int_0^{\infty} e^{-st}dt$$

$$= -\frac{1}{s}\int_{t=0}^{t=\infty} e^{-st}d(-st)$$

$$= -\frac{e^{-st}}{s}\bigg|_{t=0}^{t=\infty} = -\frac{e^{-\sigma t}}{s}e^{-j\omega t}\bigg|_{t=0}^{t=\infty}$$

$$= -\frac{e^{-\sigma\infty}}{s}e^{-j\omega\infty} + \frac{e^{-\sigma 0}}{s}e^{-j\omega 0}$$

$$\left(\left|e^{-j\omega\infty}\right| = 1 \quad e^{-\sigma\infty} = 0, \quad (if \quad \sigma > 0)\right) \; e^{-\sigma\infty} \to \infty, \quad (if \quad \sigma < 0)$$

$$= \frac{1}{s}(\cos 0 - j\sin 0) = \frac{1}{s}$$

$$\Rightarrow L(1) = L[u(t)] = \frac{1}{s} \qquad (\mathrm{Re}(s)) > 0$$

(2) $x(t) = e^{-\alpha t}u(t)$

$$L[e^{-\alpha t}u(t)] = \int_0^{\infty} e^{-\alpha t}e^{-st}dt = \int_0^{\infty} e^{-(\alpha+s)t}dt$$

Define a new complex variable. $\tilde{s} = s + \alpha$

⇨ $\int_0^{\infty} e^{-\tilde{s}t}dt$

we know $\int_0^{\infty} e^{-st}dt = \frac{1}{s} \quad Re(s) > 0$

$\Rightarrow \int_0^{\infty} e^{-\tilde{s}t}dt = \frac{1}{\tilde{s}} \quad Re(\tilde{s}) > 0$

$$\Rightarrow \int_0^{\infty} e^{-(\alpha+s)t} dt = \frac{1}{s+\alpha} \; Re(s+\alpha) > 0$$

$$\Rightarrow L[e^{-\alpha t} u(t)] = \frac{1}{s+\alpha} \; Re(s+\alpha) > 0$$

$$\Rightarrow L[e^{-\alpha t}] = \frac{1}{s+\alpha}$$

(3) $x(t) = \delta(t)$

$$L[\delta(t)] = \int_0^{\infty} \delta(t) e^{-st} dt$$

$$= e^{-st}|_{t=0} = e^{-\sigma t} e^{-j\omega t}\Big|_{t=0}$$

$$= e^{-\sigma t}(\cos\omega t - j\omega \sin\omega t)|_{t=0}$$

$$= 1$$

No constraint on s.

7.6 Convergence of the Laplace Transform

(1) To assure $\int_0^{\infty} x(t) e^{-st} dt = \int_0^{\infty} x(t) e^{-\sigma t} e^{-j\omega t} dt$ converge, $\sigma = Re(s)$ must be positive enough such that $x(t)e^{-\sigma t}$ goes to zero when t goes to positive infinite

(2) Region of absolute convergence and pole

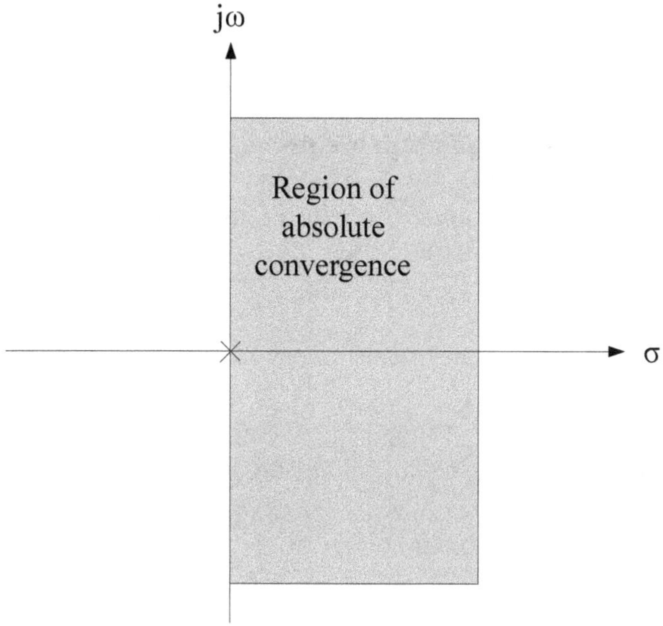

(a)

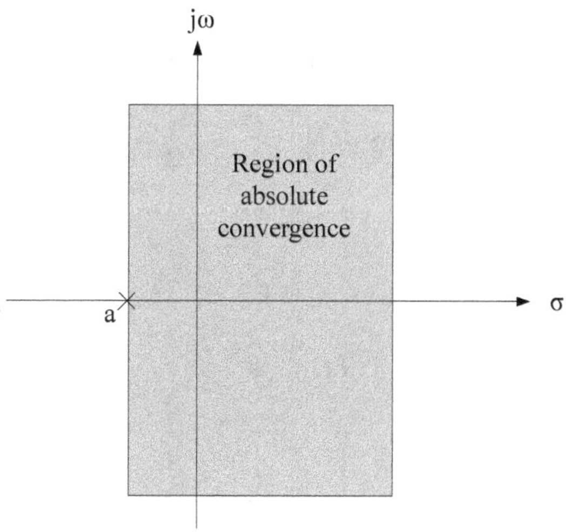

(b)

Figure. Regions of absolute convergence for the Laplace transforms of (a) u(t) and (b) $e^{-at}u(t)$

(3) How to obtain Fourier transform from Laplace transform:

$$L[x(t)] = X(s) \overset{s=j\omega}{\Rightarrow} X(j\omega) = F\big(x(t)\big)$$

7.7 Some Laplace Transform theorems

(Tools for evaluating Laplace transform based on the Laplace transforms of the basic functions)

7.7.1 Linearity

Assume

$x(t) = a_1 x_1(t) + a_2 x_2(t)$ (a_1 and a_2 are time-independent)

$X_1(s) = L[x_1(t)], X_2(s) = L[x_2(t)]$
then $X(s) = L[x(t)] = a_1 X_1(s) + a_2 X_2(s)$

HW#2-1: Assume $x(t) = a_1(t)x_1(t) + a_2(t)x_2(t), X_1(s) = L[x_1(t)]$ and $X_2(s) = L[x_2(t)]$.

(a) Is $X(s) = L[x(t)]$ equal to $a_1(t)X_1(s) + a_2(t)X_2(s)$?

(Answer: Yes or No)

(b) If $A_1(s) = L[a_1(t)]$, is it true

$$L[x(t)] = A_1(s)X_1(s) + A_2(s)X_2(s) \text{ ?}$$

(Answer: Yes or No)

Example

(1) Find $L(cos\, \omega_0\, t)$

Key to the solution: express ($cos\, \omega_0\, t$) as linear combination of $\delta(t)$, $u(t)$, and/or $e^{-\alpha t}$:

$$L[\delta(t)] = 1$$

$$L[u(t)] = \frac{1}{s}$$

$$L[e^{-\alpha t}] = \frac{1}{s+\alpha}$$

let $\alpha = j\omega_0$

$$L[e^{-j\omega_0 t}] = \frac{1}{s+j\omega_0}$$

let $\alpha = -j\omega_0$

$$L[e^{j\omega_0 t}] = L[e^{-(-j\omega_0)t}] = \frac{1}{s-j\omega_0}$$

Can we use $e^{-j\omega_0 t}$ and $e^{j\omega_0 t}$ to express $cos(\omega_0 t)$?

$$e^{-j\omega_0 t} = cos(-\omega_0 t) + j\,sin(-\omega_0 t)$$

$$= cos(\omega_0 t) - j\,sin(\omega_0 t)$$

$$e^{j\omega_0 t} = cos(\omega_0 t) + j\,sin(\omega_0 t)$$

$$e^{-j\omega_0 t} + e^{j\omega_0 t} = 2\,cos(\omega_0 t)$$

$$\Rightarrow cos(\omega_0 t) = \frac{e^{-j\omega_0 t} + e^{j\omega_0 t}}{2}$$

$$\Rightarrow L[cos(\omega_0 t)] = \frac{1}{2}[L(e^{-j\omega_0 t}) + L(e^{j\omega_0 t})]$$

$$= \frac{1}{2}\left[\frac{1}{s+j\omega_0} + \frac{1}{s-j\omega_0}\right]$$

$$= \frac{1}{2}\left[\frac{(s-j\omega_0)+(s+j\omega_0)}{(s+j\omega_0)(s-j\omega_0)}\right]$$

$$= \frac{s}{s^2+{\omega_0}^2}$$

(2) Find $L[sin\,\omega_0\,t]$

7.7.2 Transforms of Derivatives

Assume $X(s) = L[x(t)]$

Then $L\left(\frac{dx(t)}{dt}\right) = sX(s) - x(0^-)$

Proof:

(1) Definition $L\left[\frac{dx(t)}{d(t)}\right] = \int_0^\infty \frac{dx(t)}{dt} e^{-st} dt = \int_0^\infty e^{-st} dx(t)$

(2) Integration by parts:

General equation:

$$\int_a^b u(t)dv(t) = u(t)v(t)|_{t=a}^{t=b} - \int_a^b v(t)du(t)$$

(3) Use the above equation

Why? If we assume $v(t) = x(t), \quad u(t) = e^{-st}$

$$\Rightarrow \int_0^\infty v(t)du(t) = \int_0^\infty x(t)de^{-st} = -s\int_0^\infty x(t)e^{-st}\,dt$$

$$X(s) = L(x(t))$$

$$u(t) = e^{-st}, v(t) = x(t) \Rightarrow \int_0^\infty u(t)dv(t) = \int_0^\infty e^{-st}dx(t)$$

$$= [\ L\left(\frac{dx(t)}{dt}\right) \text{ from (1) }]$$

$$u(t)v(t)|_{t=0}^{t=\infty} - \int_a^b v(t)du(t)$$

$$= e^{-st}x(t)|_{t=0}^{t=\infty} + s\int_0^\infty x(t)e^{-st}dt$$

$$= e^{-st}x(t)|_{t=0}^{t=\infty} + sL[x(t)]$$

$\lim_{t\to\infty} e^{-st}x(t)$ must go to zero. Otherwise, $L[x(t)] = \int_0^\infty x(t)e^{-st}dt$ does not exist!

$$\Rightarrow e^{-st}x(t)|_{t=0}^{t=\infty} = -e^{-s\cdot 0}x(0) = -x(0)$$

use 0^- as lower limit => $x(0^-)$

$$\Rightarrow u(t)v(t)|_{t=0}^{t=\infty} - \int_0^{\infty} v(t)du(t)$$

$$= -x(0^-) + sX(s) = sX(s) - x(0^-)$$

Problem:

Assume: $X(s) = L[x(t)]$. Prove

$$L\left[\frac{d^2x(t)}{dt}\right] = s^2X(s) - sx(0^-) - x^{(1)}(0^-)$$

Problem:Express $L\left[\frac{d^{(n)}x(t)}{dt^n}\right]$ using $L[x(t)]$

Example

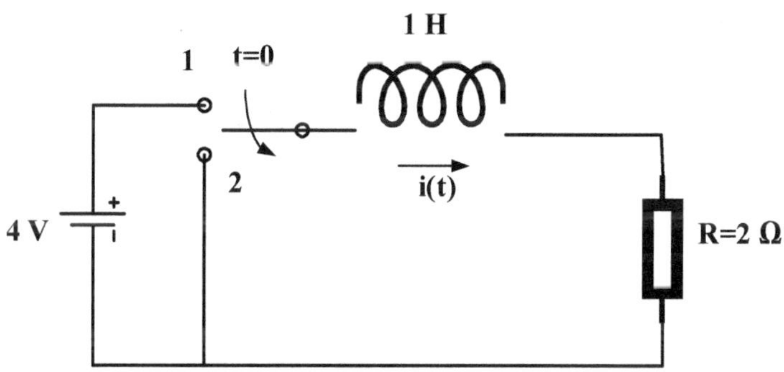

Find $i(t)$ using Laplace transform method for $t>0$

Solution:

(1) Before switching from 1 to 2 at $t=0$

$$i = \frac{4}{2} = 2A \Rightarrow i(0^-) = 2A$$

(2) System equation ($t>0$)

KVL: $L\frac{di(t)}{dt} + Ri(t) = 0\ (L = 1H)\ (R = 2\ ohm)$

$$\Rightarrow \frac{di(t)}{dt} + 2i(t) = 0$$

(3) Solve the system equation using Laplace transform

$$L\left[\frac{di(t)}{dt} + 2i(t)\right] = L\left[\frac{di(t)}{dt}\right] + 2L[i(t)]$$

$$= sI(s) - i(0^-) + 2I(s)$$

$$= (s+2)I(s) - 2 = 0$$

$$\Rightarrow I(s) = \frac{2}{s+2}$$

$$\Rightarrow i(t) = 2e^{-2t}u(t)\ A$$

7.7.3 Laplace Transform of an Integral

Assume $y(t) = \int_{-\infty}^{t} x(\lambda)d\lambda, X(s) = L[x(t)]$

Then $L\left[\int_{-\infty}^{t} x(\lambda)d\lambda\right] = \frac{X(s)}{s} + \frac{y(0^-)}{s}$

where $y(0^-) = \int_{-\infty}^{0^-} x(\lambda)d\lambda$

Proof :

$$L\left[\int_{-\infty}^{t} x(\lambda)d\lambda\right] = \int_{0^-}^{\infty} \left(\int_{-\infty}^{0^-} x(\lambda)d\lambda\right) e^{-st}dt$$

$$= \int_{0^-}^{\infty} udv = uv|_{t=0^-}^{t=\infty} - \int_{0^-}^{\infty} vdu$$

(1)

$$uv|_{t=0^-}^{t=\infty} = \left(\int_{-\infty}^{t} x(\lambda)d\lambda\right)\frac{-e^{-st}}{s}\Bigg|_{t=0^-}^{t=\infty}$$

$$= \lim_{t\to\infty}\frac{-e^{-st}}{s}\int_{-\infty}^{t} x(\lambda)d\lambda + \frac{e^{-s\cdot 0}}{s}\int_{-\infty}^{0^-} x(\lambda)d\lambda = \frac{y(0^-)}{s}$$

(2)

$$\int_{0^-}^{\infty} v du = -\int_{0^-}^{\infty} \frac{e^{-st}}{s} x(t)dt = -\frac{1}{s}\int_{0^-}^{\infty} x(t)e^{-st}dt = -\frac{1}{s}X(s)$$

(3)

$$L\left[\int_{-\infty}^{t} x(\lambda)d\lambda\right] = \frac{y(0^-)}{s} + \frac{X(s)}{s} = \frac{X(s)}{s} + \frac{y(0^-)}{s} \quad \text{Proved!}$$

Example:

Find $I(s) = L(i(t))$

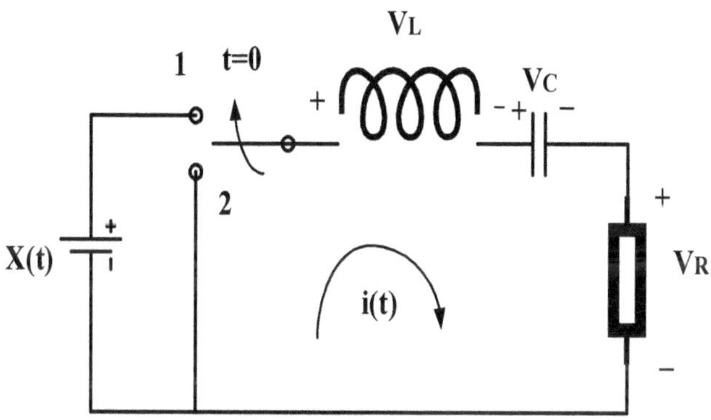

Solution:

(1) Differential equation

$$\text{KVL: } x(t) = v_L(t) + v_C(t) + v_R(t)$$

$$\left.\begin{array}{l} i(t) = C\frac{dv_C(t)}{dt} \\ dv_C(t) = \frac{1}{C}i(t)dt \\ \int_{-\infty}^{t} dv_C(\lambda) = \frac{1}{C}\int_{-\infty}^{t} i(\lambda)d\lambda \\ v_C(t) = v_C(-\infty) + \frac{1}{C}\int_{-\infty}^{t} i(\lambda)d\lambda \\ v_C(-\infty) = 0 \Rightarrow v_C(t) = \frac{1}{C}\int_{-\infty}^{t} i(\lambda)d\lambda \end{array}\right\}$$

$$v_L(t) = L\frac{di(t)}{dt}$$

$$v_C(t) = \frac{1}{C}\int_{-\infty}^{t} i(\lambda)d\lambda$$

$$v_R(t) = Ri(t)$$

(2) Laplace transform

$$X(s) = V_L(s) + V_C(s) + V_R(s)$$

$$V_L(s) = L[sI(s) - i(0^-)] = LsI(s)\ i(0^-) = zero$$

$$V_C(s) = \frac{1}{C}\left[\frac{I(s)}{s} + \frac{1}{s}\int_{-\infty}^{0^-} i(\lambda)d\lambda\right]$$

$$= \frac{1}{Cs}I(s) + \frac{1}{s}\frac{1}{c}\int_{-\infty}^{0^-} i(\lambda)d\lambda$$

$$= \frac{1}{Cs}I(s) + \frac{1}{s}v_C(0^-)$$

$$V_R(s) = RI(s)$$

$$X(s) = LsI(s) + \frac{1}{Cs}I(s) + \frac{1}{s}v_C(0^-) + RI(s)$$

$$\Rightarrow I(s) = \frac{X(s) - \frac{1}{s}v_C(0^-)}{Ls + \frac{1}{Cs} + R}$$

$$= \frac{sX(s) - v_C(0^-)}{Ls^2 + \frac{1}{C} + sR}$$

$$= \frac{sX(s) - v_C(0^-)}{L\left[s^2 + \left(\frac{R}{L}\right)s + \frac{1}{LC}\right]}$$

7.8 Complex Frequency shift (s-shift) Theorem

Assume $y(t) = x(t)e^{-\alpha t}$

$$X(s) = L[x(t)] \quad Y(s) = L[y(t)]$$

Then $Y(s) = X(s + \alpha)$

- $L[u(t)] = \frac{1}{s}, \quad L[u(t)e^{-\alpha t}] = \frac{1}{s+\alpha}$
- $L[\cos \omega_0 t] = \frac{s}{s^2+\omega_0{}^2} \quad L[\sin \omega_0 t] = \frac{\omega_0}{s^2+\omega_0{}^2}$

=> $L[\cos \omega_0 t \cdot e^{-\alpha t}]$
$= \frac{s + \alpha}{(s + \alpha)^2 + \omega_0{}^2} \quad L[\sin \omega_0 t \cdot e^{-\alpha t}] = \frac{\omega_0}{(s + \alpha)^2 + \omega_0{}^2}$

Example: Find $x(t) = L^{-1}[X(s)] = L^{-1}\frac{s+8}{s^2+6s+13}$

Solution:

$$X(s) = \frac{s + 8}{s^2 + 6s + 13} = \frac{(s + 3) + 5}{s^2 + 6s + 9 + 4}$$

$$= \frac{s + 3}{(s + 3)^2 + 2^2} + \frac{(5/2) \times 2}{(s + 3)^2 + 2^2}$$

$$x(t) = L^{-1}[X(s)] = e^{-3t} \cos 2t + \frac{5}{2}e^{-3t} \sin 2t \quad (t > 0)$$

7.8.1 Delay Theorem

Question: How do express delayed function?

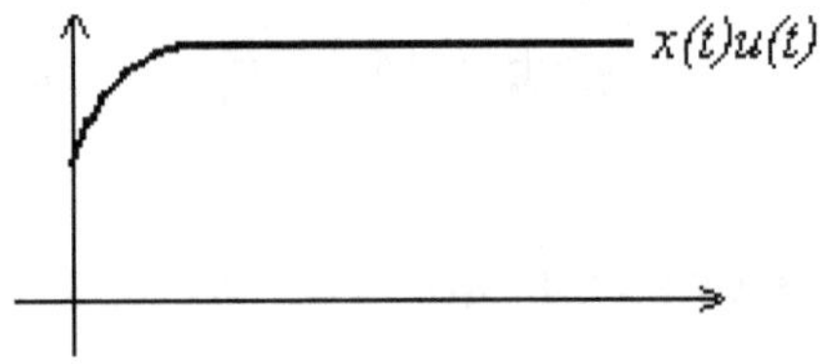

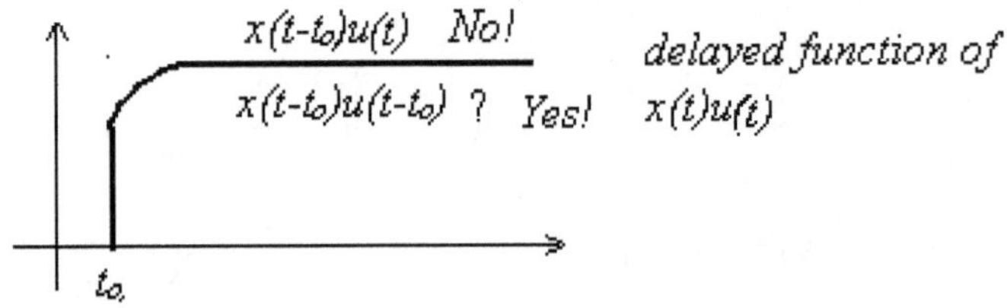

Assume $\quad L[x(t)] \equiv L[x(t)u(t)] = X(s)$

Then $\quad L[x(t-t_0)u(t-t_0)] = e^{-st_0}X(s) \;\; (t_0 > 0)$

(If $(t_0 < 0)$, it will not be a delay!)

Proof :

$$L[x(t-t_0)u(t-t_0)]$$

$$= \int_0^{\infty} x(t-t_0)u(t-t_0)e^{-st}dt$$

$$= \int_0^{t_0} x(t-t_0)u(t-t_0)e^{-st}dt + \int_{t_0}^{\infty} x(t-t_0)u(t-t_0)e^{-st}dt$$

$$= \int_{t_0}^{\infty} x(t-t_0)e^{-st}dt = \int_{t_0}^{\infty} x(t-t_0)e^{-s(t-t_0)-st_0}dt$$

$$= e^{-st_0}\int_{t_0}^{\infty} x(t-t_0)e^{-s(t-t_0)}d(t-t_0)$$

$$\overset{\tau=t-t_0}{=} e^{-st_0}\int_0^{\infty} x(\tau)e^{-s\tau}d\tau = e^{-st_0}\int_0^{\infty} x(t)e^{-st}dt$$
$$= e^{-st_0}X(s)$$

Question: will $L[x(t-t_0)u(t-t_0)] = e^{-st_0}X(s)$ be true if $t_0 < 0$?

No! (it will not be a delay)

Example: Square wave beginning at t = 0

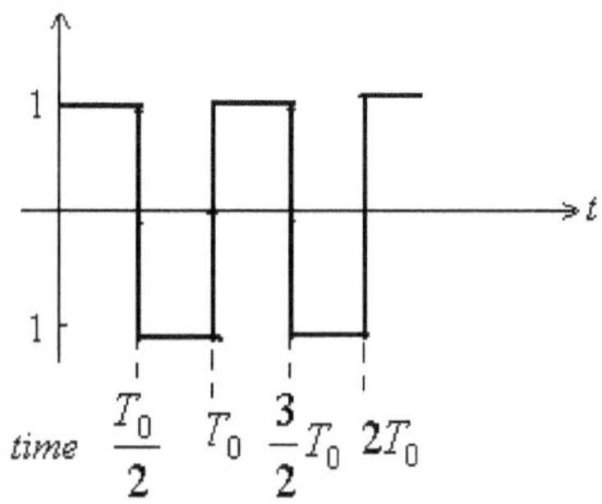

$$x_{sq}(t) = u(t) - 2u(t-\frac{T_0}{2}) + 2u(t-T_0)$$
$$-2u(t-\frac{3}{2}T_0) + 2u(t-2T_0)$$
......

$$L[x_{sq}(t)] = \frac{1}{s} - 2\frac{1}{s}e^{-\frac{T_0}{2}s} + 2\frac{1}{s}e^{-T_0 s} - 2\frac{1}{s}e^{-\frac{3T_0}{2}s} + 2\frac{1}{s}e^{-2T_0 s}$$

$$\overset{\lambda=-e^{-\frac{1}{2}T_0 s}}{=} \frac{1}{s} + 2\frac{1}{s}\lambda + 2\frac{1}{s}\lambda^2 + 2\frac{1}{s}\lambda^3 + 2\frac{1}{s}\lambda^4 + \cdots$$

$$= \frac{1}{s} + \frac{2}{s}(\lambda + \lambda^2 + \lambda^3 + \ldots)$$

$$= \frac{1}{s} + \frac{2\lambda}{s(1-\lambda)} = \frac{1}{s}\cdot\frac{1+\lambda}{1-\lambda} = \frac{1}{s}\cdot\frac{1-e^{-\frac{T_0}{2}s}}{1+e^{-\frac{T_0}{2}s}}$$

7.8.2 Convolution

Signal 1: $x_1(t)$ Signal 2 : $x_2(t)$

$$y(t) = x_1(t) * x_2(t) = \int_{-\infty}^{\infty} x_1(\lambda)x_2(t-\lambda)\,d\lambda$$

if $x_1(t) = 0. \quad \forall t < 0$

$$\Rightarrow y(t) = \int_0^{\infty} x_1(\lambda)x_2(t-\lambda)\,d\lambda$$

if $\quad x_2(t) = 0. \forall t < 0. \quad (x_2(t-\lambda) = 0 \quad \forall \lambda > t)$

$\Rightarrow y(t) = \int_0^t x_1(\lambda)x_2(t-\lambda)\,d\lambda$

Therefore, if $\quad x_1(t) = 0, \quad x_2(t) = 0 \quad \forall t < 0$

$$\Rightarrow \int_0^t x_1(\lambda)x_2(t-\lambda)\,d\lambda = \int_0^{\infty} x_1(\lambda)x_2(t-\lambda)\,d$$

$$= \int_{-\infty}^{\infty} x_1(\lambda)x_2(t-\lambda)\,d\lambda$$

$$\Rightarrow L[\int_0^t x_1(\lambda)x_2(t-\lambda)\,d\lambda] = L[\int_0^{\infty} x_1(\lambda)x_2(t-\lambda)\,d\lambda]$$

$$= L[\int_{-\infty}^{\infty} x_1(\lambda)x_2(t-\lambda)\,d\lambda]$$

$$\int_0^{\infty}[\int_0^{\infty} x_1(\lambda)x_2(t-\lambda)d\lambda]e^{-st}dt$$

$$= \int_0^{\infty} x_1(\lambda)[\int_0^{\infty} x_2(t-\lambda)e^{-st}dt]d\lambda$$

Look at

$$\int_0^{\infty} x_2(t-\lambda)e^{-st}dt = \int_0^{\infty} x_2(t-\lambda)e^{-s(t-\lambda)}e^{-s\lambda}dt$$

$$\underset{\substack{d\tau=dt \\ t=0\to\tau=-\lambda \\ t=\infty\to\tau=\infty}}{\overset{\tau=t-\lambda}{=}} e^{-s\lambda}\int_{-\lambda}^{\infty} \lambda \quad {}^{-s\tau}d\,\tau \quad t$$

$$\overset{x_2(\tau)=0\forall \quad \tau\in[-\lambda \quad 0]}{=} e^{-s\lambda}\int_0^{\infty} x_2(\tau)e^{-s\tau}d\,\tau = e^{-s\lambda}X_2(s)$$

Then

$$Y(s) = \int_0^{\infty} x_1(\lambda)e^{-s\lambda}X_2(s)d\,\lambda$$

$$= X_2(s) \int_0^{\infty} x_1(\lambda) e^{-s\lambda} d\,\lambda$$

$$= X_1(s) X_2(s)$$

7.8.3 Product

Initial Value Theorem

$$x(t) = L^{-1}[X(s)]$$

$$\Rightarrow x(0^+) = \lim_{s \to \infty} sX(s)$$

Example: A demonstration where *x*(0) is obvious

$$x(t) = e^{-\alpha t} \cos \omega_0\, t u(t)$$

It is evident: $x(0) = e^{-0} \cos \omega_0 \cdot 0 = 1$

Using Laplace transform

$$X(s) = L[x(t)] = \frac{s + \alpha}{(s + \alpha)^2 + {\omega_0}^2}$$

$$x(0) = \lim_{s \to \infty} sX(s) = \lim_{s \to \infty} \frac{s(s + \alpha)}{(s + \alpha)^2 + {\omega_0}^2}$$

$$= \lim_{s \to \infty} \frac{s^2 + \alpha s}{s^2 + 2\alpha s + \alpha^2 + {\omega_0}^2}$$

$$= \lim_{s \to \infty} \frac{\frac{d(s^2 + \alpha s)}{ds}}{\frac{d(s^2 + 2\alpha s + \alpha^2 + {\omega_0}^2)}{ds}} \quad \left(\frac{\infty}{\infty}\right)$$

$$= \lim_{s \to \infty} \frac{2s + \alpha}{2s + 2\alpha} = \lim_{s \to \infty} \frac{d(2s + \alpha)/ds}{d(2s + 2\alpha)/ds} = \lim_{s \to \infty} \frac{2}{2} = 1$$

7.8.4 Final Value Theorem

If $x(t)$ and $dx(t)/dt$ are Laplace transformable, then

$$\lim_{t\to\infty} x(t) = \lim_{s\to 0} sX(s)$$

(condition: $sX(s)$ has no poles on $j\omega - axis$ or in the right-half s-plan or $lim_{t\to\infty}\, x\,(t)$ exists)

7.8.5 Scaling

a>0: *x*(*at*) ← *a* times fast (if *a*>1) or slow (if *a*<1) as *x*(*t*)

$$X(s) \stackrel{\Delta}{=} L[x(t)]$$

What do we expect on $L[x(at)]$?

$$L[x(at)] \Rightarrow X(\frac{s}{a})?$$

$$L[x(at)] = \int_0^\infty x(at)e^{-st}dt \stackrel{a>0}{=} \frac{1}{a}\int_0^\infty x(at)e^{-\frac{s}{a}(at)}d(at)$$

$$\underset{\substack{a>0\\ t=\infty\\ \Rightarrow\tau=\infty}}{\stackrel{\tau=at}{=}} \frac{1}{a}\int_0^\infty x(\tau)e^{-\left(\frac{s}{a}\right)\tau}d\tau = \frac{1}{a}X\left(\frac{s}{a}\right)$$

7.8.6 Inversion of Rational Functions

(1) Ways to find $x(t)$ from $X(s)$

(1) $x(t) = \frac{1}{2\pi j}\int_{\sigma-\omega}^{\sigma+\omega} X(s)e^{st}dt$ (Contour Integral)

(2) Transform pair

$$\frac{1}{s} \leftrightarrow u(t)$$

$$\frac{1}{s+\alpha} \leftrightarrow e^{-\alpha t}u(t)$$

Therefore $X(s) = \frac{1}{s+\alpha} \Rightarrow x(t) = e^{-\alpha t}u(t)$

All kinds of Laplace Transform? No!

We almost only see

$$\frac{b_0 s^m + b_1 s^{m-1} + \ldots + b_{m-1}s + b_m}{s^n + a_1 s^{n-1} + \ldots a_{n-1}s + a_n} e^{-\tau s}$$

rational function X(s) Delay

$$x(t) = L^{-1}[X(s)]$$

$$y(t) = L^{-1}[X(s)e^{-\tau s}] = x(t-\tau)u(t-\tau)$$

⇨ Consider Rational Functions only!

(3) Non-proper Rational Function
proper Rational Function

Non-proper *m>=n* $(b_0 \neq 0)$
proper *m<n*

Non-proper => proper + Polynomial (using long division)

$$\frac{s^3 + 4s^2 + 6s + 7}{s^2 + 3s + 2} = s + 1 + \frac{s+5}{s^2 + 3s + 2}$$

$$s + 1$$

$$s^2+3s+2 \overline{)s^3 + 4s^2 + 6s + 7}$$

$$\overline{} \\ s^2 + 4s + 7$$

$$s^2 + 3s + s$$

$$\overline{} \\ s + 5$$

How to find the inverse Laplace transform for polynomials?

$$s^n \leftrightarrow \delta^{(n)}(t)$$

$$L^{-1}(s+1) = L^{-1}(s) + L^{-1}(1) = \delta^{(1)}(t) + \delta(t)$$

➔ consider proper rational functions only!

(4) Proper Rational Functions: Partial Fraction Expansion

➔ sum of $\frac{1}{(s+\alpha)^{n+1}} \to \frac{t^n e^{-\alpha t}}{n!} u(t)$

$$\frac{s+\alpha}{(s+\alpha)^2 + {\omega_0}^2} \to e^{-\alpha t} \cos \omega_0\, t u(t)$$

$$\frac{\omega_0}{(s+\alpha)^2 + {\omega_0}^2} \to e^{-\alpha t} \sin \omega_0\, t u(t)$$

$$1 \to \delta(t)$$

$$\frac{1}{s} \to u(t)$$

$$\frac{1}{s+\alpha} \to e^{-\alpha t}$$

Let's look at examples and then summarize!

Techniques:

Common Denominator
The specific value of s
Heaviside's Expansion
Matlab

Factorize first!
Expands second!
Find coefficients third

Example: Simple Factors

$$Y(s) = \frac{10}{s^2+10s+16}$$

Solution:

(1) Factorize and expand $Y(s) = \frac{10}{(s+8)(s+2)} = \frac{A}{s+8} + \frac{B}{s+2}$

(3) Common Denominator Methods

$$\frac{10}{(s+8)(s+2)} = \frac{A(s+2)+B(s+8)}{(s+8)(s+2)}$$

$$\Rightarrow A(s+2) + B(s+8) = 10$$

$$\Rightarrow \begin{cases} A + B = 0 \Rightarrow A = -B \\ 2A + 8B = 10 \Rightarrow -2B + 8B = 10 \end{cases}$$

$$\Rightarrow B = \frac{10}{6} = \frac{5}{3} \quad A = -5/3$$

$$Y(s) = -\frac{5}{3} \cdot \frac{1}{s+8} + \frac{5}{3} \cdot \frac{1}{s+2}$$

$$\Rightarrow y(t) = (-\frac{5}{3}e^{-8t} + \frac{5}{3}e^{-2t})u(t)$$

specific values of s

$$\frac{10}{(s+8)(s+2)} = \frac{A}{s+8} + \frac{B}{s+2}$$

$$\overset{s=0}{\Rightarrow} \frac{10}{8 \times 2} = \frac{A}{8} + \frac{B}{2}$$

$$\overset{s=2}{\Rightarrow} \frac{10}{10 \times 4} = \frac{A}{10} + \frac{B}{4}$$

Can you solve for A and B?

Heaviside Expansion

$$\frac{10}{(s+8)(s+2)} = \frac{A}{s+8} + \frac{B}{s+2}$$

$$\Rightarrow \frac{10}{s+2} = A + \frac{B(s+8)}{s+2} \overset{s=-8}{\Rightarrow} \frac{10}{-8+2} = A = -5/3$$

and $\Rightarrow \frac{10}{s+8} = \frac{A(s+2)}{s+8} + B \overset{s=-2}{\Rightarrow} B = \frac{10}{-2+8} = 5/3$

Example: Imaginary Roots

$$Y(s) = \frac{15s^2 + 25s + 20}{(s^2+1)(s+2)(s+8)}$$

Solution: what do we have:

$$Y(s) = \frac{A_1}{s+j} + \frac{A_2}{s-j} + \frac{A_3}{s+2} + \frac{A_4}{s+8}$$

Same as real roots!

$$Y(s) = \frac{A_1(s-j)+A_2(s+j)}{s^2+1} + \frac{A_3}{s+2} + \frac{A_4}{s+8}$$

$$= \frac{(A_1+A_2)s + (A_2 - A_1)j}{s^2+1} + \frac{A_3}{s+2} + \frac{A_4}{s+8}$$

A_1+A_2 must be a real number

$(-A_1+A_2)j$ must be a real number

$$Y(s) = \frac{c_1 s + c_2}{s^2+1} + \frac{A_3}{s+2} + \frac{A_4}{s+8}$$

Heaviside Expansion => A_3=1 and A_4= - 2.

$$Y(s) = \frac{15s^2 + 25s + 20}{(s^2+1)(s+2)(s+8)} = \frac{c_1 s + c_2}{s^2+1} + \frac{1}{s+2} + \frac{-2}{s+8}$$

s=1 => $\frac{15+25+20}{2\times3\times9} = \frac{c_1+c_2}{2} + \frac{1}{3} + \frac{-2}{9}$

s=2 => $\frac{15\times4+25\times2+20}{5\times4\times10} = \frac{2c_1+c_2}{5} + \frac{1}{4} + \frac{-2}{10}$

Can we solve for c_1 and c_2?

c_1=1 c_2=1

$$Y(s) = \frac{s+1}{s^2+1} + \frac{1}{s+2} - \frac{2}{s+8}$$

$$= \frac{s}{s^2+1} + \frac{1}{s^2+1} + \frac{1}{s+2} - \frac{2}{s+8}$$

=>$[\cos t + \sin t + e^{-2t} - 2e^{-8t}]u(t)$

Too complex: use MATLAB

Example: Repeated linear Factors

$$Y(s) = \frac{10s}{(s+2)^2(s+8)}$$

$$= \frac{A_1}{s+8} + \frac{A_2}{s+2} + \frac{A_3}{(s+2)^2}$$

Example:

$$Y(s) = \frac{10s}{(s+2)^3(s+8)}$$

$$= \frac{A_1}{s+8} + \frac{A_2}{s+2} + \frac{A_3}{(s+2)^2} + \frac{A_4}{(s+2)^3}$$

Example: Complex - Conjugate Factors

$$Y(s) = \frac{2s^2 + 6s + 6}{(s+2)(s^2 + 2s + 2)}$$

$$= \frac{2s^2 + 6s + 6}{(s+2)[(s+1)^2 + 1]} = \frac{A_1}{s+2} + \frac{A_2}{s-1-j} + \frac{A_3}{s-1+j}$$

Example: Repeated Quadratic Factors

$$Y(s) = \frac{s^4 + 5s^3 + 12s^2 + 7s + 15}{(s+2)(s^2+1)^2}$$

$$= \frac{A_1}{s+2} + \frac{B_1 s + c_1}{s^2+1} + \frac{B_2 s + c_2}{(s^2+1)^2}$$

* Summary of Partial–Fraction Expansion

(1) Expansion Structure:

Simple Roots (including complex conjugate)

=> $\frac{A_j}{s-\alpha_j}$ It could be complex.

Repeated Roots: m multiplicity

=> $\frac{B_1}{s-\beta_j} + \frac{B_2}{(s-\beta_j)^2} + \ldots + \frac{B_m}{(s-\beta_j)^m}$

real number or complex number

(2) Avoid complex number

For complex conjugates: $\alpha_j = a + jb$

$$\frac{A_j}{s-\alpha_j} + \frac{A_j^*}{s-\alpha_j^*} \Rightarrow \frac{cs + D}{(s-a)^2 + b^2}$$

$$\frac{B_k}{(s-\alpha_j)^k} + \frac{B_k^*}{\left(s-\alpha_j^*\right)^k} \Rightarrow \frac{cs + D}{[(s-a)^2 + b^2]^k}$$

(3) Inverse Laplace transform

$$\frac{A_j}{s-\alpha_j} \leftrightarrow A_j e^{\alpha_j t} u(t)$$

$$\frac{A_j}{(s-\alpha_j)^k} \leftrightarrow A_k \frac{t^{k-1} e^{\alpha_j t}}{(k-1)!} u(t) \quad k \geq 2$$

$$\frac{A_j}{(s-\alpha_j)^k} + \frac{B_j}{(s-\alpha_j^*)^k} \leftrightarrow \frac{t^{k-1}}{(k-1)!} [A_j e^{\alpha_j t} + B_j e^{\alpha_j^* t}] u(t)$$

$$\overset{\alpha_j = a+jb}{\underset{\alpha_j^* = a-jb}{=}} \frac{t^{k-1}}{(k-1)!} e^{at} [A_j(\cos b\,t + j \sin b\,t) + B_j(\cos b\,t - j \sin b\,t] u(t)$$

7.9 Two-port Network

A **two-port network** (a **four-terminal** or **quadripole**) is an electrical network (circuit) or device with two *pairs* of terminals to connect to external circuits. Two terminals would constitute a port if the currents applied to them satisfy the essential requirement known as the port condition: the electric current entering one terminal must equal the current emerging from the other terminal on the same port. The ports constitute interfaces where the network connects to other networks, the points where signals are applied or outputs are taken. In a two-port network, port one is often considered the input port, and port two is the output port.

The two-port network model is used in mathematical circuit analysis techniques to isolate portions of larger circuits. A two-port network is regarded as a "black box" with properties specified by a matrix of numbers. This allows the network's response to signals applied to the ports to be calculated easily without solving for all the internal voltages and currents in the network. It also allows similar circuits or devices to be compared easily. For example, transistors are often regarded as two-ports, characterized by their h-parameters (see below) which the manufacturer lists. Any linear circuit with four terminals can be regarded as a two-port network, provided it does not contain an independent source and satisfies the port conditions.

Examples of circuits analyzed as two-ports are filters, matching networks, transmission lines, transformers, and small-signal models for transistors (such as the hybrid-pi model). The analysis of passive two-port networks is an outgrowth of reciprocity theorems first derived by Lorentz.

In two-port mathematical models, the network is described by a two-by-two-square matrix of complex numbers. The common models are *z-parameters*, *y-parameters*, *h-parameters*, *g-parameters*, and *ABCD parameters*. These are all limited to linear networks since their derivation assumes that any given circuit condition is a linear superposition of various short-circuit and open-circuit conditions.

They are usually expressed in matrix notation and establish relationship between the variables.

V_1, the voltage across port 1
I_1,the current into port 1
V_2, the voltage across port 2
I_2,the current into port 2

which are shown in figure 1. The difference between the various models lies in which of these variables are regarded as the independent variables. These current and voltage variables are most useful at low-to-moderate frequencies. At high frequencies, the use of power and energy variables is more appropriate, and the two-port current–voltage approach is replaced by an approach based on scattering parameters.

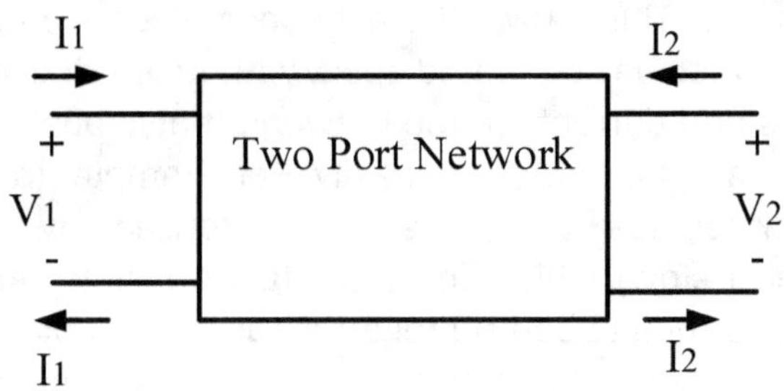

1) Z-Parameters

$$V_1 = Z_{11}I_1 + Z_{12}I_2$$

$$V_2 = Z_{21}I_1 + Z_{22}I_2$$

2) Y-Parameters

$$I_1 = Y_{11}V_1 + Y_{12}V_2$$

$$I_2 = Y_{21}V_1 + Y_{22}V_2$$

3) Hybrid-Parameters

$$V_1 = h_{11}I_1 + h_{12}V_2$$

$$I_2 = h_{21}I_1 + h_{22}V_2$$

4) ABCD-Parameters (Transmission line Parameters)

$$V_1 = A\,V_2 + B\,I_2$$

$$I_1 = C\,V_2 + D\,I_2$$

7.10 Electric Circuits Formulas

METRIC PREFIX TABLE

Metric Prefix	Symbol	Multiplier (Traditional Notation)	Expo-nential	Description
Yotta	Y	1,000,000,000,000,000,000,000,000	10^{24}	Septillion
Zetta	Z	1,000,000,000,000,000,000,000	10^{21}	Sextillion
Exa	E	1,000,000,000,000,000,000	10^{18}	Quintillion
Peta	P	1,000,000,000,000,000	10^{15}	Quadrillion
Tera	T	1,000,000,000,000	10^{12}	Trillion
Giga	G	1,000,000,000	10^{9}	Billion
Mega	M	1,000,000	10^{6}	Million
kilo	k	1,000	10^{3}	Thousand
hecto	h	100	10^{2}	Hundred
deca	da	10	10^{1}	Ten
Base	b	1	10^{0}	One
deci	d	1/10	10^{-1}	Tenth
centi	c	1/100	10^{-2}	Hundredth
milli	m	1/1,000	10^{-3}	Thousandth
micro	µ	1/1,000,000	10^{-6}	Millionth
nano	n	1/1,000,000,000	10^{-9}	Billionth
pico	p	1/1,000,000,000,000	10^{-12}	Trillionth
femto	f	1/1,000,000,000,000,000	10^{-15}	Quadrillionth
atto	a	1/1,000,000,000,000,000,000	10^{-18}	Quintillionth
zepto	z	1/1,000,000,000,000,000,000,000	10^{-21}	Sextillionth
yocto	y	1/1,000,000,000,000,000,000,000,000	10^{-24}	Septillionth

4-Band Resistor Color Code Table

BAND	COLOR	DIGIT
Band 1: 1st Digit Band 2: 2nd Digit Band 3: Multiplier (# of zeros following 2nd digit)	Black	0
	Brown	1
	Red	2
	Orange	3
	Yellow	4
	Green	5
	Blue	6
	Violet	7
	Gray	8
	White	9
Band 4: Tolerance	GOLD	± 5%
	SILVER	± 10%

5-Band Resistor Color Code Table

BAND	COLOR	DIGIT
Band 1: 1st Digit Band 2: 2nd Digit Band 3: 3rd Digit Band 4: Multiplier (# of zeros following 3rd digit)	Black	0
	Brown	1
	Red	2
	Orange	3
	Yellow	4
	Green	5
	Blue	6
	Violet	7
	Gray	8
	White	9
	GOLD	0.1
	SILVER	0.01
Band 5: Tolerance	GOLD	± 5%
	SILVER	± 10%

1: FUNDAMENTAL CIRCUITS

CHARGE

$$Q = \frac{Number\ of\ electrons}{6.25\ X\ 10^{18}\ Electrons/C}$$

Where:

Q = Charge in Coulombs (C)

Note:

1 C = Total charge possessed by 6.25×10^{18} electrons

VOLTAGE

$$V = \frac{W}{Q}$$

Where:

V = Voltage in Volts (V)

W = Energy in Joules (J)

Q = Charge in Coulombs (C)

CURRENT

$$I = \frac{Q}{t}$$

Where:

I = Current in Amperes (A)

Q = Charge in Coulombs (C)

t = Time in seconds (s)

OHM'S LAW

$$I = \frac{V}{R}$$

Where:

I = Current in Amperes (A)

V = Voltage in Volts (V)

R = Resistance in Ohms (Ω)

RESISTIVITY

$$\rho = \frac{A \cdot R}{l}$$

Where:

ρ = Resistivity in Circular Mil – Ohm per Foot (CM-Ω/ft)

A = Cross-sectional area in Circular Mils (CM)

R = Resistance in Ohms (Ω)

l = Length in Feet (ft)

Note:

CM: Area of a wire with a 0.001 inch (1 mil) diameter

CONDUCTANCE

$$G = \frac{1}{R}$$

Where:

G = Conductance in Siemens (S)

R = Resistance in Ohms (Ω)

CROSS-SECTIONAL AREA

$$A = d^2$$

Where:

A = Cross-sectional area in Circular Mils (CM)

d = Diameter in thousandths of an inch (mils)

ENERGY

$$W = P \cdot t$$

Where:

W = Energy in Joules (J). The symbol is an italic *W*.

P = Power in Watts (W). Unit is not an italic W.

t = Time in seconds (s)

Note:

1 W = Amount of power when 1 J of energy is used in 1 s

POWER

$$P = V \cdot I$$

Where:

P = Power in Watts (W)

V = Voltage in Volts (V)

I = Current in Amperes (A)
Note:
P_{true} = P in a resistor is also called a true power.

OUTPUT POWER

$$P_{OUT} = P_{IN} - P_{LOSS}$$

Where:
P_{OUT} = Output power in Watts (W)
P_{IN} = Input power in Watts (W)
P_{LOSS} = Power loss in Watts(W)

POWER SUPPLY EFFICIENCY

$$\text{Efficiency} = \frac{P_{OUT}}{P_{IN}}$$

Where:
P_{OUT} = Output power in Watts (W)
P_{IN} = Input power in Watts (W)
Efficiency = Unitless value
Note:
Efficiency expressed as a percentage:

$$\text{Efficiency \%} = \text{Efficiency} \cdot 100\%$$

2: SERIES CIRCUITS (R1, R2, ⋯, Rn)

TOTAL RESISTANCE
$$R_T = R_1 + R_2 + R_3 + \cdots + R_n$$
Where:
R_T = Total series resistance in Ohms (Ω)
R_n = Circuit's last resistor in Ohms (Ω)

KIRCHHOFF'S VOLTAGE LAW
$$V_S = V_1 + V_2 + V_3 + \cdots + V_n$$
Where:
V_S = Voltage source in Volts (V)
V_n = Circuit's last voltage drop in Volts (V)

VOLTAGE – DIVIDER
$$V_x = \left(\frac{R_x}{R_T}\right) \cdot V_S$$
Where:
V_x = Voltage drop in Ohms (Ω)
R_x = Resistance where V_x occurs in Ohms (Ω)
R_T = Total series resistance in Ohms (Ω)
V_S = Voltage source in Volts (V)

TOTAL POWER
$$P_T = P_1 + P_2 + P_3 + \cdots + P_n$$
Where:
P_T = Total power in Watts (W)
P_n = Circuit's last resistor's power in Watts (W)

3: PARALLEL CIRCUITS (R1||R2||⋯||Rn)
TOTAL RESISTANCE
$$R_T = \frac{1}{\left(\frac{1}{R_1}\right) + \left(\frac{1}{R_2}\right) + \left(\frac{1}{R_3}\right) + \cdots + \left(\frac{1}{R_n}\right)}$$
Where:
R_T = Total parallel resistance in Ohms (Ω)
R_n = Circuit's last resistor in Ohms

TOTAL RESISTANCE - TWO RESISTORS IN PARALLEL
$$R_T = \frac{R_1 \cdot R_2}{R_1 + R_2}$$
Where:
R_T = Total parallel resistance in Ohms (Ω)

TOTAL RESISTANCE - EQUAL-VALUE RESISTORS

$$R_T = \frac{R}{n}$$

Where:

R_T = Total parallel resistance in Ohms (Ω)

R - Resistor Value in Ohms (Ω)

n = Number of equal value resistors (Unitless)

UNKNOWN RESISTOR

$$R_x = \frac{R_A \cdot R_T}{R_A - R_T}$$

Where:

R_x = Unknown resistance in Ohms (Ω)

R_A = Known parallel resistance in Ohms (Ω)

R_T = Total parallel resistance in Ohms (

KIRCHHOFF'S CURRENT LAW

$$\mathrm{I}_{\mathrm{IN}(1)} + \mathrm{I}_{\mathrm{IN}(2)} + \cdots + \mathrm{I}_{\mathrm{IN}(n)} = \mathrm{I}_{\mathrm{OUT}(1)} + \mathrm{I}_{\mathrm{OUT}(2)} + \cdots + \mathrm{I}_{\mathrm{OUT}(m)}$$

Where:

n = Number of currents into node (Unitless)

m = Number of currents going out of node (Unitless)

CURRENT – DIVIDER

$$I_x = \left(\frac{R_T}{R_x}\right) \cdot I_T$$

Where:

I_x=Branch "x" current in Amperes (A)

R_T = Total parallel resistance in Ohms (Ω)

R_x = Branch "x" resistance in Ohms (Ω)

I_T = Total current in Amperes (A)

TWO-BRANCH CURRENT – DIVIDER

$$I_1 = \left(\frac{R_2}{R_1 + R_2}\right) \cdot I_T \qquad I_2 = \left(\frac{R_1}{R_1 + R_2}\right) \cdot I_T$$

Where:

I_1=Branch "1" current in Amperes (A)

R_2 = Branch "2" resistance in Ohms (Ω)

R_1 = Branch "1" resistance in Ohms (Ω)

I_T = Total current in Amperes (A)

TOTAL POWER

$$P_T = P_1 + P_2 + P_3 + \cdots + P_n$$

Where:

P_T = Total power in Watts (W)

P_n = Circuit's last resistor's power in Watts (W)

OPEN BRANCH RESISTANCE

$$R_{Open} = \frac{1}{G_{T(Calc)} - G_{T(Meas)}} \qquad G_{T(Meas)} = \frac{1}{R_{T(Meas)}}$$

Where:

R_{Open} = Resistance of open branch in Ohms (Ω)

$R_{T(Meas)}$ = Measured resistance in Ohms (Ω)

$G_{T(Calc)}$ = Calculated total conductance in Siemens (S)

$G_{T(Meas)}$ = Measured total conductance in Siemens (S)

Note:

$G_{T(Meas)}$ obtained by measuring total resistance, $R_{T(Meas)}$

4: SERIES - PARALLEL CIRCUITS

BLEEDER CURRENT

$I_{BLEEDER} = I_T - I_{RL1} - I_{RL2}$

Where:

$I_{BLEEDER}$ = Bleeder current in Amperes (A)

I_T = Total current in Amperes (A)

I_{RL1} = Load resistor 1 current in Amperes (A)

I_{RL2} = Load resistor 2 current in Amperes (A)

THERMISTOR BRIDGE OUTPUT

$$\Delta V_{OUT} \cong \Delta R_{Therm} \left(\frac{V_S}{4 \cdot R} \right)$$

Where:

ΔV_{OUT} = Change in output voltage in Volts (V)

ΔR_{Therm} = Change in thermal resistance in Ohms (Ω)

V_S = Voltage source in Volts (V)

R = Resistance value in Ohms (Ω)

UNKNOWN RESISTANCE IN A WHEATSTONE BRIDGE

$$R_X = R_V \left(\frac{R_2}{R_4} \right)$$

Where:

R_X = Unknown resistance in Ohms (Ω)

R_V = Variable resistance in Ohms (Ω)

R_2 = Resistance 2 in Ohms (Ω)

R_4 = Resistance 4 in Ohms (Ω)

5: MAGNETISM AND ELECTROMAGNETISM

MAGNETIC FLUX DENSITY

$$B = \frac{\phi}{A}$$

Where:

B = Magnetic flux density in Tesla (T)

ϕ = Flux in Weber (Wb) (Greek letter Phi)

A = Cross-sectional area in square meters (m^2)

Note:

Tesla (T) equals a Weber per square meter (Wb/m^2)

RELATIVE PERMEABILITY

$$\mu_r = \frac{\mu}{\mu_0}$$

Where:

μ_r = Relative permeability (Unitless) (Greek letter Mu)

μ = Permeability in Webers per Ampere-turn meter (Wb/At·m)

μ_0 = Vacuum permeability in Webers per Ampere-turn · meter (Wb/At·m)

Note:

$\mu_0 = 4\pi \times 10^{-7}$ Wb/ At·m

RELUCTANCE

$$R = \frac{l}{\mu \cdot A}$$

Where:

R = Reluctance in Ampere-turn per Weber (At/Wb)

l = Length of magnetic path in meters (m)

μ = Permeability in Weber per Ampere-turn · meter (Wb/At · m)

A = Cross-sectional area in meters squares (m^2)

MAGNETOMOTIVE FORCE

$$F_m = N \cdot I$$

Where:

F_m = Magnetomotive force (mmf) in Ampere-turn (At)

N = Number of Turns of wire (t)

I = Current in Amperes (A)

MAGNETIC FLUX

$$\phi = \frac{F_m}{R}$$

Where:

ϕ = Flux in Weber (Wb)

F_m = Magnetomotive force in Ampere-turn (At)

R = Reluctance in Ampere-turn per Weber (At/Wb)

MAGNETIC FIELD INTENSITY

$$H = \frac{F_m}{l}$$

Where:

H = Magnetic field intensity in Amperes-turn per meter (At/m)

F_m = Magnetomotive force in Ampere-turn (At)

l = Length of material in meters (m)

INDUCED VOLTAGE

$$v_{ind} = B \cdot l \cdot v$$

Where:

v_{ind} = Induced voltage in Volts (V)

B= Magnetic flux density in Tesla (T)

l = Length of the conductor exposed to the magnetic field in meters (m)

v = Relative velocity in meters per second (m/s)

Note:

Tesla (T) equals a Weber per square meter (Wb/m^2)

FARADAY'S LAW

$$v_{ind} = N\left(\frac{d\phi}{dt}\right)$$

Where:

v_{ind} = Induced voltage in Volts (V)

N = Number of turns of wire in the coil (Unitless)

$\frac{d\phi}{dt}$ = Rate of change of magnetic field with respect to the coil in Webers per second (Wb/s)

6: ALTERNATE CURRENT & INDUCTORS

ALTERNATE CURRENT FREQUENCY & PERIOD

$$f = \frac{1}{T} \qquad T = \frac{1}{f}$$

Where:

f = Frequency in Hertz (Hz)

T = Period in Seconds (s)

Note:

1 Hertz = 1 cycle per 1 second

PEAK TO PEAK VOLTAGE

$$V_{pp} = 2 \cdot V_p$$

Where:

V_{pp}=Peak to peak voltage in Volts (V)

V_p = Peak voltage in Volts (V)

ROOT MEAN SQUARE (RMS) VOLTAGE

$$V_{rms} = 0.707 \cdot V_p$$

Where:

V_{rms} = Root mean square voltage in Volts (V)

V_p = Peak voltage in Volts (V)

HALF-CYCLE AVERAGE VOLTAGE

$$V_{avg} = 0.637 \cdot V_p$$

Where:

V_{avg} = Half-cycle average voltage in Volts (V)

V_p = Peak voltage in Volts (V)

RADIAN & DEGREE CONVERSION

$$Rad = \left(\frac{\pi\ rad}{180^0}\right) \cdot (Degrees)$$

$$Degrees = \left(\frac{180^0}{\pi\, rad}\right) \cdot (Rad)$$

Where:

Rad = Number of radians in Rad (rad)

Degrees = Number of degrees in Degrees (^{0})

Note:

π = 3.1416 (Greek letter Pi)

GENERATOR OUTPUT FREQUENCY

$$f = (number\ of\ pole\ pairs) \cdot (rps)$$

Where:

f = Frequency in Hertz (Hz)

Number of pole pairs = Number of pole pairs (Unitless)

rps = Revolutions per second in Revolutions per Second (rps)

PEAK TO PEAK CURRENT

$$I_{pp} = 2 \cdot I_p$$

Where:

I_{pp} = Peak to peak current in Amperes (A)

I_p = Peak current in Amperes (A)

ROOT MEAN SQUARE (RMS) CURRENT

$$I_{rms} = 0.707 \cdot I_p$$

Where:

I_{rms} = Root mean square current in Amperes (A)

I_p = Peak current in Amperes (A)

HALF-CYCLE AVERAGE CURRENT

$$I_{avg} = 0.637 \cdot I_p$$

Where:

I_{avg} = Half-cycle average current in Amperes (A)

I_p = Peak current in Amperes (A)

SINE WAVE GENERAL FORMULA

$$y = A \cdot sin\theta$$

Where:

y = Instantaneous voltage or current value at angle θ in Volts or Amperes (V or A) (Greek letter Theta)

A = Maximum voltage or current value in Volts or Amperes (V or A)

θ = Angle where given instantaneous voltage or current value exists

SINE WAVE LAGGING THE REFERENCE

$$y = A \cdot \sin(\theta - \phi)$$

Where:

y = Instantaneous voltage or current value at angle θ in Volts or Amperes (V or A)

A = Maximum voltage or current value in Volts or Amperes (V or A)

θ = Angle where given instantaneous voltage or current value exists

ϕ = Angle sine wave is shifted right (lagging) of reference (Greek letter Phi)

ANGULAR VELOCITY

$$\omega = 2 \cdot \pi \cdot f$$

Where:

ω = Angular velocity in Radians per second (rad/s) (Small Greek letter omega)

f = Frequency in Hertz (Hz)

Note:

π = 3.1416

SINE WAVE VOLTAGE

$$v = V_p \cdot sin(2 \cdot \pi \cdot f \cdot t)$$

Where:

v = Sinusoidal voltage in Volts (V)

V_p = Peak voltage in Volts (V)
f = Frequency in Hertz (Hz)
t = Time in Seconds (s)
Note:
π = 3.1416

PULSE WAVEFORM AVERAGE VALUE

$$v_{avg} = \text{baseline} + (\text{duty cycle}) \cdot (\text{amplitude})$$

Where:
v_{avg} = Pulse waveform average value in Volts (V)
baseline = Baseline in Volts (V)
duty cycle = Percent duty cycle in Percent/100% (Unitless)
Amplitude = Amplitude in Volts (V)

SINE WAVE LEADING THE REFERENCE

$$y = A \cdot \sin(\theta + \phi)$$

Where:
y = Instantaneous voltage or current value at angle θ in Volts or Amperes (V or A)
A = Maximum voltage or current value in Volts or Amperes (V or A)
θ =Angle where given instantaneous voltage or current value exists
ϕ = Angle sine wave is shifted left (leading) of reference

PHASE ANGLE

$$\phi = \omega \cdot t$$

Where:
ϕ = Angle sine wave is shifted in Radians (rad)
ω = Angular velocity in Radians per second (rad/s)
t = Time in Seconds (s)

DUTY CYCLE

$$\text{Percent duty cycle} = \left(\frac{t_w}{T}\right) \cdot 100\%$$

Where:
Percent duty cycle = Percent duty cycle in Percentage (%)
t_w = Pulse width in Seconds (s)
T = Period in Seconds (s)
F = Frequency in Hertz (Hz)

INDUCTORS

INDUCED VOLTAGE

$$v_{ind} = L\left(\frac{di}{dt}\right)$$

Where:
v_{ind} = Induced voltage in Volts (V)
L = Inductance in Henries (H)
$\frac{di}{dt}$ = Time rate of change of the current in Amperes per second (A/s)

INDUCTANCE OF A COIL

$$L = \frac{N^2 \cdot \mu \cdot A}{l}$$

Where:
L =Inductance of a coil in Henries (H)
N =Number of turns of wire (Unitless)
μ=Permeability in Henries per meter (H/m)
A = Cross-sectional area in Meter squared (m^2)
l = Core length in Meters (m)
Notes:
Permeability in H/m is equal to Wb/At·m
Non-magnetic core = Permeability of a vacuum, μ_0.
$\mu_0 = 4\pi \times 10^{-7}$ H/m

RL TIME CONSTANT

$$\tau = \frac{L}{R}$$

Where:
τ = RL time constant in Seconds (s) (Greek letter Tau)
L = Inductance in Henries (H)
R = Resistance in Ohms (Ω)

GENERAL EXPONENTIAL VOLTAGE FORMULA

$$v = V_F + (V_i - V_F)e^{-Rt/L}$$

Where:
v = Instantaneous voltage at time, t, in Volts (V)
V_F = Voltage final value in Volts (V)
V_i =Voltage initial value in Volts (V)
R = Resistance in Ohms (Ω)
t = Time in Seconds (s)
L = nductance in Henries (H)

INDUCTOR ENERGY STORAGE

$$W = \frac{1}{2} \cdot L \cdot I^2$$

Where:
W = Energy in Joules (J)
L = Inductance in Henries (H)
I = Current in Amperes (A)

TOTAL INDUCTANCE - SERIES

$$L_T = L_1 + L_2 + L_3 + \cdots + L_n$$

Where:
L_T = Total series inductance in Henries (H)
L_n = Circuit's last inductor in Henries (H)

TOTAL INDUCTANCE – PARALLEL

$$L_T = \frac{1}{\left(\frac{1}{L_1}\right) + \left(\frac{1}{L_2}\right) + \left(\frac{1}{L_3}\right) + \cdots + \left(\frac{1}{L_n}\right)}$$

Where:
L_T = Total parallel inductance in Henries (H)
L_n = Circuit's last inductor in Henries (H)

RL CIRCUIT CURRENT INCREASE AND DECREASE FOR GIVEN NUMBER OF TIME CONSTANTS

# of Time Constants	Approx % of Final Current	Approx % of Initial Charge
1	63	37
2	86	14
3	95	5
4	98	2
5	99 Considered 100%	1 Considered 0%

GENERAL EXPONENTIAL CURRENT FORMULA

$$i = I_F + (I_i - I_F)e^{-Rt/L}$$

Where:
i = Instantaneous current at time, t, in Amperes (A)
I_F = Current final value in Amperes (A)
I_i=Current initial value in Amperes (A)
R = Resistance in Ohms (Ω)
t = Time in Seconds (s)
L = Inductance in Henries (H)

INDUCTIVE REACTANCE

$$X_L = 2 \cdot \pi \cdot f \cdot L$$

Where:
X_L = Inductive reactance in Ohms (Ω)
f = Frequency in Hertz (Hz)
L = Inductance in Henries (H)
Note:
π = 3.1416 (Greek letter "Pi")

INDUCTOR REACTIVE POWER

$$P_r = V_{rms} \cdot I_{rms} = \frac{V_{rms}^2}{X_L} = I_{rms}^2 \cdot X_L$$

Where:
P_r = Reactive Power in Watts (W)

V_{rms} = Voltage rms in Volts (V)
I_{rms} = Current rms in Amperes (A)
X_L =Inductive reactance in Ohms (Ω)

7: RL CIRCUITS

SERIES RL CIRCUIT IMPEDANCE IN RECTANGULAR FORM

$$\boldsymbol{Z} = R + jX_L$$

Where:
Z = Impedance in Ohms (Ω)
R = Resistance in Ohms (Ω)
X_L = Inductive reactance in Ohms (Ω)
Note:
Bold letters represent complete phasor quantities. For example, "**Z**" in the formula above

VOLTAGE IN RECTANGULAR FORM

$$\boldsymbol{V_s} = V_R + jV_L$$

Where:
V_s = Voltage in Volts (V)
V_R = Resistor voltage in Volts (V)
V_L = Inductor voltage in Volts (V)

INDUCTOR TRUE POWER

$$P_{true} = (I_{rms})^2 \cdot R_W$$

Where:
P_{true} = True Power in Watts (W)
I_{rms} = Current rms in Amperes (A)
R_W = Winding resistance in Ohms (Ω)

COIL QUALITY FACTOR $Q = \frac{X_L}{R_W}$

Where:
Q= Coil quality factor (Unitless)
X_L = Inductive reactance in Ohms (Ω)
R_W = Winding resistance of the coil or the resistance in series with the coil in Ohms (Ω)

Note:
Circuit Q and the coil Q are the same when the resistance is only the coil winding resistance

IMPEDANCE IN POLAR FORM

$$\boldsymbol{Z} = \sqrt{R^2 + X_L^2} \cdot \angle tan^{-1}\left(\frac{X_L}{R}\right)$$

Where:
Z = Impedance in Ohms (Ω)
R = Resistance in Ohms (Ω)
X_L = Inductive reactance in Ohms (Ω)
Note:
$\sqrt{R^2 + X_L^2}$ = Magnitude
$\frac{X_L}{R}$ = Phase Angle

VOLTAGE IN POLAR FORM

$$\boldsymbol{V_s} = \sqrt{V_R^2 + V_L^2} \cdot \angle tan^{-1}\left(\frac{V_L}{V_R}\right)$$

Where:
V_s = Voltage in Volts (V)
V_R = Resistor voltage in Volts (V)
V_L = Inductor voltage in Volts (V)

LEAD CIRCUIT
ANGLE BETWEEN VOLTAGE IN & OUT

$$\phi = tan^{-1}\left(\frac{R}{X_L}\right)$$

Where:
ϕ = Angle between voltage in and out in Degrees (^{0})
R = Resistance in Ohms (Ω)
X_L = Inductive reactance in Ohms (Ω)

OUTPUT VOLTAGE MAGNITUDE

$$V_{out} = \left(\frac{X_L}{\sqrt{R^2 + X_L^2}}\right) \cdot V_{in}$$

Where:
V_{out} = Voltage output in Volts (V)

X_L = Inductive reactance in Ohms (Ω)
R = Resistance in Ohms (Ω)

LAG CIRCUIT
ANGLE BETWEEN VOLTAGE IN & OUT

$$\phi = -tan^{-1}\left(\frac{X_L}{R}\right)$$

Where:
ϕ = Angle between voltage in and out in Degrees (^{0})
X_L = Inductive reactance in Ohms (Ω)
R = Resistance in Ohms (Ω)

OUTPUT VOLTAGE MAGNITUDE

$$V_{out} = \left(\frac{R}{\sqrt{R^2 + X_L^2}}\right) \cdot V_{in}$$

Where:
V_{out} = Output voltage in Volts (V)
R = Resistance in Ohms (Ω)
X_L = Inductive reactance in Ohms (Ω)
V_{in} = Input voltage in Volts (V)

PARALLEL RL CIRCUIT
TOTAL 2-COMPONENT IMPEDANCE

$$\boldsymbol{Z} = \left(\frac{R \cdot X_L}{\sqrt{R^2 + X_L^2}}\right) \cdot \angle tan^{-1}\left(\frac{R}{X_L}\right)$$

Where:
Z = Total 2-component impedance in Ohms (Ω)
R = Resistance in Ohms (Ω)
X_L = Inductive reactance in Ohms (Ω)

CURRENT IN POLAR FORM

$$\boldsymbol{I_{tot}} = \sqrt{I_R^2 + I_L^2} \cdot \angle - tan^{-1}\left(\frac{I_L}{I_R}\right)$$

Where:
$\mathbf{I_{tot}}$ = Total current in Amperes (A)
I_R =Resistor current in Amperes (A)
I_L =Inductor current in Amperes (A)

TOTAL ADMITTANCE

$$\boldsymbol{Y} = G - jB_L$$

Where:
Y = Total admittance in Siemens (S)
G = Conductance in Siemens (S)
B_L = Inductive Susceptance in Siemens (S)
Note:

$$\boldsymbol{Y} = \frac{1}{Z\angle \pm \theta} = Y\angle \mp \theta$$

$$\boldsymbol{G} = \frac{1}{R\angle 0^0} = G\angle 0^0$$

$$\boldsymbol{B_L} = \frac{1}{X_L\angle 90^0} = B_L\angle - 90^0 = -jB_L$$

CURRENT IN RECTANGULAR FORM

$$\boldsymbol{I_{tot}} = I_R - jI_L$$

Where:
$\mathbf{I_{tot}}$ = Total current in Amperes (A)
I_R =Resistor current in Amperes (A)
I_L = Inductor current in Amperes (A)

PARALLEL TO SERIES FORM CONVERSION

$$R_{eq} = Z \cdot cos\theta \qquad X_{L(eq)} = Z \cdot sin\theta$$

Where:
R_{eq} = Resistance in Ohms (Ω)
Z = Impedance in Ohms (Ω)
X_L =Inductive reactance in Ohms (Ω)
θ = Angle where given instantaneous voltage or current value exists

POWER
RL CIRCUIT REACTIVE POWER

$$P_r = I_{tot}^2 \cdot X_L$$

Where:

P_r = Reactive power in Volt-Ampere Reactive (VAR)
I_{tot} = Total current in Amperes (A)
X_L = Inductive reactance in Ohms (Ω)

8: CAPACITORS

CAPACITANCE

$$C = \frac{Q}{V}$$

Where:
C = Capacitance in Farads (F)
Q = Charge in Coulombs (C)
V = Voltage in Volts (V)

ENERGY STORED IN A CAPACITOR

$$W = \frac{1}{2} \cdot C \cdot V^2$$

Where:
W = Energy in Joules (J)
C = Capacitance in Farads (F)
V = Voltage in Volts (V)

DIELECTRIC CONSTANT (RELATIVE PERMITTIVITY)

$$\varepsilon_R = \frac{\varepsilon}{\varepsilon_0}$$

Where:
ε_R = Dielectric constant (Unitless) (Greek letter Epsilon)
ε=Absolute permittivity of a material in Farads per meter (F/m)
ε_0=Absolute permittivity of a vacuum in Farads permeter (F/m)
Note:
ε_0 = 8.85 x 10^{-12} F/m

CAPACITANCE

$$C = \frac{A \cdot \varepsilon_r \cdot (8.85 \text{ x } 10^{-12} \text{ F/m})}{d}$$

Where:
C = Capacitance in Farads (F)
A =Plate area in Meters squared (m^2)
ε_R = Dielectric constant (Unitless)
d = Plate separation in Meters(m)
Note:
If d is in mils, 1 mil = 2.54 x 10^{-5} meters

SERIES CAPACITORS

TOTAL CHARGE

$$Q_T = Q_1 = Q_2 = Q_3 = \cdots = Q_n$$

Where:
Q_T = Total charge in Coulombs (C)
Q_n = Circuit's last capacitor charge in Coulombs (C)

TOTAL CAPACITANCE

$$C_T = \frac{1}{\left(\frac{1}{C_1}\right) + \left(\frac{1}{C_2}\right) + \left(\frac{1}{C_3}\right) + \cdots + \left(\frac{1}{C_n}\right)}$$

Where:
C_T = Total series capacitance in Farads (F)
C_n = Circuit's last capacitor's capacitance in Farads (F)

TOTAL CAPACITANCE - TWO CAPACITORS

$$C_T = \frac{C_1 \cdot C_2}{C_1 + C_2}$$

Where:
C_T = Total series capacitance in Farads (F)

VOLTAGE ACROSS A CAPACITOR

$$V_x = \left(\frac{C_T}{C_x}\right) \cdot V_T$$

Where:
V_x = Voltage drop in Volts (V)
C_T = Total series capacitance in Farads (F)
C_x=Capacitor x's capacitance in Farads (F)
V_T=Total voltage in Volts (V)

TOTAL CAPACITANCE - EQUAL-VALUE CAPACITORS

$$C_T = \frac{C}{n}$$

Where:

C_T = Total series capacitance in Farads (F)

n = Number of equal-value capacitors (Unitless)

PARALLEL CAPACITORS TOTAL CHARGE

$$Q_T = Q_1 + Q_2 + Q_3 + \cdots + Q_n$$

Where:

Q_T = Total charge in Coulombs (C)

Q_n = Circuit's last capacitor charge in Coulombs (C)

TOTAL CAPACITANCE - EQUAL-VALUE CAPACITORS

$$C_T = \mathrm{n} \cdot \mathrm{C}$$

Where:

C_T = Total series capacitance in Farads (F)

n = Number of equal-value capacitors (Unitless)

CAPACITORS IN DC CIRCUITS RC TIME CONSTANT

$$\tau = \mathrm{R} \cdot \mathrm{C}$$

Where:

τ = Time constant in Seconds (s)

R = Resistance in Ohms (Ω)

C = Capacitance in Farads (F)

TOTAL CAPACITANCE

$$C_T = C_1 + C_2 + C_3 + \cdots + C_n$$

Where:

C_T = Total series capacitance in Farads (F)

C_n= Circuit's last capacitor's capacitance inFarads (F)

RC CIRCUIT CURRENT INCREASES AND DECREASES FOR A GIVEN NUMBER OF TIME CONSTANTS

# of Time Constants	Approx % of Final Current	Approx % of Initial Charge
1	63	37
2	86	14
3	95	5
4	98	2
5	99 Considered 100%	1 Considered 0%

GENERAL EXPONENTIAL VOLTAGE FORMULA

$$v = V_F + (V_i - V_F)e^{-t/\tau}$$

Where:

v = Instantaneous voltage at the time, t, in Volts (V)

V_F =Voltage final value in Volts (V)

V_i =Voltage initial value in Volts (V)

t = Time in Seconds (s)

τ = Time constant in Seconds (s)

CHARGING TIME TO A SPECIFIED VOLTAGE

$$t = -R \cdot C \cdot ln\left(1 - \frac{v}{V_F}\right)$$

Where:

t = Time in Seconds (s)

R =Resistance in Ohms (Ω)

C=Capacitance in Farads (F)

v=Specified voltage level in Volts (V)

V_F = Final voltage level in Volts (V)

Note:

Assumes V_i = 0 Volts

GENERAL EXPONENTIAL CURRENT FORMULA

$$i = I_F + (I_i - I_F)e^{-t/\tau}$$

Where:
i=Instantaneous current at the time, t, in Amperes (A)
I_F=Current final value in Amperes (A)
I_i=Current initial value in Amperes (A)
t= Time in Seconds (s)
τ = Time constant in Seconds (s)

DISCHARGING TIME TO A SPECIFIED VOLTAGE

$$t = -R \cdot C \cdot ln\left(\frac{v}{V_i}\right)$$

Where:
t = Time in Seconds (s)
R = Resistance in Ohms (Ω)
C = Capacitance in Farads (F)
v = Specified voltage level in Volts (V)
V_i = Initial voltage level in Volts (V)
Note:
Assumes V_F = 0 Volts

CAPACITORS IN AC CIRCUITS
INSTANTANEOUS CAPACITOR CURRENT

$$i = C \cdot \left(\frac{dv}{dt}\right)$$

Where:
i=Instantaneous current in Amperes (A)
C = Capacitance in Farads (F)
$\frac{dv}{dt}$ = Instantaneous rate of change of the voltage
across the capacitor in Volts per second (V/s)

CAPACITOR REACTIVE POWER

$$P_r = V_{rms} \cdot I_{rms} = \frac{V_{rms}^2}{X_C} = I_{rms}^2 \cdot X_C$$

Where:
P_r = Reactive Power in Volt-Ampere Reactive (VAR)
V_{rms} = Voltage RMS in Volts (V)
I_{rms} = Current RMS in Amperes (A)
X_C = Capacitive reactance in Ohms (Ω)

CAPACITIVE REACTANCE

$$X_C = \frac{1}{2 \cdot \pi \cdot f \cdot C}$$

Where:
X_C = Capacitive reactance in Ohms (Ω)
f = Frequency in Hertz (Hz)
C = Capacitance in Farads (F)
Note:
π = 3.1416 (Greek letter "Pi")

SWITCHED-CAPACITORS CIRCUITS
AVERAGE CURRENT

$$I_{1(avg)} = \frac{C \cdot (V_1 - V_2)}{T}$$

Where:
$I_{1(avg)}$ = Instantaneous current in Amperes (A)
C = Capacitance in Farads (F)
V_1 = Voltage 1 in Volts (V)
V_2 = Voltage 2 in Volts (V)
T = period in Seconds (s)

9: RC CIRCUITS
RC SERIES CIRCUITS
IMPEDANCE IN RECTANGULAR FORM

$$\boldsymbol{Z} = R - jX_C$$

Where:
Z = Impedance in Ohms (Ω)
R = Resistance in Ohms (Ω)
X_C = Capacitive reactance in Ohms (Ω)

OHM'S LAW

$$\boldsymbol{V} = \boldsymbol{I} \cdot \boldsymbol{Z}$$

Where:
I = Current in Amperes (A)
Z = Impedance in Ohms (Ω)
V = Voltage in Volts (V)

VOLTAGE IN RECTANGULAR FORM

$$\boldsymbol{V_s} = V_R - jV_C$$

Where:
$\mathbf{V_s}$ = Voltage in Volts (V)
V_R = Resistor voltage in Volts (V)
V_C = Capacitor voltage in Volts (V)

LEAD CIRCUIT
ANGLE BETWEEN VOLTAGE IN & OUT

$$\phi = tan^{-1}\left(\frac{X_C}{R}\right)$$

Where:
ϕ = Angle between the voltage in and out in Degrees (^{0})
X_C = Capacitive reactance in Ohms (Ω)
R = Resistance in Ohms (Ω)

EQUIVALENT RESISTANCE

$$R = \frac{T}{C} = \frac{1}{f \cdot C}$$

Where:
R = Equivalent resistance in Ohms (Ω)
T = period in Seconds (s)
C = Capacitance in Farads (F)
f = Frequency in Hertz (Hz)

IMPEDANCE IN POLAR FORM

$$\boldsymbol{Z} = \sqrt{R^2 + X_C^2} \cdot \angle - tan^{-1}\left(\frac{X_C}{R}\right)$$

Where:
Z = Impedance in Ohms (Ω)
R = Resistance in Ohms (Ω)
X_C = Capacitive reactance in Ohms (Ω)

VOLTAGE IN POLAR FORM

$$\boldsymbol{V_s} = \sqrt{V_R^2 + V_C^2} \cdot \angle - tan^{-1}\left(\frac{V_C}{V_R}\right)$$

Where:
$\mathbf{V_s}$ = Voltage in Volts (V)
V_R = Resistor voltage in Volts (V)
V_C = Capacitor voltage in Volts (V)

OUTPUT VOLTAGE MAGNITUDE

$$V_{out} = \left(\frac{R}{\sqrt{R^2 + X_C^2}}\right) \cdot V_{in}$$

Where:
V_{out} = Voltage output in Volts (V)
R = Resistance in Ohms (Ω)
X_C = Capacitive reactance in Ohms (Ω)

LAG CIRCUIT
ANGLE BETWEEN VOLTAGE IN & OUT

$$\phi = -tan^{-1}\left(\frac{R}{X_C}\right)$$

Where:
ϕ = Angle between the voltage in and out in Degrees (^{0})
R = Resistance in Ohms (Ω)
X_C = Capacitive reactance in Ohms (Ω)

RC PARALLEL CIRCUITS
TOTAL 2-COMPONENT IMPEDANCE

$$\boldsymbol{Z} = \left(\frac{R \cdot X_C}{\sqrt{R^2 + X_C^2}}\right) \cdot \angle - tan^{-1}\left(\frac{R}{X_C}\right)$$

Where:
Z = Total 2-component impedance in Ohms (Ω)
R = Resistance in Ohms (Ω)
X_C = Capacitive reactance in Ohms (Ω)

OHM'S LAW

$$\boldsymbol{I} = \boldsymbol{V} \cdot \boldsymbol{Y}$$

Where:
I = Current in Amperes (A)
V = Voltage in Volts (V)
Y = Admittance in Siemens (S)

CURRENT IN RECTANGULAR FORM

$\boldsymbol{I_{tot}} = I_R + jI_C$

Where:
$\mathbf{I_{tot}}$ = Total current in Amperes (A)
I_R =Resistor current in Amperes (A)
I_C =Capacitor current in Amperes (A)

PARALLEL TO SERIES FORM CONVERSION

$R_{eq} = Z \cdot cos\theta$ $\quad X_{C(eq)} = Z \cdot sin\theta$

Where:
R_{eq} =Resistance in Ohms (Ω)
Z=Impedance in Ohms (Ω)
X_C=Capacitive reactance in Ohms (Ω)
θ =Angle where given instantaneous voltage or current value exists

OUTPUT VOLTAGE MAGNITUDE

$$V_{out} = \left(\frac{X_C}{\sqrt{R^2 + X_C^2}}\right) \cdot V_{in}$$

Where:
V_{out} = Voltage output in Volts (V)
X_C = Capacitive reactance in Ohms (Ω)
R = Resistance in Ohms (Ω)

TOTAL ADMITTANCE

$\boldsymbol{Y} = G + jB_C$

Where:
Y =Total admittance in Siemens (S)
G =Conductance in Siemens (S)
B_C =Capacitive susceptance in Siemens (S)

Note:

$\boldsymbol{Y} = \frac{1}{Z\angle\pm\theta} = Y\angle\mp\theta$

$\boldsymbol{G} = \frac{1}{R\angle0^0} = G\angle0^0$

$\boldsymbol{B_C} = \frac{1}{X_C\angle-90^0} = B_C\angle90^0 = +jB_C$

CURRENT IN POLAR FORM

$$\boldsymbol{I_{tot}} = \sqrt{I_R^2 + I_C^2} \cdot \angle tan^{-1}\left(\frac{I_C}{I_R}\right)$$

Where:
$\mathbf{I_{tot}}$ = Total current in Amperes (A)
I_R =Resistor current in Amperes (A)
I_C =Capacitor current in Amperes (A)

RC SERIES –PARALLEL CIRCUITS PHASE ANGLE

$$\theta = \left(\frac{\Delta t}{T}\right) \cdot 360^0$$

Where:
R_{eq} = Resistance in Ohms (Ω)
Z= Impedance in Ohms (Ω)
X_C=Capacitive reactance in Ohms (Ω)
θ=Angle where given instantaneous voltage or current value exists

POWER

APPARENT POWER

$P_a = I^2 \cdot Z$

Where:
P_a=Apparent power in Volt ampere (VA)
I =Current in Amperes (A)
Z= Impedance in Ohms (Ω)

POWER FACTOR

$PF = cos\theta$

Where:
PF = Power Factor (Unitless)
θ = Phase angle in Degrees (^{0})

OSCILLATOR AND FILTER
OSCILLATOR OUTPUT FREQUENCY

$$f_r = \frac{1}{2 \cdot \pi \cdot \sqrt{6} \cdot R \cdot C}$$

Where:

f_r=Output frequency in Hertz (Hz)
R=Resistance in Ohms (Ω)
C=Capacitance in Farads (F)
Note:
π= 3.1416

10: RLC CIRCUITS AND PASSIVE FILTERS
RLC SERIES CIRCUITS
TOTAL REACTANCE

$$X_{tot} = |X_L - X_C|$$

Where:

X_{tot} = Total reactance in Ohms (Ω)
X_L=Inductive reactance in Ohms (Ω)
X_C= Capacitive reactance in Ohms (Ω)

TOTAL IMPEDANCE IN POLAR FORM

$$\boldsymbol{Z} = \sqrt{R^2 + (X_L - X_C)^2} \cdot \angle \pm tan^{-1}\left(\frac{X_{tot}}{R}\right)$$

Where:

Z =Total impedance in Ohms (Ω)
R= Resistance in Ohms (Ω)
X_L =Inductive reactance in Ohms (Ω)
X_C =Capacitive reactance in Ohms (Ω)
X_{tot} = Total reactance in Ohms (Ω)
Note:
When $X_L > X_C$, the angle is positive
When $X_C > X_L$, the angle is negative

TRUE POWER

$$P_{true} = V \cdot I \cdot cos\theta$$

Where:

P_{true} = True power in Watts (W)
V = Voltage in Volts (V)
I = Current in Amperes (A)
θ = Phase angle in Degrees (^{0})

FILTER CUTOFF FREQUENCY

$$f_c = \frac{1}{2 \cdot \pi \cdot R \cdot C}$$

Where:

f_c = Cutoff frequency in Hertz (Hz)
R = Resistance in Ohms (Ω)
C = Capacitance in Farads (F)
Note:
π = 3.1416

TOTAL IMPEDANCE IN RECTANGULAR FORM

$$\boldsymbol{Z} = R + jX_L - jX_C$$

Where:

Z = Total impedance in Ohms (Ω)
R = Resistance in Ohms (Ω)
X_L = Inductive reactance in Ohms (Ω)
X_C = Capacitive reactance in Ohms (Ω)

RESONANT FREQUENCY

$$f_r = \frac{1}{2 \cdot \pi \cdot \sqrt{L \cdot C}}$$

Where:

f_r = Resonant frequency in Hertz (Hz)
L = Inductance in Henries (H)
C = Capacitance in Farads (F)
Note:
At resonance, $X_L = X_C$ and the j terms cancel π = 3.1416

RLC PARALLEL CIRCUITS
TOTAL CURRENT

$$\boldsymbol{I}_{tot} = \sqrt{I_R^2 + (I_C - I_L)^2} \cdot \angle tan^{-1}\left(\frac{I_{CL}}{I_R}\right)$$

Where:

$\boldsymbol{I}_{tot}$ = Total current in Amperes (A)
I_R= Resistor current in Amperes (A)

I_C= Capacitor current in Amperes (A)
I_L= Inductor current in Amperes (A)
I_{CL}= Total current into the L and C branches in Amperes (A)

RLC PARALLEL RESONANCE
RESONANT FREQUENCY - IDEAL

$$f_r = \frac{1}{2 \cdot \pi \cdot \sqrt{L \cdot C}}$$

Where:
f_r = Resonant frequency in Hertz (Hz)
L = Inductance in Henries (H)
C = Capacitance in Farads (F)
Note:
At resonance, $X_L = X_C$ and $Z_r = \infty$
π = 3.1416

CURRENT AND PHASE ANGLE

$$I_{tot} = \frac{V_S}{Z_r}$$

Where:
I_{tot} = Total current in Amperes (A)
V_S = Voltage source in Volts (V)
Z_r= Impedance at resonance in Ohms (Ω)

RESONANT FREQUENCY - PRECISE

$$f_r = \frac{\sqrt{1-(R_W^2\, C/L)}}{2 \cdot \pi \cdot \sqrt{L \cdot C}}$$

Where:
f_r= Resonant frequency in Hertz (Hz)
R_W = Winding resistance in Ohms (Ω)
C = Capacitance in Farads (F)
L = Inductance in Henries (H)
Note:
π = 3.1416

RLC SERIES – PARALLEL CIRCUITS

SERIES-PARALLEL TO PARALLEL CONVERSION
EQUIVALENT INDUCTANCE

$$L_{eq} = L \cdot \left(\frac{Q^2+1}{Q^2}\right)$$

Where:
L_{eq} = Equivalent inductance in Henries (H)
L = Inductance in Henries (H)
Q = Coil quality factor (Unitless)

EQUIVALENT PARALLEL RESISTANCE

$$R_{p(eq)} = R_W \cdot (Q^2+1)$$

Where:
$R_{p(eq)}$ = Equivalent parallel resistance in Ohms (Ω)
R_W= Winding resistance in Ohms (Ω)
Q= Coil quality factor (Unitless)

NON-IDEAL TANK CIRCUIT TOTAL IMPEDANCE AT RESONANCE

$$Z_R = R_W \cdot (Q^2+1)$$

Where:
Z_R = Total impedance in Ohms (Ω)
R_W = Resistance in Ohms (Ω)
Q = Coil quality factor (Unitless)

SPECIAL TOPICS
RESONANT CIRCUIT BANDWIDTH
BANDWIDTH

$$BW = f_2 - f_1$$

Where:
BW = Bandwidth in Hertz (Hz)
f_2 = Upper critical frequency at $Z=0.707 \cdot Z_{max}$ in Hertz (Hz)
f_1 = Lower critical frequency at $Z=0.707 \cdot Z_{max}$ in Hertz (Hz)

BANDWIDTH AND QUALTIY FACTOR

$$BW = \frac{f_r}{Q}$$

Where:
BW = Bandwidth in Hertz (Hz)
f_r = Center (resonant) frequency in Hertz (Hz)
Q = Coil quality factor (Unitless)

PASSIVE FILTERS

POWER RATIO IN DECIBELS

$$dB = 10 \cdot log\left(\frac{P_{out}}{P_{in}}\right)$$

Where:
dB = Power ratio in decibels (dB)
P_{out} = Output power in Watts (W)
P_{in} = Input power in Watts (W)

OVERALL QUALITY FACTOR WITH AN EXTERNAL LOAD

$$Q_O = \frac{R_{p(tot)}}{X_{L(eq)}}$$

Where:
Q_O = Overall quality factor (Unitless)
$R_{p(tot)}$= Total parallel equivalent resistance in Ohms (Ω)
X_L = Inductive reactance in Ohms (Ω)

CENTER (RESONANT) FREQUENCY

$$f_r = \frac{f_1 + f_2}{2}$$

Where:
f_r = Center (resonant) frequency in Hertz (Hz)
f_1 = Lower critical frequency at $Z=0.707 \cdot Z_{max}$ in Hertz (Hz)
f_2 = Upper critical frequency at $Z=0.707 \cdot Z_{max}$ in Hertz (Hz)

VOLTAGE RATIO IN DECIBELS

$$dB = 20 \cdot log\left(\frac{V_{out}}{V_{in}}\right)$$

Where:
dB = Power ratio in decibels (dB)
V_{out} = Output voltage in Volts (V)
V_{in} = Input voltage in Volts (V)

LOW-PASS & HIGH-PASS FILTERS

RC FILTERS

$$f_c = \frac{1}{2 \cdot \pi \cdot R \cdot C}$$

Where:
f_C = Filter critical frequency in Hertz (Hz)
R = Resistance in Ohms (Ω)
C = Capacitance in Farads (F)
Note:
π = 3.1416
at f_C, $V_{out} = (0.707) \cdot V_{in}$

SERIES RESONANT BAND-PASS FILTER

$$BW = \frac{f_0}{Q}$$

Where:
BW = Bandwidth in Hertz (Hz)
f_0 = Center frequency in Hertz (Hz)
Q = Coil quality factor (Unitless)

RL FILTERS

$$f_c = \frac{1}{2 \cdot \pi \cdot (L/R)}$$

Where:
f_c = Filter critical frequency in Hertz (Hz)
L = Inductance in Henries (H)
R = Resistance in Ohms (Ω)
Note:
π = 3.1416
At f_C, $V_{out} = (0.707) \cdot V_{in}$

11: GENERAL INFORMATION

AREA AND VOLUMES

AREAS

CIRCLE AREA

$A = \pi \cdot r^2$

Where:

A=Circle area in meters squared (m^2)

r = Radius in meters (m)

Note:

π = 3.1416

RECTANGULAR AND POLAR FORMS

RECTANGULAR FORM

$A + jB$

Where:

A = Coordinate value on the real axis (Horizontal Plane)

j = j operator

B= Coordinate value on imaginary axis (Vertical Plan)

Note:

The "j operator" prefix indicates designated coordinate value is on the imaginary axis.

COMPLEX PLANE AND RECTANGULAR FORM PHASOR

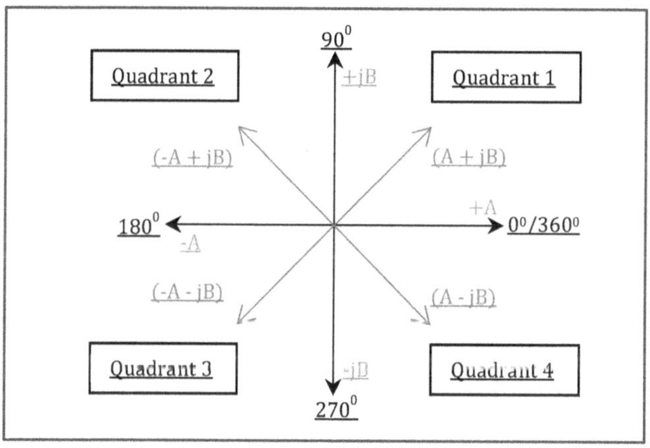

POLAR FORM

$C\angle \pm \theta$

Where:

C = Phasor magnitude

θ = Phasor angle relative to the positive real axis

COMPLEX PLANE AND POLAR FORM PHASOR

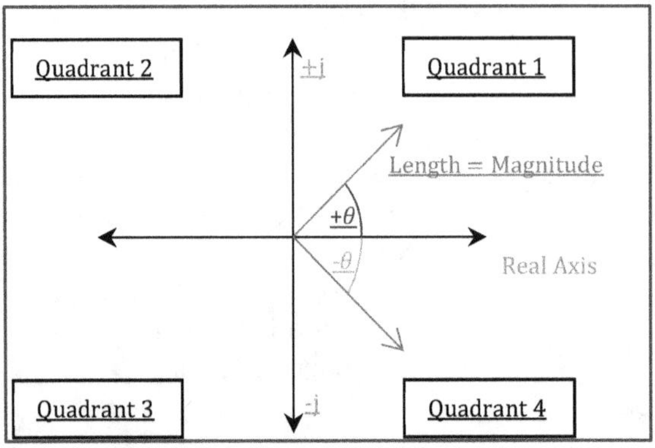

RECTANGULAR TO POLAR CONVERSION

$\pm A \pm jB = C\angle \pm \theta$

Where:
A = Coordinate value on the real axis (Horizontal Plane)
j = j operator
B = Coordinate value on imaginary axis (Vertical Plan)
C = Phasor magnitude
θ = Phasor angle relative to the positive real axis

Note:
To calculate C:
$C = \sqrt{A^2 + B^2}$
To calculate θ in Quadrants 1 and 4 (A is positive):
$\theta = \tan^{-1}\left(\frac{\pm B}{A}\right)$
Use +B for +B values, -B for –B values

To calculate θ in Quadrants 2 and 3 (A is negative):
$\theta = \pm 180^0 \mp tan^{-1}\left(\frac{B}{A}\right)$
Use $+180^0 - tan^{-1}$ for +B values
Use $-180^0 + tan^{-1}$ for –B values

POLAR TO RECTANGULAR CONVERSION

$C\angle\theta = A + jB$
Where:

C = Phasor magnitude
θ = Phasor angle relative to the positive real axis
A = Coordinate value on real axis (Horizontal Plane)
j = j operator
B = Coordinate value on imaginary axis (Vertical Plan)

Note:
To calculate A:
$A = C \cdot cos\theta$

To calculate B:
$B = C \cdot sin\theta$

1. For the circuit below, calculate the following.
   a. Total circuit impedance.
   b. Total circuit current,
   c. True power.
   d. Reactive power
   e. Apparent power.

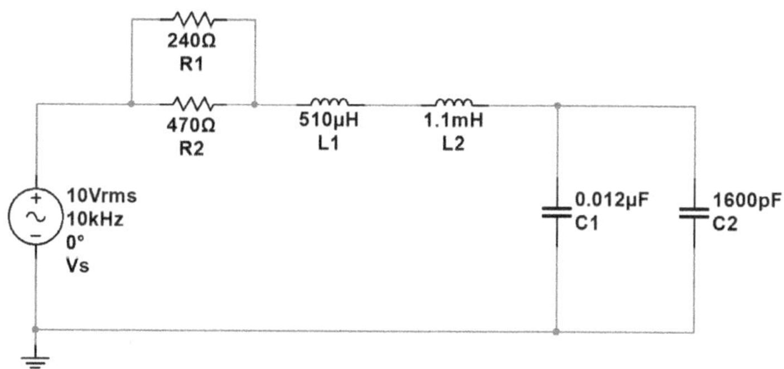

Solution:
a.

$$R_T = R_1//R_2 = \frac{240 * 470}{240 + 470} = 158.873\Omega$$

$$L_T = L_1 + L_2 = 510 * 10^{-6} + 1.1 * 10^{-3} = 1.61\ mH$$

$$X_L = 2\pi f L_T = 2\pi * 10 * 10^3 * 1.61 * 10^{-3} = 101.159\ \Omega$$

$$C_T = C_1 + C_2 = 0.012 * 10^{-6} + 1600 * 10^{-12} = 1.36 * 10^{-8} F$$

$$X_C = 1/2\pi f C_T = 1/(2\pi * 10 * 10^3 * 1.36 * 10^{-8}) = 1170.26\ \Omega$$

$$Z_T = R_T + j(X_L - X_C) = 158.873 + j(101.159 - 1170.26)$$

$$= 1080.841\angle -81.547°\ \Omega$$

b.

$$I_T = \frac{V_S}{Z_T} = \frac{10}{1080.841\angle -81.547°} = 9.252\angle 81.547°\ mA$$

c,d and e.

$$S = VI^* = 10 * 9.252 * 10^{-3}\angle -81.547° = 0.09252\angle -81.547°$$

$$= 0.0136 - j0.09151\ VA$$

True power

$$P = 13.6\ mW$$

Reactive power

$$Q = -91.51\ mVAR$$

Apparent power

$$S = 92.52\ mVA$$

2. For the circuit below, calculate the following.

   a. The resonant frequency.
   b. The Bandwidth.
   c. The upper and lower critical frequencies.

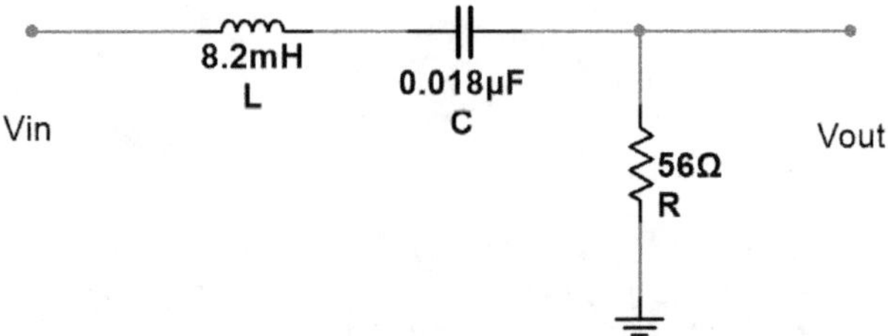

Solution:

a. The resonant frequency.

$$f_r = \frac{1}{2\pi\sqrt{LC}} = \frac{1}{2\pi\sqrt{8.2 * 10^{-3} * 0.018 * 10^{-6}}} = 13.1\ kHz$$

b. The Bandwidth.

$$Q = \frac{X_L}{R} = \frac{2\pi * 13.1\ kHz * 8.2\ mH}{56} = 12$$

$$BW = \frac{f_r}{Q} = \frac{13.1}{12} = 1.0916\ kHz$$

c. The upper and lower critical frequencies.

$$f_1 = f_r - \frac{BW}{2} = 13.1 - \frac{1.0916}{2} = 12.5542\ kHz$$

$$f_2 = f_r + \frac{BW}{2} = 13.1 + \frac{1.0916}{2} = 13.6458\ kHz$$

Chapter 8

Communication Engineering

Communication engineering is a field of electrical engineering that focuses on designing, developing, and implementing communication systems. It involves using various technologies to transmit and receive information between two or more points.

Communication systems can be broadly classified into two categories: analog and digital. Analog systems use continuous signals, while digital systems use discrete signals. Some examples of communication systems include telephone networks, radio and television broadcasting, satellite communications, and internet communication.

Communication engineering is essential for modern society, enabling people and businesses to communicate effectively over long distances. It plays a vital role in various industries, such as telecommunications, broadcasting, and internet communication. The field is constantly evolving, with new technologies and techniques being developed to improve the performance and reliability of communication systems.

8.1 Communication

Conveying or transmission of information from one place and/or time to another.

The principle objective of a communication system is to transmit information signals from one point to another.

The information signals may result from a voice message, a TV picture, or a meter reading, or may take on a variety of other formats depending on the specific application. Communication Engineering involves analyzing, designing, and fabricating an operating system that performs the communication objective.

8.2 Electrical Communication System

Figure 8.1 shows a commonly used model for a single-link communication system. Although it suggests a system for communication between two remotely located points, this block diagram also applies to remote sensing systems, such as radar or sonar, where the system input and output may be located at the same site.

Regardless of the particular application and configuration, all information transmission systems involve three major subsystems :
The transmitter, the channel, and the receiver.

The system uses electrical signals for information transmission. It is a reliable and economical communication technology for transmitting information over long distances. The function of each element is listed below:

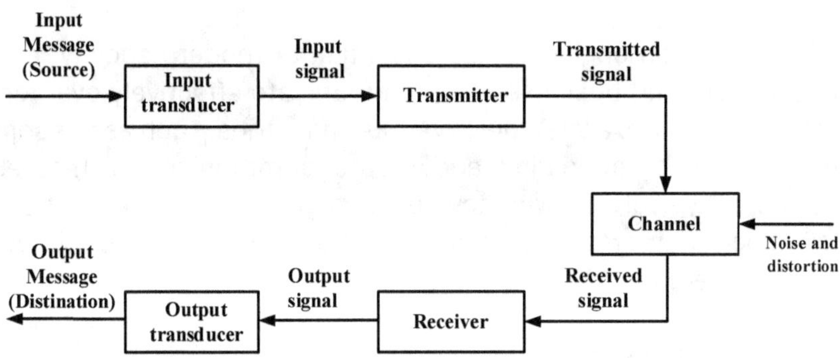

Figure 8.1

Source

Originates the message (e.g., human voice, TV picture, teletype message, or data).

Input Transducer

The wide variety of possible sources of information results in many different forms of messages. Messages may be analog or digital. A transducer must convert the message produced by a source to a form suitable for the particular type of communication system employed. For example, speech waves are converted by a microphone to voltage variations in electrical communications. Such a converted message is referred to as the message signal. The input transducer converts the physical message into an electrical signal called a "baseband signal."

Transmitter

Modifies the baseband signal for efficient transmission.
The purpose of the transmitter is to couple the message to the channel. Modulating a carrier wave with the signal from the input transducer is often necessary. Modulation is the systematic variation of some carrier attribute,such as amplitude, phase, or frequency, by a function of the message signal.

There are several reasons for using a carrier and modulating it. Important ones are

- For ease of radiation.
- To reduce noise and interference.
- For channel assignment.
- For multiplexing or transmission of several messages over a single channel, and
- To overcome equipment limitations.

Channel

It is a medium such as wire or coaxial cable.The channel can have many different forms; the most familiar is the channel between the transmitting antenna of a commercial radio station and the receiving antenna of a radio. In this channel, the transmitted signal propagates to the receiving antenna through the atmosphere or free space. Other forms of channels are

- Transmission lines (such as open two-wire systems and coaxial cables).
- Optical fiber channels.
- Guided electromagnetic-wave channels.

All channels have one thing in common: the signal undergoes degradation from transmitter to receiver. This degradation results from noise and other undesired signals or interference but may also include other distortion effects, such as fading signal levels, multiple transmission paths, and filtering.

Receiver

Reprocesses the signal from the channel by undoing the signal modifications made at the transmitter and the channel. The receiver's function is to extract the desired message from the received signal at the channel output and to convert it to a form suitable for the output transducer. Although amplification may be one of the first operations performed by the receiver, where the received signal may be extremely weak, the main function of the receiver is to demodulate the received signal.

Often, the receiver output is desired to be a scaled, possibly delayed, version of the message signal at the modulator input.

Output transducer

converts the electrical signal to its original form - the message. The output transducer completes the communication system. The device converts the electric signal at its input into the form desired by the system user. The most common output transducer is a loudspeaker. Many other examples include tape recorders, personal computers, meters, and cathode–ray tubes.

Destination

The unit to which the message is communicated. The transmission of information using electrical signals (current or voltage) by varying these signals with time (e.g., changing the amplitude, phase, or frequency of a sinusoidal signal). At the transmission channel, the transmitted signal suffers from the following:

- Attenuation: increases as the length of the channel increases.
- Distortion: Caused by transmitting equipment and can be

corrected at the receiver by an equalizer.

- Interference: caused by some other existing signals.
- Additive noise: Random and unpredictable electric signalgen-erated naturally from cases external and internal.

8.3 Signal–to–Noise Ratio (SNR)

The ratio of the signal power (S) to the noise power (N)

$$SNR = \frac{S}{N}$$

$$SNR(dB) = 10\ \log_{10}\left(\frac{S}{N}\right)$$

8.4 Analog and Digital Messages

i. Digital messages: Constructed with a finite number of symbols (e.g., 26 letters, binary information).

ii. Analog messages: characterized by data whose value var-iesover a continuous range (e.g., the temperature, pres-sure speech).

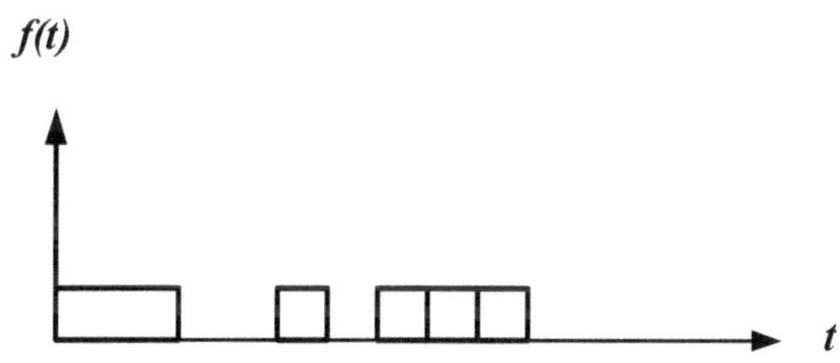

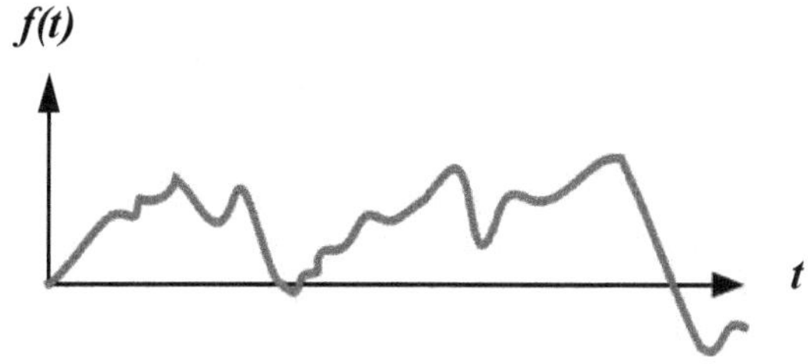

In comparison, the digital message is easier to extract and has high noise immunity than the analog one.

Bandwidth (B): The range of frequencies it can transmit with reasonable fidelity (e.g., from 0 Hz d.c. to 5 kHz) yields bandwidth = 5kHz.

$$W = 2\pi B$$

W is the bandwidth (BW) in rad/sec, while the bandwidth is in Hertz.

Basic transmitter components:

Encoder: chooses the best form for the signal to optimize its detection at the input.

Modulator: produces a varying signal (carrier) proportional to the information signal.

Basic receiver components:

Demodulator: Performs the inverse operation of the modulator to recover the signal in its original form.

Decoder: performs the inverse operation of the encoder to make the best decision that the message was indeed to send.

8.5 Signals and Systems Analysis

Signals are time-varying quantities such as voltages or currents. A system combines devices and networks (subsystems) chosen to perform the desired function. Because of the sophistication of modern communication systems, a great deal of analysis and experimentation with trial subsystems occurs before building the desired system. Thus, the communications engineer's tools are mathematical models for signals and systems.

The signal serves to start some action. In communication systems, the signal is an electrical voltage $v(t)$ or current $i(t)$.

8.5.1 Signal classification

- A signal is deterministic if it could be expressed

mathematically as a function of time (e.g., $f(t) = cos\omega t$). Otherwise, it is said to be random (e.g., noise).

- A signal is said to be periodic if it repeats itself after a fixed length of time (i.e., $f(t) = f(t+T)$) when $-\infty < t < \infty$). Otherwise, it is said to be non-periodic.
- A signal is said to be powerful if it has a finite value of average power, while it is an energy signal if it has a finite value of energy.

Signal powerwatt

$$P_{av} = \frac{1}{T}\int_{0}^{T} |f(t)|^2 dt$$

Signal energy

$$E = \int_{-\infty}^{\infty} |f(t)|^2 dt$$

at R =1 Ω.

Notes:

1. For a given value of resistance R.

For $f(t)$ is voltage

$$P_{av} = \frac{1}{T.R}\int_{0}^{T} |f(t)|^2 dt$$

$$E = \frac{1}{R}\int_{-\infty}^{\infty} |f(t)|^2 dt$$

and for $f(t)$ is current

$$P_{av} = \frac{R}{T}\int_{0}^{T} |f(t)|^2 dt$$

$$E = R\int_{-\infty}^{\infty} |f(t)|^2 dt$$

2. A periodic signal is almost a power signal.
3. The non-periodic signal is almost an energy signal.

4. Some signals are neither power nor energy (increasing ordivergent signals like $tan\ t, 1/t, e^{a'}$ etc.).
5. The power of the energy signal is zero, while the energy of the power signalis infinite.

Example: Classify the following signals:

1-$f(t) = cos\ 3\ t$
2-$f(t) = e^{-2|t|}$
3-$f(t) = e^{+2t}$
Solution:

$$1 - f(t) = cos\ 3\ t$$

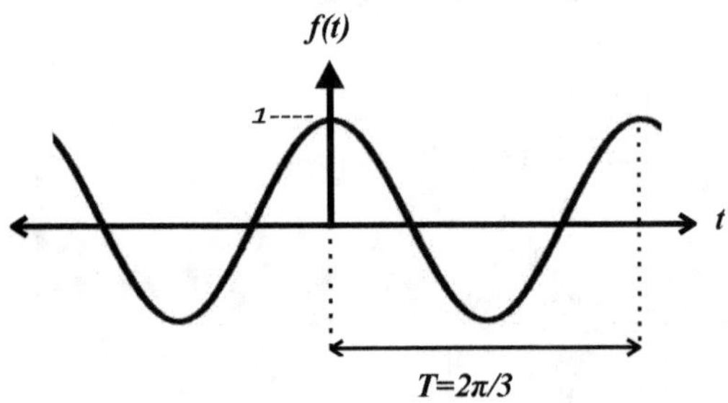

- Deterministic
- Periodic
- Energy or power?

Test energy

$$E = R \int_{-\infty}^{\infty} |f(t)|^2 dt = \infty$$

not energy signal
Test Power;

$$P_{av} = \frac{R}{T} \int_{0}^{T} |f(t)|^2 dt = 1/2 \text{ watt}$$

∴ power signal
2-$f(t) = e^{-2|t|}$

- Deterministic
- Non-periodic

Text power:

8.5.2 Sampling

We can create discrete-time signals by sampling continuous-time signals at regular length intervals. If we multiply a continuous-time signal x(t) with a train of delta functions, we have a sampled signal:

$$x(t)\sum_{-\infty}^{\infty}\delta(t-nT_s)=\sum_{n=-\infty}^{\infty}x(t)\delta(t-nT_s)$$

$$=\sum_{n=-\infty}^{\infty}x(nT_S)\delta(t-nT_s)$$

T_s is the sampling period or interval, and the sampling frequency is f_s $=1/T_s$. If x(t) is band-limited, we can prevent overlap in the frequency domain by selecting T_s.

$$T_S \leq \frac{1}{2W}$$

Where W is the signal's bandwidth, this is the Nyquist-Shannon Sampling theorem. We refer to $f_S = 2W$ as the Nyquist frequency since it is the highest frequency at which a signal can contain energy and remain compatible with the sampling theorem. The sampled signal $x_s(t)$ is, therefore

$$x_s(t)=x(t)\sum_{n=-\infty}^{\infty}\delta\,(t-nT_s)$$

We take the Fourier transform of this and apply the frequency convolution property.

$$X_s(\omega)=\frac{1}{2\pi}[X(\omega)] * \omega_S \sum_{n=-\infty}^{\infty}\delta(\omega-n\omega_s)$$

$$=\frac{1}{T_S}\sum_{n=-\infty}^{\infty}X(\omega) * \delta(\omega-n\omega_s)$$

$$=\frac{1}{T_S}\sum_{n=-\infty}^{\infty}X(\omega-n\omega_s)$$

Where $\omega_s = 2\pi / T_s$.

Example: For frequencies above 4 kHz, the spectrum of a signal is zero. To sample the signal, find the maximum time spacing between samples.

Solution

$$W = 4\ kHz$$

The Nyquist rate is

$$f_s = 2W = 8\ kHz$$

The Nyquist interval

$$T_s = \frac{1}{f_s} = \frac{1}{8 \times 10^3} = 125\ \mu s$$

Example: A voice signal is band-limited to 2.5 kHz. Find the Nyquist rate and the Nyquist interval for the signal.

$$W = 2.5\ kHz$$

The Nyquist rate is

$$f_s = 2W = 5\ kHz$$

The Nyquist interval

$$T_s = \frac{1}{f_s} = \frac{1}{5 \times 10^3} = 0.2\ ms$$

Multiple chose questions

1. ___ is the full form of the abbreviation TQM.

A. Team quality management
B. Total quality management
C. Total quality manager
D. Total quality management
Ans: D

2. The slowest transmission speeds are those of
A. Coaxial cable
B.Twisted pair wire
C. Fiber-optic cable
D. Microwave
Ans: B

3. AM radio uses frequency division multiplexing. 117 AM radio stations can be present at 535 kHz to 1705 kHz. What is the maximum audio bandwidth of an AM radio station to allow 117 simultaneous AM stations in an area?

A) 5 kHz B) 10 kHz C) 15 kHz D) 20 kHz

Solution

The total bandwidth allotted for AM radio is:

$$1705 - 535 = 1170 \text{ kHz}$$

If 117 stations are needed, each station will be allotted a bandwidth of:

(1170 kHz)/(117 Stations)=117 stations

We know the bandwidth for AM is two times the baseband bandwidth. Therefore, the maximum baseband (audio) bandwidth is 5 kHz. The answer is A

4. A sinusoid of frequency 1 kHz is transmitted using amplitude modulation (AM) with a carrier frequency of 800 kHz. What frequencies are contained in the AM signal
A) 800 kHz
B) 799, 801 kHz
C) 799, 800, 801 kHz
D) 800+n kHz, where n=0,1,2,3...
Ans.: C

5. Which of the following statements is true?

1. A single parity bit allows for error detection, not error correction.
2. Odd parity cannot be used for an even number of data bits.
3. Parity bits lower the bandwidth required to transmit data.

A) 1 B) 2 C) 3 D) None of the above
The answer is A.

6. Which of the following statements is true?

1. One parity bit for every seven data bits allows error correction.
2. Even and odd parity refers to the number of ones in a data segment. It does not refer to the total number of data bits.
3. Parity bits lower the bandwidth required to transmit data.

A) 1 B) 2 C) 3 D) None of the above
The answer is B.

7. Which of the following statements is true?
1. One parity bit for every seven data bits allows error correction.
2. Odd parity cannot be used for an even number of data bits.
3. The parity bits would be additional data to transmit, requiring additional bandwidth.

A) 1 B) 2 C) 3 D) None of the above
The answer is C.

8.6 Information Theory and Coding

Information theory and coding are two closely related fields in computer science and telecommunications that deal with the representation, transmission, and storage of information reliably and efficiently. Let's explore each of these concepts further:

8.6.1 Concept of Probability

If an experiment has A1, A2, A3... An outcome then:

$$p(A_i) = \lim_{N\to\infty} \frac{n(A_i)}{N}$$

Where n(Ai): Number of times event Ai occurs if the experiment is repeated N times. Note that:

$$1 \geq p(A_i) \geq 0 \quad and \sum_{i=1}^{n} P(A_i) = 1$$

If $p(Ai) = 1$, then A_i is a certain event.

Example: A simple language consists of only two symbols, **A** and **B,** produced in a long continuous sequence.

$$BAAABBBAAA\ AABBBAAAAB$$

$$p(A) = \frac{12}{20}, p(B) = \frac{8}{20}$$

Joint Probability: *If we have two experiments,* ***A*** *&* ***B****, experiment* ***A*** *has* $\boldsymbol{A_1, A_2, A_3 \ldots A_n}$ *outcomes, and experiment* ***B*** *has* $\boldsymbol{B_1, B_2, B_3 \ldots B_n}$ *outcomes* $p(A_i, B_j)$ *Joint probability that event* ***Ai*** *occurs from experiment* ***A*** *and event* ***Bj*** *from experiment* **B**. note that:

$$\sum_{i=1}^{n}\sum_{j=1}^{m} p(A_i, B_j) = 1$$

$p(A_i, B_j)$ is written in matrix form as follows

$$p(A_i, B_j) = \begin{bmatrix} B_1 & B_2 & \ldots & B_m \\ \ldots & \ldots & \ldots & \ldots \\ \ldots & \ldots & \ldots & \ldots \\ \ldots & \ldots & \ldots & \ldots \end{bmatrix} \begin{matrix} A_1 \\ A_2 \\ \ldots \\ A_n \end{matrix}$$

Note that

$$\sum_{i=1}^{n} p(A_i, B_j) = p(B_j)$$

$$\sum_{j=1}^{m} p(A_i, B_j) = p(A_i)$$

Example: From the previous example, assume that the **21**st letter is **A** to have **20** pairs.

Then

$$p(A, A) = \frac{9}{20}\ , p(A, B) = \frac{3}{20}\ ,$$

$$p(B, A) = \frac{3}{20}\ ,$$

$$and\ p(B, B) = \frac{5}{20}$$

Or in matrix form

$$p(x, y) = \begin{bmatrix} \frac{9}{20} & \frac{3}{20} \\ \frac{3}{20} & \frac{5}{20} \end{bmatrix}$$

8.6.2 Conditional Probability

Two experiments, **A & B,** with their outcomes, affect each other. p(Ai/Bj) = Conditional probability of A_i given that B_j has already occurred in the experiment.

p(B_j/A_i)= Conditional probability of B_j given that A_i has already occurred in the experiment

Note that: $p(A_i\ , B_j) = p(A_i).p(B_j/A_i\ )$

$$p(B_j\ , A_i) = p(B_j)\ .p(A_i/B_j\ )$$

Both $p(A_i/B_j)$ and $p(B_j/A_i)$ can be written in matrix form

$$p(A_i/B_j) = \begin{bmatrix} B_1 & B_2 & \dots & B_m \\ \dots & \dots & \dots & \dots \\ \dots & \dots & \dots & \dots \\ \dots & \dots & \dots & \dots \end{bmatrix} \begin{matrix} A_1 \\ A_2 \\ \dots \\ A_n \end{matrix}$$

Note that

$$\sum_{i=1}^{n} p(A_i/B_j) = 1$$

and

$$\sum_{j=1}^{m} p(B_j/A_i) = 1$$

Example: *From the previous example, we have*

$$p(A, A) = \frac{9}{12} \;, p(A, B) = \frac{3}{8} \;,$$

$$p(B, A) = \frac{3}{12} \;,$$

$$and\; p(B, B) = \frac{5}{8} \;,$$

on matrix form:

$$p(x/y) = \begin{bmatrix} \frac{9}{12} & \frac{3}{8} \\ \frac{3}{12} & \frac{5}{8} \end{bmatrix}$$

8.6.3 Statistical Independence

If A_i does not affect the probability of B_j then they are called independent, and $P(A_i\,/B_j\,) = \; P(A_i)$ and $P(B_j\,/A_i\,) = \; P(B_j)$ and $P(A_i/B_j) = \; P(A_i).P(B_j)$

Example: *The two experiments, **A & B,** have the joint probability matrix*

$$p(A,B) = \begin{bmatrix} 0.1 & 0.25 \\ 0 & 0.2 \\ 0.25 & 0.2 \end{bmatrix}$$

find: $p(A), p(B), p(A/B), p(B/A)$

Solution:

$$p(A_1) = \sum_{j=1}^{2} p(A_1, B_j) = 0.1 + 0.25 = 0.35$$

$$p(A_2) = \sum_{j=1}^{2} p(A_2, B_j) = 0 + 0.2 = 0.2$$

$$p(A_3) = \sum_{j=1}^{2} p(A_3, B_j) = 0.25 + 0.2 = 0.45$$

So $p(A) = [0.35 \quad 0.2 \quad 0.45]$

We note that: $p(A_1) + p(A_2) + p(A_3) = 1$

To find *p(B) we have:*

$$p(B_1) = \sum_{i=1}^{3} p(A_i, B_1) = 0.1 + 0 + 0.25 = 0.35$$

$$p(B_2) = \sum_{i=1}^{3} p(A_i, B_2) = 0.25 + 0.2 + 0.2 = 0.65$$

Also note that : $p(B_1) + p(B_2) = 1$

$$p(A_i/B_j) = p(A_i\,, B_j)/p(B_j)$$

$$p(B_j/A_i) = p(A_i\,, B_j)/p(A_i)$$

$$p\left(\frac{A}{B}\right) = \begin{bmatrix} \frac{0.1}{0.35} & \frac{0.25}{0.65} \\ 0 & \frac{0.2}{0.65} \\ \frac{0.25}{0.35} & \frac{0.2}{0.65} \end{bmatrix} = \begin{bmatrix} \frac{2}{7} & \frac{5}{13} \\ 0 & \frac{4}{13} \\ \frac{5}{7} & \frac{4}{13} \end{bmatrix}$$

$$p(B_j/A_i) = p(A_i\,,B_j)/p(A_i)$$

$$p\left(\frac{B}{A}\right) = \begin{bmatrix} \frac{0.1}{0.35} & \frac{0.25}{0.35} \\ 0 & \frac{0.2}{0.2} \\ \frac{0.25}{0.45} & \frac{0.2}{0.45} \end{bmatrix} = \begin{bmatrix} \frac{2}{7} & \frac{5}{7} \\ 0 & 1 \\ \frac{5}{9} & \frac{4}{9} \end{bmatrix}$$

8.7 Information Theory

This subject deals with information and data transmission from one point to another. The block diagram of any communication system is shown below:

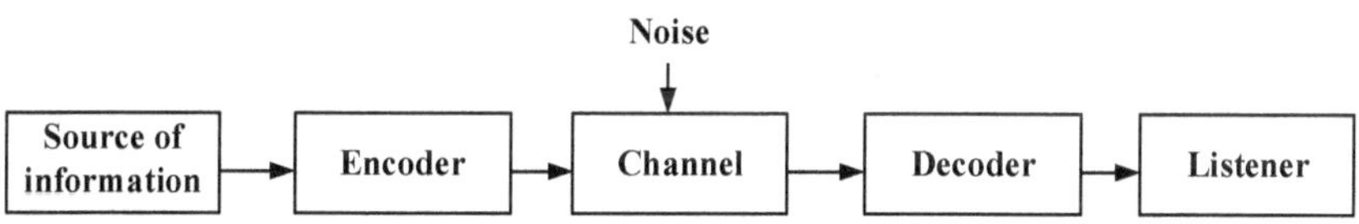

8.7.1 Self-Information

Suppose that the source of information produces a finite set of messages $x_1, x_2, x_3, \dots x_n$ *with probabilities* $p(x_1), p(x_2), p(x_3), \dots\dots p(x_n)$ *such that*

$$\sum_{i=1}^{n} p(x_i) = 1$$

The amount of information gained from knowing that the source produces the symbol (x_i) is *related to* $p(x_i)$ *as follows:*

1) Information is zero if $p(x_i) = 1$ (certain event).
2) Informationincreases as $p(x_i)$ decrease to zero.
3) Information is a positive quantity.

The function that relates p(xi) with information of xi is :

$$I(x_i) = -\log_a p(x_i)$$

The units of $I(x_i)$ depend on (a).

i. if a=2, then $I(x_i)$ has the units of bits.
ii. if a=e=2.718, $I(x_i)$ has the units of nat.
iii. if a=10, $I(x_i)$ has the units of hartly.

Source Entropy $H(x)$: If $I(x_i),\ i = 1,2,3 \ldots n$ is different for the source producing unequal probability symbols, then the statistical average of $\boldsymbol{I}(x_i)$ will give the average amount of uncertainty associated with the source x. This average is called source entropy and is denoted by $H(x)$. This is given by:

$$H(x) = \sum_{i=1}^{n} p(x_i)\, I(x_i) \qquad (bits/symbol)$$

$$H(x) = -\sum_{i=1}^{n} p(x_i)\, \log_a p(x_i) \qquad (bits/symbol)$$

Example: Find the entropy of the source producing the symbols:

$$p(x) = [0.25\ 0.1\ 0.15\ 0.5]$$

Solution:

$$H(x) = -\sum_{i=1}^{n} p(x_i)\, \log_a p(x_i)$$

$$= -\frac{1}{ln2}[0.25\ln(0.25) + 0.1\ln(0.1) + 0.15\ln(0.15) + 0.5\ln(0.5)]$$

$$= 1.7427\ bits/symbol$$

8.7.2 Source Entropy Rate R(X)

The average amount of information produced per second.

In general, If the n symbols $x_1, x_2, x_3, \dots x_n$ are equiprobable, so

$$p(x_i) = \frac{1}{n} \; then \; H(x) = H(x)|_{max.} = \log_2 n$$

$$R(X) = H(x) \times rate\ of\ producing\ symbols\ \ (bits/sec)$$

or

$$R(X) = \frac{H(x)}{\bar{\tau}}$$

$$\bar{\tau} = \sum_{i=1}^{n} \tau_i \, p(x_i)$$

Where $\tau_i = time\ duration\ of\ x_i$.

Example: A source produces dots and dashes with p (dot) = 0.65, If the time duration of a dot is 200 msec and that for a dash is 800 msec. Find the average source entropyrate.

Solution:

$$p(dot) = 0.65\,, so\ p(dash) = 1 - 0.65 = 0.35$$

$$(x) = -\sum_{i=1}^{n} p(x_i)\ \log_a\ p(x_i)$$

$$= -\frac{1}{ln2}[0.65 \ln(0.65) + 0.35 \ln(0.35)] = 0.9341\ bits/symbol$$

$$\tau_{dot} = 200\ msec\,, \tau_{dash} = 800\ msec$$

$$\bar{\tau} = \sum_{i=1}^{n} \tau_i \, p(x_i) = 200 \times 0.65 + 800 \times 0.35 = 410\ msec$$

$$R(X) = \frac{H(x)}{\bar{\tau}} = \frac{0.9341}{410} = 2.2782\ bits/sec$$

8.8 Modulation

Modulation refers to modifying a carrier signal to encode information for transmission over a communication channel. It is a fundamental concept in telecommunications used in various communication systems, including radio, television, telephone networks, and wireless communication.

The purpose of modulation is to transfer information, such as voice, video, or data, efficiently and reliably from a source to a destination. It allows multiple signals to share the same channel without interference and enables long-distance communication by overcoming signal degradation and noise.

Modulation involves modifying one or more properties of the carrier signal, a high-frequency signal capable of propagating over long distances. The carrier signal's characteristics, such as amplitude, frequency, or phase, are altered by the input signal or data to be transmitted.

Here are three common types of modulation:

i. Amplitude Modulation (AM): In AM, the amplitude of the carrier signal varies in proportion to the instantaneous amplitude of the input signal. The input signal is typically an audio signal. AM is commonly used in broadcasting for AM radio transmissions.

ii. Frequency Modulation (FM): FM involves varying the carrier signal frequency based on the input signal. The instantaneous frequency of the carrier signal changes according to the amplitude of the input signal. FM is widely used in radio broadcasting and applications such as two-way radios and analog television.

iii. Phase Modulation (PM): PM is a modulation technique where the phase of the carrier signal is varied to encode the information. The phase shift of the carrier signal is proportional to the input signal. PM is used in various digital communication systems, such as GSM (Global System for Mobile Communications), to transmit data.

8.8.1 Amplitude Modulation

Amplitude Modulation (AM) is a modulation technique used in analog communication systems, particularly in broadcasting for AM radio transmissions. In AM, the amplitude of a high-frequency carrier signal is varied in proportion to the instantaneous amplitude of the modulating signal, which typically represents audio or voice signals.

Here's how the process of Amplitude Modulation works:

i. Carrier Signal: The carrier signal is a high-frequency sinusoidal waveform that serves as the carrier wave for transmitting the modulating signal. It typically has a much higher frequency than the modulating signal.

ii. Modulating Signal: The modulating signal is the input signal that carries the information to be transmitted. In AM, the modulating signal is usually an audio signal representing sound waves or voice.

iii. Modulation Process: The modulating signal is used to modulate the amplitude of the carrier signal. The instantaneous amplitude of the carrier wave is varied based on the amplitude of the modulating signal at that moment.

iv. Sidebands: The modulation process creates sidebands, which are additional frequency components around the carrier frequency. In AM, the sidebands contain the information from the modulating signal. The upper sideband (USB) and lower sideband (LSB) are mirror images of each other concerning the carrier frequency.

v. Frequency Spectrum: The frequency spectrum of an AM signal consists of the carrier frequency and the two sidebands. The carrier frequency is at the center, and the sidebands are symmetrically located above and below it.

vi. Demodulation: At the receiver end, the demodulation process extracts the original modulating signal from the AM signal. This is typically achieved using a diode or an envelope detector, which removes the carrier wave and retrieves the modulating signal.

Advantages and Disadvantages of AM:

Advantages:

i. AM is relatively simple to implement and doesn't require complex circuitry.
ii. It can cover long distances and provide good signal propagation, especially in the medium-wave and long-wave frequency bands.
iii. AM receivers are less complex and less expensive compared to FM receivers.

Disadvantages:

i. AM is susceptible to noise and interference, which can affect the quality of the received signal.
ii. Its limited bandwidth results in lower audio fidelity than other modulation techniques like FM.
iii. AM signals are prone to multipath interference, where the signal arrives at the receiver through multiple paths, causing fading and distortion.

Despite its limitations, AM remains used for various applications, including broadcasting, aviation communications, and two-way radio systems.

8.8.2 Frequency Modulation

Frequency Modulation (FM) is a modulation technique widely used in analog and digital communication systems. It is commonly employed in FM radio broadcasting, two-way radio communication, and certain applications in television transmission. FM modulates the frequency of a carrier signal based on the instantaneous amplitude of the modulating signal.

Let's explore how Frequency Modulation works:

i. **Carrier Signal**: Like AM, FM utilizes a high-frequency carrier signal acting as the carrier wave. The carrier signal has a much higher frequency than the modulating signal.

ii. **Modulating Signal**: The modulating signal carries the information to be transmitted and is typically an audio signal representing sound waves or voice. The modulating signal affects the frequency of the carrier wave.

iii. **Modulation Process**: In FM, the instantaneous frequency of the carrier signal is varied based on the amplitude of the modulating signal. As the modulating signal changes in amplitude, it causes the carrier frequency to shift up and down accordingly.

iv. **Frequency Deviation**: The frequency deviation range determines how much the carrier frequency varies in response to the modulating signal. The magnitude of frequency deviation is proportional to the amplitude of the modulating signal.

v. **Sidebands**: Frequency modulation generates sidebands around the carrier frequency. Unlike AM, FM produces infinite sidebands, forming a continuous spectrum. However, the energy in FM signals primarily concentrates around the carrier frequency and the immediate sidebands.

vi. **Bandwidth:** FM signals typically require a wider bandwidth than AM signals to accommodate the frequency variations caused by the modulating signal. The bandwidth of an FM signal depends on the modulation index, which measures the maximum frequency deviation divided by the frequency of the modulating signal.

vii. **Demodulation**: At the receiver end, a demodulator or FM receiver detects the changes in the frequency of the received FM signal. The receiver extracts the original modulating signal by recovering the variations in frequency and reconstructing the audio or data signal.

Advantages and Disadvantages of FM:

Advantages:

i. FM is less susceptible to noise and interference than AM, resulting in better signal quality.
ii. It provides higher fidelity and better audio quality due to its wider bandwidth.
iii. FM signals exhibit the capture effect, where a stronger FM signal can suppress weaker interfering signals.

Disadvantages:

i. FM signals require a wider bandwidth than AM, limiting the channels available in a given frequency range.
ii. FM receivers are typically more complex and expensive compared to AM receivers.
iii. FM signals do not propagate well over long distances or through obstacles compared to AM signals.

Despite these limitations, FM modulation is widely used for broadcasting music and voice signals in FM radio stations and has found applications in various communication systems that require high-quality audio transmission. Additionally, digital modulation techniques like frequency-shift keying (FSK) and phase-shift keying (PSK) can be built on the foundation of FM to transmit digital data.

8.8.3 Phase Modulation

Phase Modulation (PM) is a modulation technique used in analog and digital communication systems. It involves varying the phase of a carrier signal in proportion to the instantaneous amplitude of the modulating signal. PM is widely used in digital communication systems, including GSM (Global System for Mobile Communications), combined with other data-transmission modulation schemes.

Here's how Phase Modulation works:

i. Carrier Signal: Similar to AM and FM, PM utilizes a carrier signal, a high-frequency sinusoidal waveform. The carrier signal serves as the basis for carrying the modulating signal.
ii. Modulating Signal: The modulating signal, often representing digital data or analog information, influences the phase of the carrier signal. The instantaneous amplitude of the modulating signal determines the phase shift of the carrier wave.
iii. Modulation Process: In PM, the phase of the carrier signal is varied based on the amplitude of the modulating signal. As the modulating signal's amplitude changes, the carrier wave's phase shifts accordingly.
iv. Phase Shift and Deviation: The amount of phase shift or deviation depends on the amplitude of the modulating signal. The phase deviation is directly proportional to the amplitude of the modulating signal.
v. Phase Spectrum: The PM signal exhibits a phase spectrum that represents the changes in phase over time. The instantaneous phase of the PM signal encodes the information carried by the modulating signal.
vi. Demodulation: At the receiver end, a demodulator or PM receiver recovers the original modulating signal from the PM signal. To extract the information, the receiver measures the phase difference between adjacent signal samples.

Advantages and Disadvantages of PM:

Advantages:

i. PM is less affected by amplitude variations and noise than AM, making it more resilient to signal degradation.
ii. It allows for efficient use of the frequency spectrum since it does not require a wide bandwidth like FM.

iii. PM is compatible with other modulation techniques, such as frequency modulation, enabling combined modulation schemes for various applications.

Disadvantages:

i. PM is more complex to implement and requires precise phase detection and demodulation techniques.
ii. In some scenarios, it is sensitive to phase distortions caused by channel impairments, making it more susceptible to signal quality degradation.
iii. PM signals can be more challenging to amplify and transmit efficiently than simpler modulation schemes.

While PM is not as widely used as AM or FM in traditional analog broadcasting, it plays a significant role in digital communication systems. It is a key component of various digital modulation schemes, including phase-shift keying (PSK) and quadrature phase-shift keying (QPSK), commonly used in wireless communication, satellite communication, and data transmission over digital networks.

8.9 Communication Systems

A group of objects that can interact harmoniously and combine in a manner intended to achieve a desired objective.

Linear system:

The system that satisfies the following: If **$g_1(t)$** is the output when **$f_1(t)$** is the inputand **$g_2(t)$** is the output when **$f_2(t)$** is the input, then, when the input is **$f_1(t)+ f_2(t)$,** the output would be **$a_1\, g_1(t)+ a_2\, g_2(t)$.**

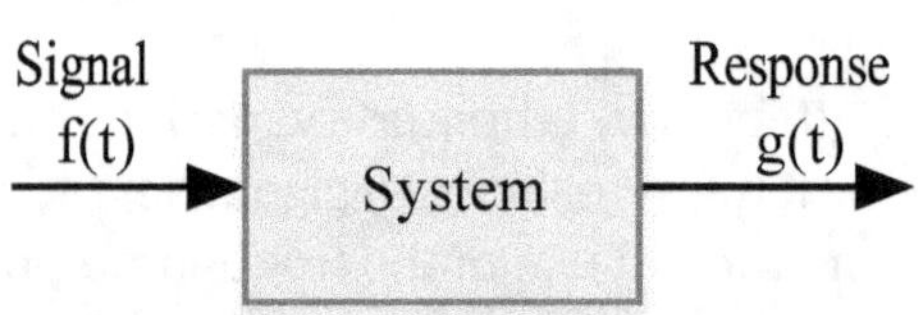

System Impulse Response:

When an input signal is applied to asystem, then the output would be;

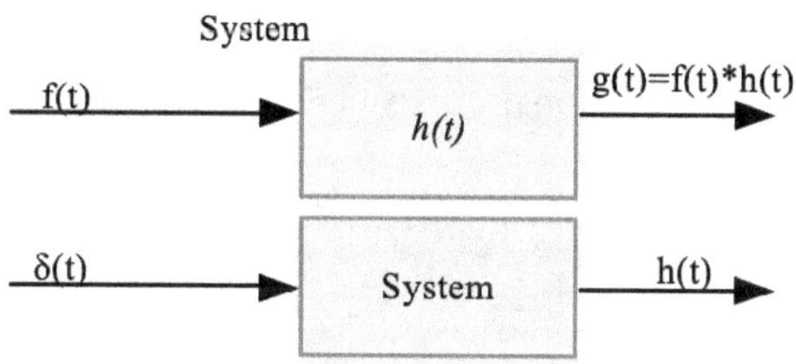

(2-19)

$$g(t) = f(t) * h(t) = \int_{-\infty}^{\infty} f(\tau)h(t-\tau)d\tau$$

Where *h(t)* is called the impulse response of the system, h(t) is found by applying δ(t) at the system input, where the output would be h(t),

$g(t) = h(t)$ when the input is δ(t).

Example: Determine the impulse response of the system shown below.

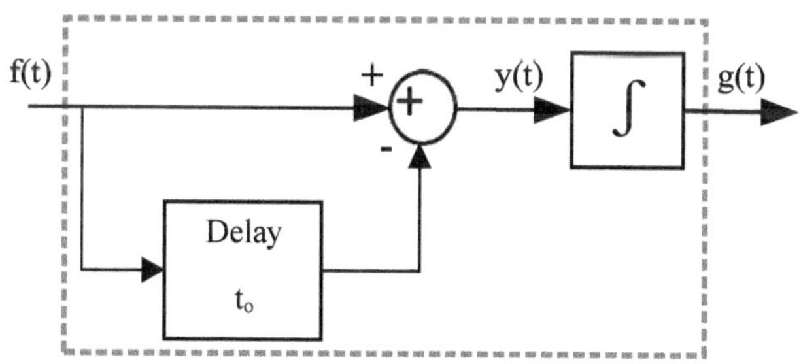

Solution

Let $f(t) = \delta(t)$

System Transfer Function:

It is the FT of the system impulse response

$$H(\omega) = F[f(t)] \quad \ldots$$

Using the time convolution property of Fourier transform, we have:

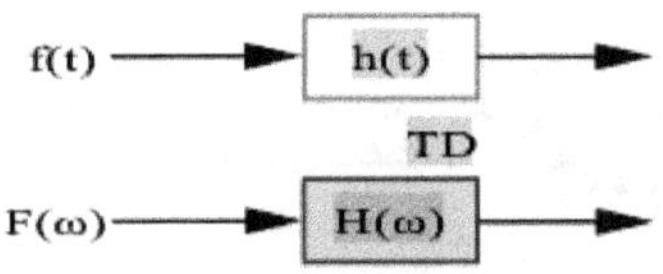

$$g(t) = f(t) \otimes h(t)$$

$$G(\omega) = F(\omega)H(\omega)$$

$$G(\omega) = F(\omega).\, H(\omega \quad \ldots$$

The amplitude and phase spectrum for g(t) would be:

$$|G(\omega)| = |F(\omega)|.\, |H(\omega)| \quad \ldots$$

$$\theta_g(\omega) = \theta_f(\omega) + \theta_h(\omega) \quad \ldots$$

Equivalent Frequency Transfer Function:

Cascaded Series:

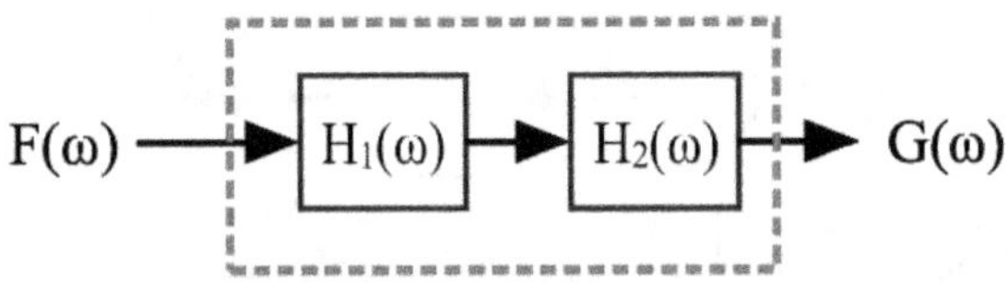

$$H_{eq}(\omega) = H_1(\omega)H_2(\omega)$$

Parallel:

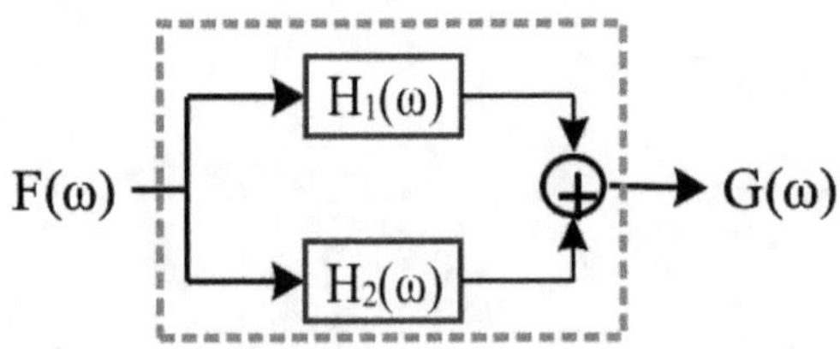

$$H_{eq}(\omega) = H_1(\omega) + H_2(\omega)$$

Feedback:

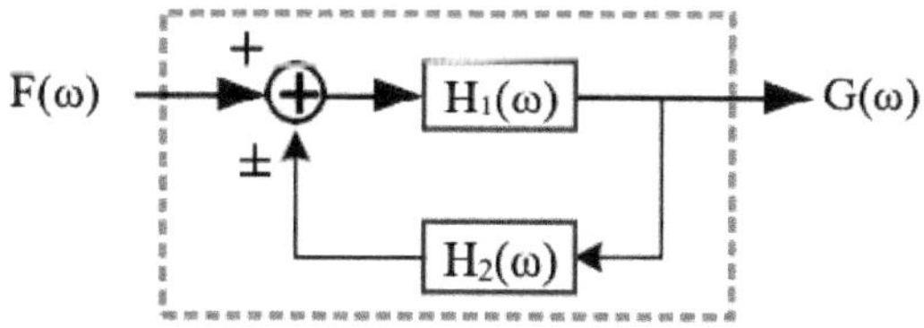

$$H_{eq}(\omega) = \frac{H_1(\omega)}{1 \pm H_1(\omega) H_2(\omega)}$$

Example: Calculate the energy of the input and output signal for the system shown below:

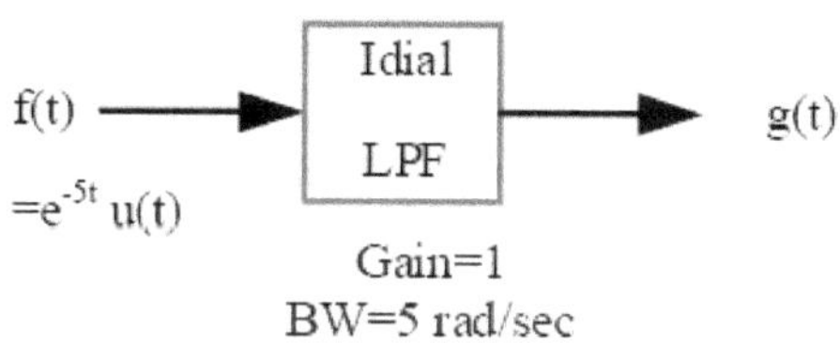

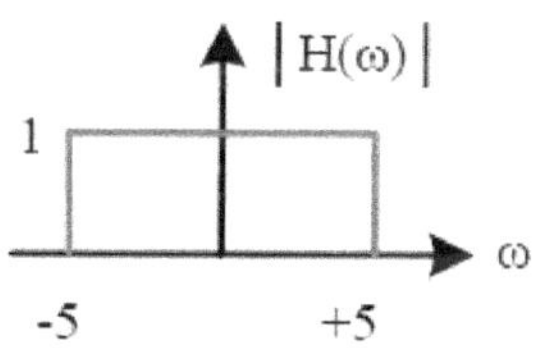

Solution:

For the input signal

$$E_f = \int_{-\infty}^{\infty} |f(t)|^2 dt$$

$$= \int_{0}^{\infty} e^{-10t}\, dt = 0.1 \ joule$$

For the output signal

$$E = \int_{-\infty}^{\infty} |G(\omega)|^2 d\omega / 2\pi = \int_{-\infty}^{\infty} |F(\omega)|^2 |H(\omega)|^2 d\omega / 2\pi$$

$$e^{-at}u(t) \Leftrightarrow \frac{1}{a+j\omega} \Rightarrow F[e^{-5t}u(t)] = \frac{1}{5+j\omega}$$

$$E_g = \frac{1}{2\pi}\int_{-5}^{5} \frac{1}{25+\omega^2} d\omega = 0.05\ joule$$

8.10 Amplitude Modulation

The general sinusoidal signal can be written as:

$$\Phi(t) = a(t)cos[\omega_C t + \gamma(t)]$$

Amplitude Frequency Phase

Angle

In amplitude modulation (AM), $a(t)$ is changed in proportion to the message signal. Frequency is constant, phase (t) = 0.

8.10.1 Types of AM

- Double-Sideband, suppressed Carrier (AM/DSB-SC).
- 2- Double-Sideband, Large Carrier (AM/DSB-LC) [AM].
- Single-sideband, suppressed carrier (AM/SSB-SC) [SSB].
- 4- Vestigial –sideband (AM/VSB).

AM/DSB-SC

The AM/DSB-SC signal, assuming proportionality constant =1, is given by:

$$\boxed{\Phi(t)_{DSB/SC} = f(t)cos\omega_c t}$$

Modulated Signal Modulating Signal Carrier Signal

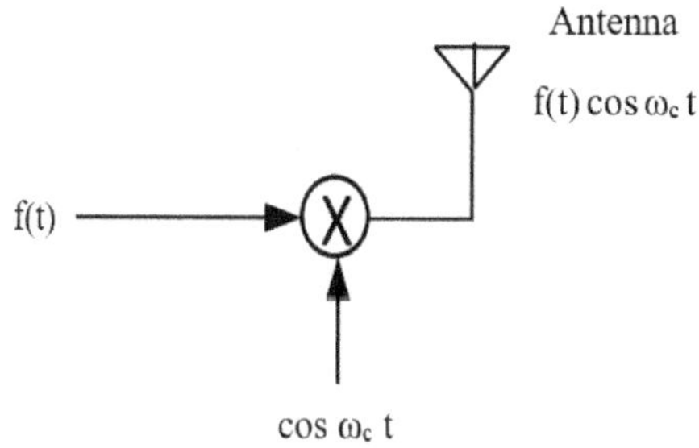

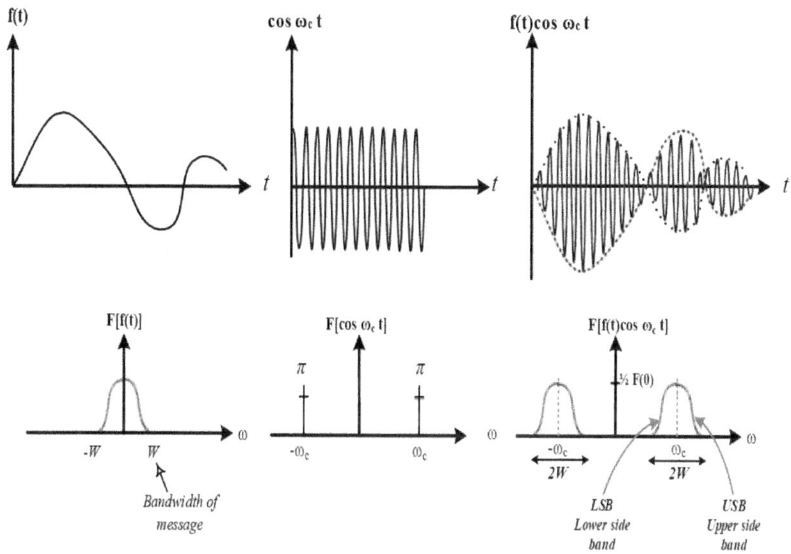

and the spectrum is:

$\Phi(\omega)_{DSB/SC} = 0.5\ F(\omega\ \ \omega_C) + 0.5\ F(\omega + \omega_C)$

Notes:

- No carrier term is present (carrier is suppressed)
- $$\text{BW}_{DSB/SC} = 2W$$

Where W is the bandwidth of the message (modulating signal), i.e., the bandwidth is doubled.

- Above process (multiplication) is called "Frequency conversion," "frequency mixing," or Heterodyning.

DSB-SC Receiver

To detect (demodulate) the DSB-SC signal, we multiply it again by $\cos\omega_c t$ as follows:

$$\Phi(t)cos\omega_C t = f(t)cos^2\omega_C t$$
$$= \frac{1}{2} f(t)\cos 2\omega_C t$$
$$F[\Phi(t)cos\omega_c t] = F(\omega) + \frac{1}{4} F(\omega - 2\omega_c) + \frac{1}{4} F(\omega + 2\omega_c)$$

Then using LPF of bandwidth W **rad/sec,** we obtain the original signal.

$\Phi(t)$=f(t) cos ω_c t

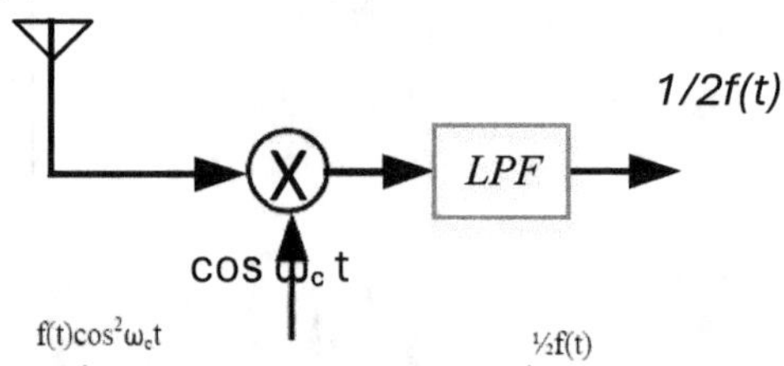

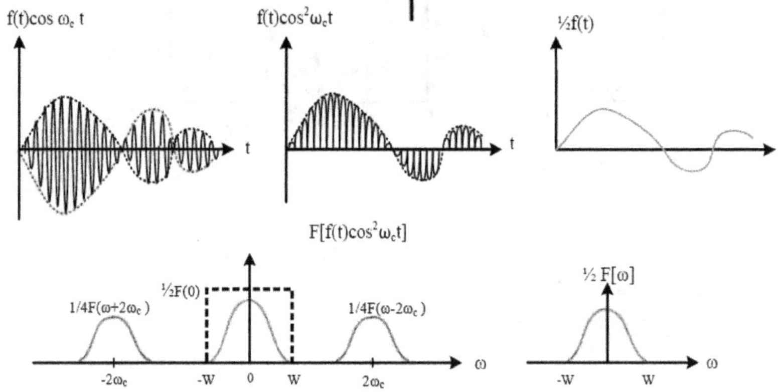

Modulator Implementation:

a) Using Electromechanical Switch

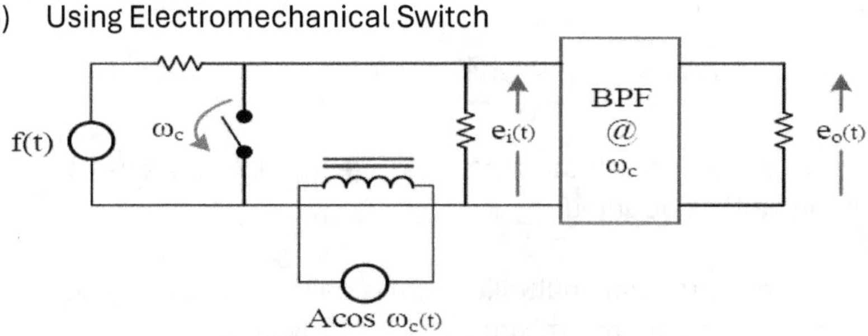

b) Using Diodes and Switches

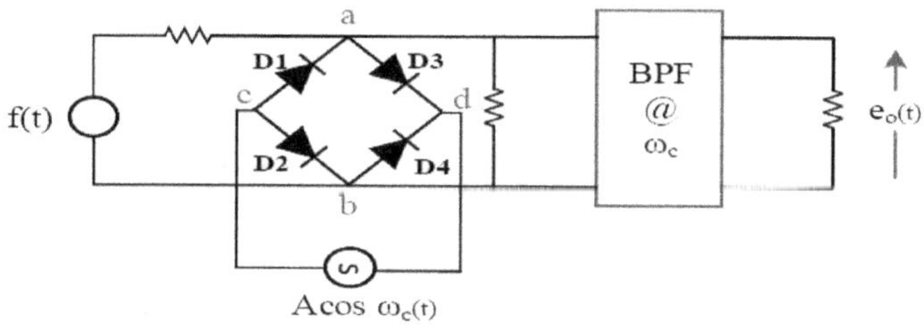

AM/DSB-LC [Standard AM]

For broadcast systems (many Rx for each Tx), it is more economical to obtain less expensive receivers. For such a case, a larger signal is transmitted with the AM/DSB-SC to eliminate the need for a local oscillator in Rx.

The AM/DSB-LC signal [AM signal] is given by

$$\Phi AM = f(t) cos\omega c\ t + Ac\ cos\omega c\ t$$
$$= [f(t) + Ac\ ] cos\omega c\ t$$

$\Phi AM(t)$ =0.5 $F(\omega + \omega c\ )$ +0.5 $F(\omega - \omega c\ ) + \pi Ac\ \delta(\omega - \omega c\ ) + \pi Ac\ \delta(\omega - \omega c\ )$

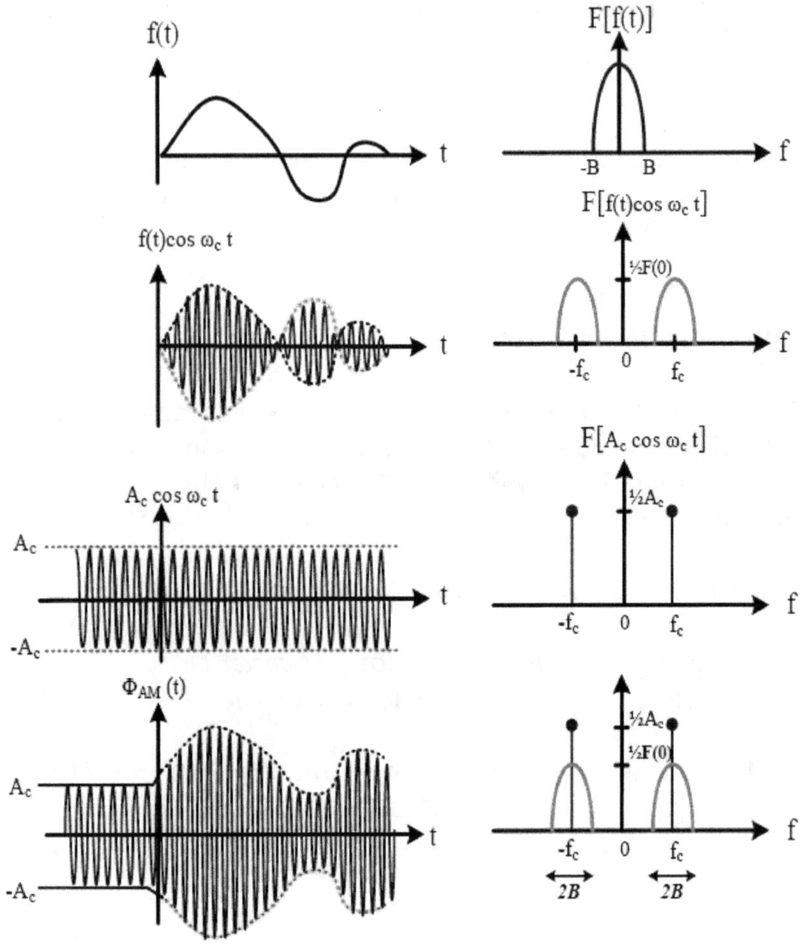

$$m = -\frac{f(t)_{min}}{A_c}$$

$m \leq 1$ *f or envelope detection*

And modulation depth as:

$D = m \times 100\%$

$$\Phi AM = [Ac + f(t)]cos\omega c\, t = Ac\, [1 + mcos\omega mt]cos\omega c\, t$$

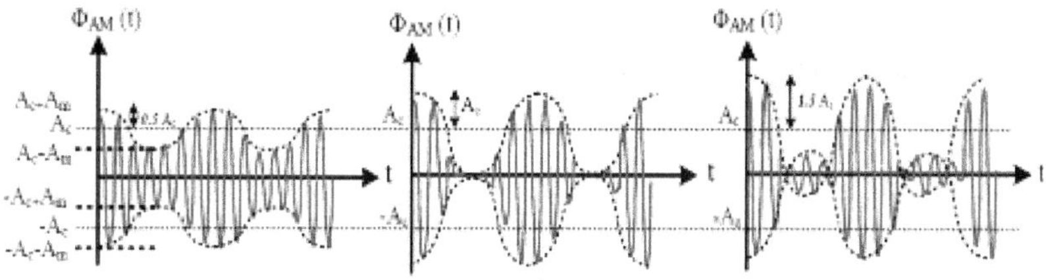

m=0.5 | m=1 | m=1.5

Correct modulation | Critical modulation | Over modulation

$$m = \frac{A_{max} - A_{min}}{A_{max} + A_{min}}$$

Where:

$Amax = Ac + Am$

$Amin = Ac - Am$

Sideband and Carrier Powers:

$\Phi AM = Ac\, cos\omega c\, t + f(t) cos\omega c\, t$

Carrier Power

$$Pc = Ac^2/2$$

Sideband Power

$$Ps = f(t)^2/2$$

$$Pt = Pc + Ps = 0.5[Ac^2 + f(t)^2)]$$

$$\eta = \frac{Ps}{Pt} \times 100\% = \frac{m^2}{m^2 + 2}$$

and

$$\frac{Pc}{Pt} = \frac{2}{m^2 + 2}$$

Example: A transmitter transmits an AM/DSB single tone-modulating signal given by $3\, cos(2\pi 10000t) volt$ with a carrier signal given by $10\, cos(2\pi 10^6 t) volt$, Find:

1- Modulation depth.

2- USB & LSB frequencies.

3- Amplitude of sideband frequencies.

4- Efficiency of Transmission.

Solution:
Am =3 V, Ac =10 V, fm =103 Hz, fc =106 Hz

1- $D = \frac{A_m}{A_c} \times 100\% = \frac{3}{10} \times 100\% == 30\%$

2- USB frequency is $fc + fm = 10^6 + 10^3 = 1.001\, MHz$,
LSB frequency is $fc - fm = 10^6 - 10^3 = 0.999\, MHz$

3- Amp. of USB & LSB is

$$\frac{mAc}{4} = \frac{0.3 \times 10}{4} = 0.75\, V$$

4- $\eta = \frac{Ps}{Pt} \times 100\% = \frac{m^2}{m^2+2} = \frac{0.3^2}{0.3^2+2} = 4.3\ \%$

AM/SSB

AM/SSB, or Amplitude Modulation with Single Sideband, is another amplitude modulation (AM) variant used in analog communication systems. It is designed to transmit the modulating signal with a reduced bandwidth by eliminating one of the sidebands.

In standard AM, the modulating signal is used to vary the amplitude of a carrier wave, resulting in the generation of the upper sideband (USB) and lower sideband (LSB) around the carrier frequency. These sidebands contain duplicate information, making the transmission less efficient regarding bandwidth utilization.

AM/SSB eliminates one of the sidebands (either USB or LSB) along with the carrier wave, transmitting only the remaining sideband. This is achieved through a process called modulation with suppressed carrier. The process involves the following steps:
The modulating signal is mixed (multiplied) with a carrier wave.
The resulting product is passed through a bandpass filter that selects either the USB or the LSB.
The filtered signal, containing only the selected sideband, is then amplified and transmitted.

At the receiver end, the AM/SSB signal is demodulated by mixing it with a local oscillator of the same frequency as the carrier. This process produces the original modulating signal, which can be further processed and played back.

The main advantage of AM/SSB over standard AM is its efficient bandwidth utilization. Eliminating one of the sidebands and the carrier achieves significant bandwidth reduction compared to standard AM. This makes it suitable for long-distance communication, especially in limited bandwidth.

AM/SSB is commonly used in applications such as long-range radio communication, amateur radio, and shortwave broadcasting. It efficiently utilizes the limited available frequency spectrum while maintaining good signal quality over long distances.

AM/VSB (Vestigial sideband)

AM/VSB, or Amplitude Modulation with Vestigial Sideband, is a modulation technique used in analog communication systems. It is a amplitude modulation (AM) variant that aims to conserve bandwidth while maintaining compatibility with standard AM receivers. In standard AM, the modulating signal is used to vary the amplitude of a carrier wave. This modulation process produces two sidebands, an upper sideband (USB) and a lower sideband (LSB), symmetrically positioned around the carrier frequency. These sidebands contain modulating signal information.

In AM/VSB, the sidebands are not transmitted fully. Instead, one sideband is significantly attenuated while the other is transmitted with some reduced amplitude. This reduction in amplitude is typically achieved by using a bandpass filter with a specific frequency response, known as a vestigial sideband filter.

The vestigial sideband filter removes a portion of one sideband, either the upper or the lower while allowing a portion to pass. The filter's design ensures that most of the modulating signal information is contained in the transmitted sideband, while the attenuated sideband helps recover the carrier signal at the receiver.

The advantage of AM/VSB over standard AM is that it requires less bandwidth since only a portion of one sideband is transmitted. This allows more channels to be accommodated within the available frequency spectrum. Additionally, the transmitted power is more efficiently utilized since one sideband is suppressed.

AM/VSB has been widely used in analog television broadcasting, specifically in the National Television Systems Committee (NTSC) standard used in North America and Japan. The NTSC television signal uses vestigial sideband modulation to transmit video and audio information.

It's important to note that with the advent of digital modulation techniques and the transition to digital television broadcasting, AM/VSB has become less prevalent and is being replaced by more advanced modulation schemes such as quadrature amplitude modulation (QAM) and orthogonal frequency-division multiplexing (OFDM).

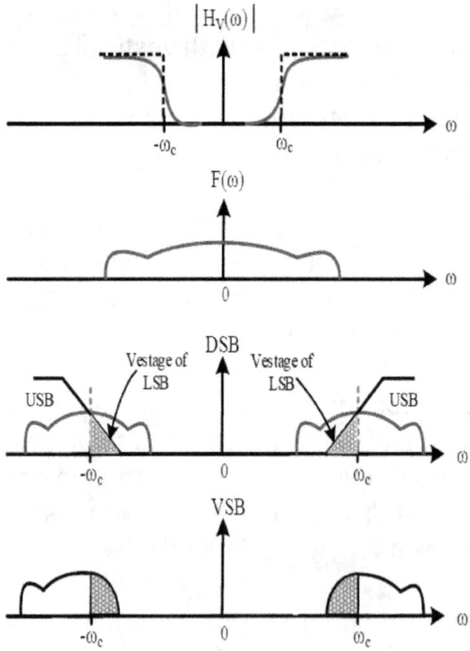

8.11 Angle modulation: (FM/PM)

Angle Modulation is a type of CW modulation in which the angle of a sinusoidal signal (either frequency or phase) is in proportion to the message where the amplitude is constant. It gives a high degree of noise immunity by band expansion; it is also widely used in high-fidelity music broadcasts and has a constant envelope.

$\emptyset(t) \; = \; A cos\, \theta(t)$ "Angle Modulation
$\emptyset(t) \; = \; f(t) cos\,(\omega c\, t \; + \; \theta o\,)$ "Amplitude Modulation"

Where $(\omega c\, t \; + \; \theta o\,)$ represents the angle of carrier frequency
Angle modulation refers to a category of modulation techniques used in communication systems where the phase or frequency of a carrier wave is varied in proportion to the modulating signal. The two primary forms of angle modulation are frequency modulation (FM) and phase modulation (PM).

8.11.1 Frequency Modulation (FM)

FM involves varying the instantaneous frequency of the carrier wave in response to the modulating signal. The amplitude of the carrier remains constant while the frequency deviates around its nominal value. The frequency deviation is directly proportional to the amplitude of the modulating signal. FM is less susceptible to noise and offers improved signal quality compared to amplitude modulation (AM). It is widely used in FM radio broadcasting, two-way radio communication, and audio signal transmission.

For FM

Where kf is constant $(rad/sec/volt)$

$$\omega i\,(t) \; = d\theta(i)/dt = \; \omega c \; + \; kf f(t)$$

$$\theta(t) \; = \; \omega c\, t \; + \; kf \int \; f(\propto) d \; \propto$$

$$\emptyset FM\,(t) \; = \; Ac\, cos\,[\,\omega c\, t \; + \; kf \int \; f(\propto) d \; \propto]$$

8.11.2 Phase Modulation (PM)

PM involves changing the phase of the carrier wave based on the modulating signal. The amplitude and frequency of the carrier remain constant while the phase continuously shifts. The amplitude of the modulating signal determines the degree of phase shift. PM is often used with FM and is the underlying principle behind many digital modulation techniques, such as phase-shift keying (PSK) and quadrature amplitude modulation (QAM).

For PM

$$\theta(t) = \omega c\, t + \theta o + kpf(t)$$

Where $kp$is constant (rad/volt)

$$\theta(t) = \omega c\, t + kpf(t)$$

$$\emptyset PM(t) = Ac\, cos[\omega c\, t + kpf(t)]$$

The instantaneous frequency $\omega i\ (t)$ is:

$$\omega i\ (t) = d\theta(i)/dt = \omega c\, t + kpf'(t)$$

Angle modulation techniques have several advantages over amplitude modulation. They provide a better signal-to-noise ratio, which results in improved resistance to noise and interference. Angle modulation is also more bandwidth-efficient than AM since the transmitted signal does not vary in amplitude significantly.

The demodulation of angle-modulated signals requires specialized receivers designed for FM or PM detection. These receivers employ frequency discriminators or phase detectors to recover the original modulating signal from the received signal.

Angle modulation is widely utilized in various communication systems, including broadcast radio, television, wireless, satellite, and digital data transmission. Its robustness, superior audio quality, and efficient bandwidth use make it a preferred choice for many applications in modern communication technology.

Chapter 9

Digital Signal Processing

Digital signal processing (DSP) is a branch of electrical engineering that deals with the processing of digital signals using various mathematical algorithms and techniques. DSP is used in various applications, including speech and audio processing, image and video processing, control systems, and telecommunications. The primary goal of DSP is to analyze, filter, and manipulate digital signals to improve their quality, extract useful information, or compress them for storage or transmission. DSP algorithms can be implemented using software or hardware and can be optimized for various performance criteria, such as speed, accuracy, and memory usage. Some of the key techniques used in DSP include:

Fourier transform: This mathematical technique converts a signal from the time domain to the frequency domain. It is used to analyze the frequency content of a signal, and it is commonly used in signal filtering and compression.

Digital filtering: This involves the application of digital filters to a signal to remove unwanted noise or interference or to extract specific features of the signal.

Sampling and quantization: These convert analog signals to digital signals. Sampling involves measuring the signal at regular intervals, while quantization involves representing each sample with a finite number of bits.

Signal compression involves reducing the data required to represent a signal while preserving its important features. Compression techniques, such as audio and video compression, are commonly used in multimedia applications.

Adaptive filtering: This involves adjusting the filter parameters based on the signal's characteristics. Adaptive filtering is used in applications where the signal characteristics may change over time.

DSP has numerous applications, including speech and audio processing, image and video processing, biomedical signal processing, control systems, and telecommunications. It is critical in modern technologies such as smartphones, digital cameras, and streaming media. As digital signal processing evolves, new techniques and applications will emerge.

9.1 Some Fundamental Sequences

Three simple discrete-time signals frequently represent and describe more complicated signals. These are the ***unit sample***, the ***unit step***, and the ***exponential***. The unit sample,

Impulse Response:

denoted by $\delta(n)$, is defined by $\delta(n) = \begin{cases} 1 & n = 0 \\ 0 & otherwise \end{cases}$

Figure 9.1 shows the generation of the unit impulse discrete function.

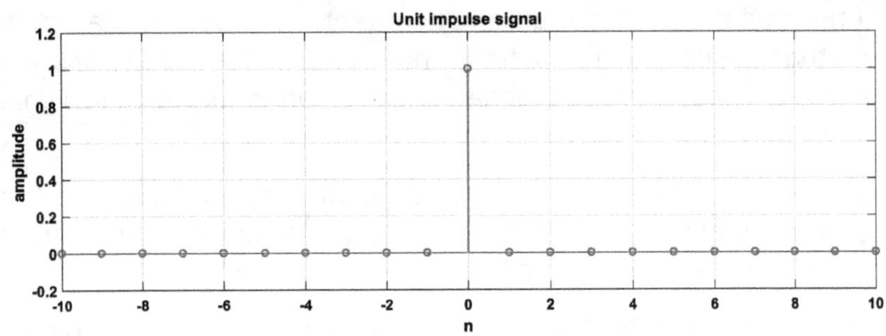

Figure 9.1 Generation of unit impulse discrete function.

The unit step, denoted by $u(n)$, is defined by

$$u(n) = \begin{cases} 1 & n \geq 0 \\ 0 & otherwise \end{cases}$$

and is related to the unit sample by

$$u(n) = \sum_{k=-\infty}^{n} \delta(n)$$

Figure 9.2 shows the generation of the unit impulse discrete function.

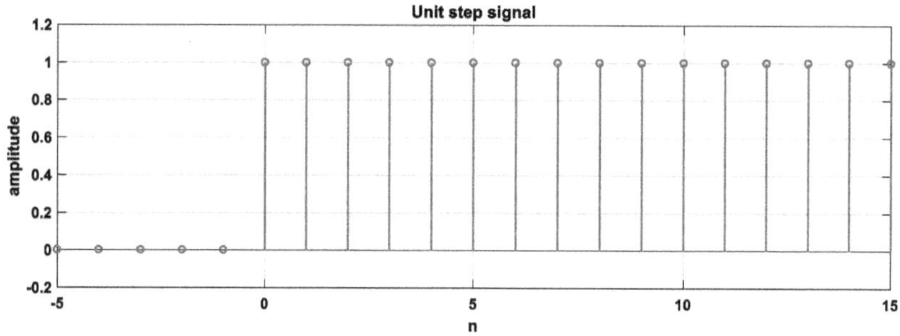

Figure 9.2 Generation of unit step discrete function.

Similarly, a unit sample is written as a difference of two steps:

$$\delta(n) = u(n) - u(n-1)$$

Finally, an exponential sequence is defined by

$$x(n) = a^n$$

Where *a* is a real or complex number. Of particular interest is the exponential sequence that formed when $a = e^{-j\omega_o}$ Where ω_o is a real number. In this case, *x*(*n*) is a complex exponential

$$e^{jn\omega_0} = \cos(n\omega_0) + j\sin(n\omega_0)$$

An infinite-length sequence *x*(*n*) is said to be *left-sided* if, for some integer n_o, $x(n) = 0$ for all $n > n_o$. An example of a left-sided sequence is

$$x(n) = u(n_0 - n) = \begin{cases} 1 & n \le n_0 \\ 0 & n > n_0 \end{cases}$$

Which is a time-reversed and delayed unit step. An infinite-length signal that is neither right-sided nor left-sided, such as the complex exponential, is called a *two-sided sequence*.

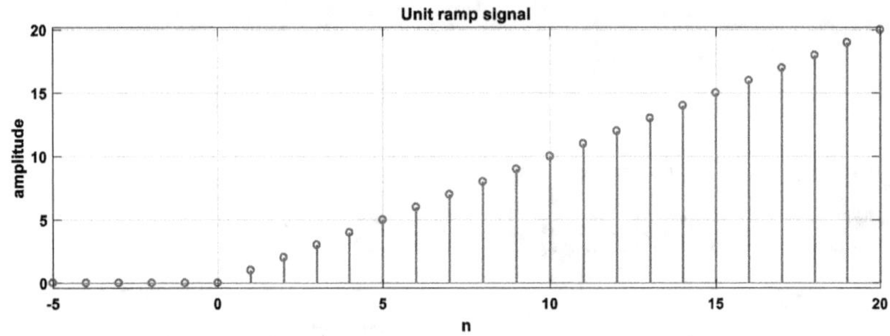

Figure 9.3 Generation of unit ramp discrete function.

Figure 9.4 shows the generation of exponential sequence a <1 discrete function.

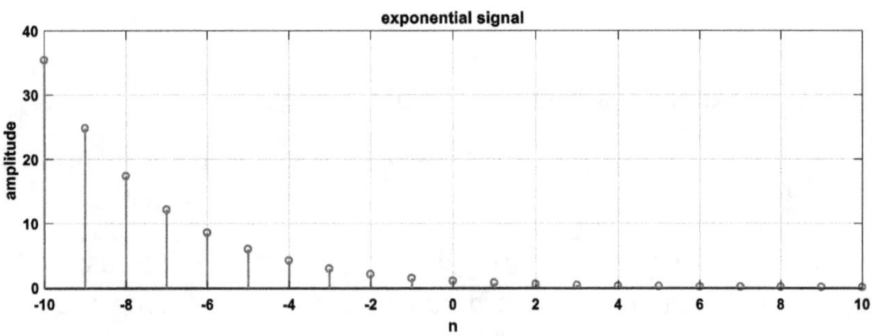

Figure 9.4 shows the generation of exponential sequence a <1 discrete function.

Figure 9.5 shows the generation of exponential sequence a >1 discrete function.

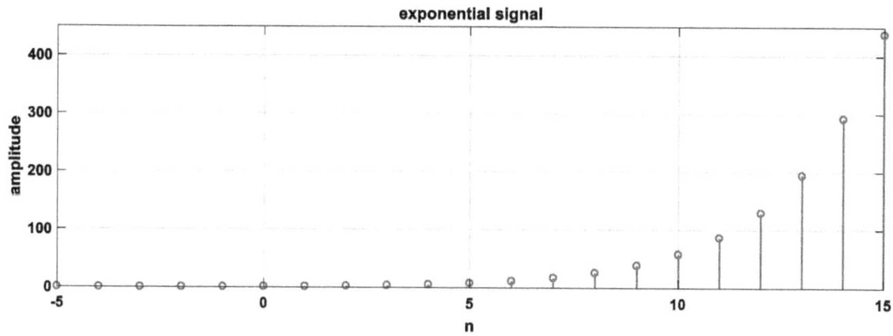

Figure 9.5 Generation of exponential sequence a>1 discrete function.

Figure 9.6 shows the generation of the exponential sequence a=1 discrete function.

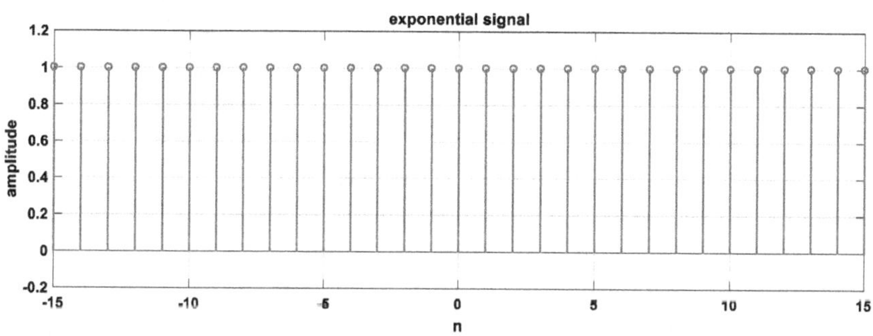

Figure 9.6 Generation of exponential sequence a=1 discrete function.

Figure 9.7 shows the generation of the discrete sine function.

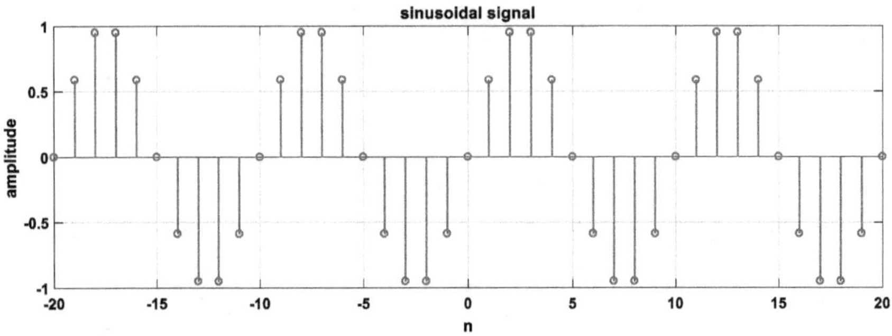

Figure 9.7 Generation of sine discrete function.

9.2 Z-Transform

The z-transform is useful for manipulating discrete data sequences and has acquired a new significance in formulating and analyzing discrete-time systems. It is used extensively today in applied mathematics, digital signal processing, control theory, population science, and economics. These discrete models are solved with different equations in a manner that is analogous to solving continuous models with differential equations. The role played by the z-transform in the solution of differential equations corresponds to that represented by the Laplace transforms in the solution of differential equations. In this chapter, the z-transform representation will present, also the region of convergence (ROC), the Properties of the z-Transform, and the Inverse z-Transform.

9.2.1 Z-Transform Representation

In the previous chapters, we saw that the discrete-time Fourier transform (DTFT) of a sequence $x(n)$ Is equal to the sum:

$$X(e^{j\omega}) = \sum_{n=-\infty}^{\infty} x(n)\, e^{-jn\omega}$$

However, for this series to converge, the signal must be summable.

Then, the z-transform is a generalization of the DTFT that allows one to deal with such sequences and is defined as follows:

Definition: The z-transform of a discrete-time signal x(n) is characterized by:

$$X(z) = \sum_{n=-\infty}^{\infty} x(n)\, z^{-n}$$

In other words, the relation between $X(e^{j\omega})$ and $X(z)$ is:

$$X(e^{j\omega}) = X(z)|_{z=e^{j\omega}}$$

where $z = e^{j\omega}$ is a complex variable.

According to the fact that for most situations, the digital signal $x(n)$ is the causal sequence, that is, $x(n) = 0$ for $n < 0$, then $X(z)$ will be:

$$X(z) = \sum_{n=0}^{\infty} x(n)\, z^{-n}$$

Thus, the definition in the above equation is referred to as a one-sided z-transform or a unilateral transform.

Example

Prove that the convolution in the time domain implies convolution in the z-domain, i.e.,

$$\boldsymbol{Z\big(x_1(n) * x_2(n)\big) = Z\big(x_1(n)\big)\, Z\big(x_2(n)\big) = X_1(z)X_2(z)}$$

Solution:

According to the definition of convolution, we have

$$x(n) = x_1(n) * x_2(n) = \sum_{k=0}^{\infty} x_1(n-k)x_2(k)$$

Taking z-transform

$$X(n) = Z(x_1(n) * x_2(n)) = Z\left(\sum_{k=0}^{\infty} x_1(n-k)x_2(k)\right)$$

$$= \sum_{n=0}^{\infty}\left(\sum_{k=0}^{\infty} x_1(n-k)x_2(k)\right)z^{-n}$$

It was written as

$$X(n) = \sum_{k=0}^{\infty} x_2(k)\, z^{-k} \,.\, \sum_{n=0}^{\infty} x_1(n-k)\, z^{-(n-k)}$$

Let $n - k = m$, and then the second summation can be written as

$$X(n) = \sum_{k=0}^{\infty} x_2(k)\, z^{-k} \,.\, \sum_{m=-k}^{\infty} x_1(m)\, z^{-m}$$

Using the causality of both sequences, the second summation can be started from $m = 0$ instead of $m = -k$. Therefore, using the definition of z-transform, we have

$$X(n) = X_1(n)\,.X_2(n)$$

Hence, proved.

9.2.2 Region of Convergence (ROC)

In the above Equation, all the values of z that make the summation exist (the sum converged) form a region of convergence in the z-transform domain, while all other values of *z* outside the region of convergence will cause the summation to diverge. The region of convergence is defined based on the particular sequence $x(n)$ Being applied. Note that we deal only with the unilateral *z*-transform. We are restricted to the causal sequence when performing inverse z-transform (which we shall study later).

Example

Given the sequence $x(n) = u(n)$. Find the z-transform of $x(n)$.0

Solution:

The z-transform is given by

$$X(z) = \sum_{n=0}^{\infty} u(n)\, z^{-n} \;\; = \sum_{n=0}^{\infty} (z^{-1})^n$$

It is an infinite geometric series that converges to

$$X(z) = \frac{z}{z-1}$$

with a condition $|z^{-1}| < 1.$

The region of convergence for all values of z is given as $|z| > 1$

Example

Considering the exponential sequence $x(n) \;= a^n \; u(n)$

Find the z-transform of the sequence $x(n)$.

Solution:

From the definition of the z-transform, it follows that.

$$X(z) = \sum_{n=0}^{\infty} a^n\, u(n)\, z^{-n} \;\; = \sum_{n=0}^{\infty} (az^{-1})^n$$

This geometric series will converge for $|az^{-1}| < 1$

$$X(z) = \frac{z}{z-a} \;\; , \quad for \quad |z| > |a|$$

The z-transform for common sequences is summarized in Table 9.1.

Table 9.1 Table of z-transform pairs (for causal sequences)

No.	Signal $x(n), \quad n \geq 0$	z-Transform $Z\big(x(n)\big) = X(z)$	Region of Convergence
1	$x(n)$	$\sum_{n=0}^{\infty} x(n)\, z^{-n}$	
2	$\delta(n)$	1	Entire z-plane
3	$a\, u(n)$	$\frac{az}{z-1}$	$\lvert z\rvert > 1$
4	$n\, u(n)$	$\frac{z}{(z-1)^2}$	$\lvert z\rvert > 1$
5	$n^2\, u(n)$	$\frac{z(z+1)}{(z-1)^3}$	$\lvert z\rvert > 1$
6	$a^n\, u(n)$	$\frac{z}{z-a}$	$\lvert z\rvert > \lvert a\rvert$
7	$e^{-na}\, u(n)$	$\frac{z}{z-e^{-a}}$	$\lvert z\rvert > e^{-a}$
8	$n\, a^n\, u(n)$	$\frac{az}{(z-a)^2}$	$\lvert z\rvert > \lvert a\rvert$
9	$\sin(an)\, u(n)$	$\frac{z\sin(a)}{z^2 - 2z\cos(a) + 1}$	$\lvert z\rvert > \lvert 1\rvert$
10	$\cos(an)\, u(n)$	$\frac{z\left(z - \cos(a)\right)}{z^2 - 2z\cos(a) + 1}$	$\lvert z\rvert > \lvert 1\rvert$
11	$a^n \sin(bn)\, u(n)$	$\frac{[a\sin(b)]\, z}{z^2 - [2a\cos(b)]z + a^2}$	$\lvert z\rvert > \lvert a\rvert$
12	$a^n \cos(bn)\, u(n)$	$\frac{z\,[z - a\cos(b)]}{z^2 - [2a\cos(b)]z + a^2}$	$\lvert z\rvert > \lvert a\rvert$
13	$e^{-an} \sin(bn)\, u(n)$	$\frac{[e^{-a}\sin(b)]\, z}{z^2 - [2e^{-a}\cos(b)]z + e^{-2a}}$	$\lvert z\rvert > e^{-a}$
14	$e^{-an} \cos(bn)\, u(n)$	$\frac{z\,[z - e^{-a}\cos(b)]}{z^2 - [2e^{-a}\cos(b)]z + e^{-2a}}$	$\lvert z\rvert > e^{-a}$

15	$2\lvert A\rvert\lvert P\rvert^n\, cos(n\theta + \phi)\; u(n)$ where P and A are complex constants defined by $P = \lvert P\rvert\angle\theta, A = \lvert A\rvert\angle\phi$	$\frac{Az}{z-P} + \frac{A^*z}{z-P^*}$	

Example

Find the z-transform for each of the following sequences $x(n)$.

a) $x(n) = 2.5\, u(n)$
b) $x(n) = 2sin(\frac{\pi n}{4})$
c) $x(n) = (0.8)^n\, u(n)$
d) $x(n) = (0.8)^n\, sin(\frac{\pi n}{4})\, u(n)$
e) $x(n) = e^{-0.5n} cos\left(\frac{\pi n}{4}\right) u(n)$

Solution:

a) $x(n) = 2.5\, u(n)$

From Table 9.1

$$X(z) = Z[x(n)] = Z[2.5\, u(n)] = 2.5\, \frac{z}{z-1}$$

b) $x(n) = 2sin\left(\frac{\pi n}{4}\right)$

From Table 9.1

$$X(z) = Z[x(n)] = Z\left[2sin\left(\frac{\pi n}{4}\right)\right] = 2\, \frac{z\, sin\left(\frac{\pi}{4}\right)}{z^2 - 2z\, cos\left(\frac{\pi}{4}\right) + 1}$$

b) $x(n) = (0.8)^n\, u(n)$

$$X(z) = Z[x(n)] = Z[(0.8)^n u(n)] = \frac{z}{z-0.8}$$

$$x(n) = (0.8)^n\, sin\left(\frac{\pi n}{4}\right)\, u(n)$$

$$X(z) = Z[x(n)] = Z\left[(0.8)^n sin\left(\frac{\pi n}{4}\right) u(n)\right]$$

$$= \frac{0.8\, z\, sin\left(\frac{\pi}{4}\right)}{z^2 - 2 * 0.8\, z\, cos\left(\frac{\pi}{4}\right) + 0.8^2}$$

a) $x(n) = e^{-0.5n} cos\left(\frac{\pi n}{4}\right) u(n)$

$$X(z) = Z[x(n)] = Z\left[e^{-0.5n} cos\left(\frac{\pi n}{4}\right) u(n)\right]$$

$$= \frac{z\left[z - e^{-0.5}\, cos\left(\frac{\pi}{4}\right)\right]}{z^2 - 2 * e^{-0.5} z\, cos\left(\frac{\pi}{4}\right) + (e^{-0.5})^2}$$

9.2.3 Properties of the z-Transform

In this section, we study some important properties of the z-transform. These properties are widely used in driving the z-transform functions of difference equations and solving the system output responses of linear digital systems with constant system coefficients.
For the following

$$Z\{f[n]\} = \sum_{n=0}^{n=\infty} f[n] z^{-n} = F(z)$$

$$Z\{g_n\} = \sum_{n=0}^{n=\infty} g_n z^{-n} = G(z)$$

- ***Linearity***:

$Z\{af_n + bg_n\} = aF(z) + bG(z)$.

and ROC is $R_f \cap R_g$ which follows from the definition of z-transform.

- ***Time Shifting***

If we have $f[n] \Leftrightarrow F(z)$ then $f[n-n_0] \Leftrightarrow z^{-n_0}F(z)$

The ROC of $Y(z)$ is the same as $F(z)$ except that there are possible pole additions or deletions at $z = 0$ or $z = \infty$.

Proof:

Let $y[n] = f[n-n_0]$then

$$Y(z) = \sum_{n=-\infty}^{\infty} f[n-n_0]\, z^{-n}$$

Assume $k = n- n_0$ then $n=k+n_0$, substituting in the above equation we have:

$$Y(z) = \sum_{k=-\infty}^{\infty} f[k]\, z^{-k-n_0} = z^{-n_0}F[z]$$

- ***Multiplication by an Exponential Sequence***

Let $y[n] = z_0^n f[n]$ then $Y(z) = X\left(\frac{z}{z_0}\right)$

The consequence is pole and zero locations are scaled by z_0. If the ROC of $FX(z)$ is $r_R < |z| < r_L$, then the ROC of $Y(z)$ is

$$r_R < |z/z_0| < r_L, \text{ i.e., } |z_0|r_R < |z| < |z_0|rL$$

Proof:

$$Y(z) = \sum_{n=-\infty}^{\infty} z_0^n x[n] z^{-n} = \sum_{n=-\infty}^{\infty} x[n]\left(\frac{z}{z_0}\right)^{-n} = X\left(\frac{z}{z_0}\right)$$

The consequence is the pole, and zero locations are scaled by z_0. If the ROC of $X(z)$ is $rR<|z|< rL$, then the ROC of $Y(z)$ is

$rR < |z/z_0| < rL$, *i.e.*, $|z_0|rR < |z| < |z_0|rL$

- ***Differentiation of X(z)***

If we have $f[n] \Leftrightarrow F(z)$ then $nf[n] \xleftrightarrow{Z} -z\frac{dF(z)}{z}$ and ROC = R_f

Proof:

$$F(z) = \sum_{n=-\infty}^{\infty} f[n]z^{-n}$$

$$-z\frac{dF(z)}{dz} = -z\sum_{n=-\infty}^{\infty} -n\,f[n]z^{-n-1} = \sum_{n=-\infty}^{\infty} -n\,f[n]z^{-n}$$

$$-z\frac{dF(z)}{dz} \xleftrightarrow{Z} nf[n]$$

- **Conjugation of a Complex Sequence**

If we have $f[n] \Leftrightarrow F(z)$ then $f^*[n] \xleftrightarrow{Z} F^*(z^*)$ and ROC = R_f

Proof:

Let $y[n] = f^*[n]$, then

$$Y(z) = \sum_{n=-\infty}^{\infty} f^*[n]z^{-n} = \left(\sum_{n=-\infty}^{\infty} f[n][z^*]^{-n}\right)^* = F^*(z^*)$$

Time reversal

If we have $f[n] \Leftrightarrow F(z)$ then $f^*[-n] \xleftrightarrow{Z} F^*(1/z^*)$

Let $y[n] = f^*[-n]$, then

$$Y(z) = \sum_{n=-\infty}^{\infty} f^*[-n]z^{-n} = \left(\sum_{n=-\infty}^{\infty} f[-n][z^*]^{-n}\right)^*$$

$$= \left(\sum_{k=-\infty}^{\infty} f[k](1/z^*)^{-k}\right)^* = F^*(1/z^*)$$

If the ROC of $F(z)$ is $r_R < |z| < r_L$, then the ROC of $Y(z)$ is

$r_R < |1/z^*| < r_L$ *i.e.,* $\frac{1}{r_R} > |z| > \frac{1}{r_L}$

When the time reversal is without conjugation, it is easy to show

$$f[-n] \xleftrightarrow{Z} F(1/z)$$

and ROC is $\frac{1}{r_R} > |z| > \frac{1}{r_L}$

A comprehensive summary of the *z*-transform properties is shown in Table 9.2.

Table 9.2 Summary of *z*-transform properties

No.	Property	Sequence	z-transform
1	Addition	$x_1(n) + x_2(n)$	$X_1(z) + X_2(z)$
2	Constant multiple	$c\,x(n)$	c X(z)
3	Linearity	$ax_1(n) + bx_2(n)$	$aX_1(z) + bX_2(z)$
4	Delay unit step	u(n-m)	$\frac{Z^{1-m}}{Z-1}$

5	Time delay shift	$x(n-m)$ U(n − m)	$Z^{-m}\,X(z)$
6	Forward one tap	*x(n+1)*	Z(X(z)- $x(0)$)
7	Forward m tap	*x(n+1)*	$Z^{m}\left(X(z) - \sum_{i=0}^{m-1} x_i\,Z^{-i}\right)$
8	Complex translation	$e^{n}\,x(n)$	X(ze^{-1})
9	Frequency scale	$b^{n}\,x(n)$	X(zb^{-1})
10	Differentiation	$n\,x(n)$	-Z X'(z)
11	Conjugation	$x^{*}(n)$	$X^{*}(z^{*})$
12	Time reversal	$x(-n)$	$X(z^{-1})$
13	Integration	$\frac{1}{n}\,x(n)$	$-\int \frac{X(z)}{Z}\,dz$
14	Discrete-time convolution	$x_1(n) * x_2(n)$	$X_1(z)\,X_2(z)$
15	Initial time	$x(0)$	$\lim_{n\to\infty} X(z)$
16	Final value	$\lim_{n\to\infty} x(n)$	$\lim_{n\to 1}(Z-1)X(z)$

Example

Find the *z*- transform of $x(n) = 3n + 2 \times 3^{n}$.

Solution:

From the linearity property

$$Z\,[\,x(n)] = Z\,[3n + 2 \times 3^{n}] = 3Z\,[n] + 2\,Z\,[3^{n}]$$

$$Z\{n\} = \frac{z}{(z-1)^2}$$

and

$$Z\{3^n\} = \frac{z}{(z-3)}$$

Therefore

$$Z\,[x(n)] = \frac{3z}{(z-1)^2} + \frac{2z}{(z-3)}$$

Example

Find the z-transform of each of the following sequences:

(a) ***x*(*n*) = 2*ⁿu*(*n*) + 3(½)*ⁿu*(*n*)**

(b) ***x*(*n*) = cos(5*n*)*u*(*n*).**

Solution:

(a) Because $x(n)$ is a sum of two sequences of the form $\alpha^n u(n)$, using the linearity property of the z-transform, and referring to Table 9.2, the z-transform pair

$$X(z) = \frac{1}{1-2z^{-1}} + \frac{3}{1-\frac{1}{2}z^{-1}} = \frac{4-\frac{13}{2}z^{-1}}{(1-2z)\left(1-\frac{1}{2}z^{-1}\right)}$$

(b) For this sequence, we write

$x(n) = \cos(5n)\, u(n) = ½(e^{j5n} + e^{-j5n})\, u(n)$

Therefore, the *z*-transform is

$$X(z) = \frac{1}{2}\frac{1}{1-e^{j5n}z^{-1}} + \frac{1}{2}\frac{1}{1-e^{-j5n}z^{-1}}$$

With a region of convergence $|z| > 1$. Combining the two terms, we have

$$X(z) = \frac{1-z^{-1}\cos 5}{1-2z^{-1}\cos 5 + z^{-2}}$$

Example

Find the z-transform of the sequence given by

$x(n) = u(n) - (0.85)^n u(n)$

Solution

$X(z) = Z[x(n)] = Z[u(n)] - Z[(0.85)^n u(n)]$

$$= \frac{z}{z-1} - \frac{z}{z-0.85}$$

9.2.4 Inverses z-transform

The z-transform of the sequence x(n)and the inverse z-transform of the function x(z) are defined as, respectively,

$$\boldsymbol{x(n) = \frac{1}{2\pi j}\oint X(z) z^{n-1} dz}$$

The circular symbol on the integral sign denotes a closed counter in the complex plane.

The *z*-transform is a useful tool in linear systems analysis. However, just as important as techniques for finding the z-transform of a sequence are methods that may be used to invert the z-transform and recover the sequence ***x(n)*** from ***X(z)***. Three possible approaches are described below. At least three methods may obtain the inverse z-transform:

1. Partial fraction expansion and look-up table
2. Power series expansion
3. Residue method

Partial fraction expansion and a look-up table

Now we are ready to deal with the inverse z-transform using the partial fraction

Expansion and look-up table. The general procedure is as follows:

1. Eliminate the negative powers of z for the z-transform function X (z)

2. Determine the rational function X(z)\z (assuming it is proper), and apply the partial fraction expanded function X(z)\z using the formula in Table 9.1.

3. Multiply the expanded function X(z)\z by z on both sides of the equation to obtain X(z).

4. Apply the inverse z- transform using Table 9.1

The partial fraction format and the formula for calculating the constants are listed in Table 9.2.

For *z*-transforms that are rational functions of *z*,

$$X(z) = \frac{\sum_{k=0}^{q} b(k)z^{-1}}{\sum_{k=0}^{p} a(k)z^{-1}} = C\,\frac{\prod_{k=0}^{q}(1-\beta_k z^{-1})}{\prod_{k=1}^{p}(1-\beta_k z^{-1})}$$

A simple approach to finding the inverse *z*-transform is to perform a partial fraction expansion of *X*(*z*). Assuming that p > *q* and that all of the roots in the denominator are simple, $\alpha_i \neq \alpha_k$ for $i \neq k$, *X*(*z*) expanded as follows:

$$X(z) = \sum_{k=1}^{p} \frac{A_k}{1-\alpha_k z^{-1}}$$

for some constants A_k for $k = 1,2, \ldots p$. The coefficients A_k may be found by multiplying both sides of the above equation by $(1 - \alpha_k z^{-1})$ and setting $z = \alpha_k$. The result is

$$A_k = [(1 - \alpha_k z^{-1})X(z)]$$

If $p \leq q$, the partial fraction expansion must include a polynomial in z^{-1} of order $(p\text{-}q)$. The coefficients of this polynomial are found by long division (i.e., by dividing the numerator polynomial by the denominator). For multiple-order poles, the expansion must be modified. For example, if $X(z)$ has a second-order pole at $z = \alpha_k$, the expansion will include two terms,

$$\frac{B_1}{1 - \alpha_k z^{-1}} + \frac{B_2}{(1 - \alpha_k z^{-1})^2}$$

where B_1 and B_2 are given by

$$B_1 = \alpha_k \left[\frac{d}{dz}(1 - \alpha_k z^{-1})^2 \, X(z)\right]$$

$$B_2 = [(1 - \alpha_k z^{-1})^2 \, X(z)]$$

Example

Find the inverse of the following z-transform

$$\boldsymbol{X(z) = \frac{1}{(1 - z^{-1})(1 - 0.8z^{-1})}}$$

Solution:

Multiplying the numerator and the denominator by $\mathbf{z^2}$ we get

$$\boldsymbol{X(z) = \frac{1}{(1 - z^{-1})(1 - 0.8z^{-1})} \times \frac{z^2}{z^2} = \frac{z^2}{(z - 1)(z - 0.8)}}$$

Dividing both sides by $\boldsymbol{z}$, we have

$$\frac{X(z)}{z} = \frac{z}{(z-1)(z-0.8)}$$

We notice that the right-hand side of the above equation is a proper rational polynomial of z. Also, we see that the denominator of the right-hand side has distinct poles. Therefore, we right into partial fractions as,

$$\frac{X(z)}{z} = \frac{A}{(z-1)} + \frac{B}{(z-0.8)}$$

To find out the unknown constants A and B, we use:

$$A = \left[(z-1) \times \frac{X(z)}{z}\right]_{z=1} = \left[(z-1) \times \frac{z}{(z-1)(z-0.8)}\right]_{z=1}$$
$$= \left[\frac{z}{(z-0.8)}\right]_{z=1} = \frac{1}{(1-0.8)} = 5$$

$$B = \left[(z-0.8) \times \frac{X(z)}{z}\right]_{z=0.8} = \left[(z-0.8) \times \frac{z}{(z-1)(z-0.8)}\right]_{z=0.8}$$
$$= \left[\frac{z}{(z-1)}\right]_{z=0.8} = \frac{0.8}{(0.8-1)} = -4$$

Substituting the values we have,

$$\frac{X(z)}{z} = \frac{5}{(z-1)} + \frac{(-4)}{(z-0.8)}$$

Or it can be written as (by multiplying both sides by z)

$$X(z) = \frac{5z}{(z-1)} - \frac{4}{(z-0.8)}$$

Taking the inverse z-transform of both sides and using Table 9.1 , we have

$$x(n) = Z^{-1}(X(z)) = 5Z^{-1}\left(\frac{z}{(z-1)}\right) - 4Z^{-1}\left(\frac{z}{(z-0.8)}\right)$$
$$= 5\,u(n) - 4(0.8)^n\,u(n)$$

Example Suppose that a sequence *x*(*n*) has a *z*-transform

$$X(z) = \frac{4 - \frac{7}{4}z^{-1} + \frac{1}{4}z^{-2}}{1 - \frac{3}{4}z^{-1} + \frac{1}{8}z^{-2}}$$

Solution:

$$X(z) = \frac{4 - \frac{7}{4}z^{-1} + \frac{1}{4}z^{-2}}{1 - \frac{3}{4}z^{-1} + \frac{1}{8}z^{-2}} = \frac{4 - \frac{7}{4}z^{-1} + \frac{1}{4}z^{-2}}{(1 - \frac{1}{2}z^{-1})(1 - \frac{1}{4}z^{-1})}$$

With a region of convergence $|z| > ½$. Because $p = q = 2$, and the two poles are simple, the partial fraction expansion has the form

$$X(z) = K_1 + \frac{K_2}{(1 - \frac{1}{2}z^{-1})} + \frac{K_3}{(1 - \frac{1}{4}z^{-1})}$$

The constant K_1 is found by long division and is equal to 2 .K2 and K3 are equal to 3 and -1, respectively.

Thus, the complete partial fraction expansion becomes

$$X(z) = 2 + \frac{3}{(1 - \frac{1}{2}z^{-1})} - \frac{1}{(1 - \frac{1}{4}z^{-1})}$$

Finally, because the region of convergence is the exterior of the circle $|z| > 1$, $x(n)$ is the right-sided sequence

$$x(n) = 2\delta(n) + 3(0.5)^n u(n) - (0.25)^n u(n)$$

Power Series

The z-transform is a power series expansion,

$$X(z) = \sum_{n=}^{\infty} x(n) z^{-n} = \cdots + x(-2)z^2 + x(-1)z^1 + x(0) + x(1)z^{-1} + x(2)z^{-2} + \cdots$$

Where the sequence values $x(n)$ are the coefficients of z^{-n} in the expansion, therefore, if we can find the power series expansion for $X(z)$, the sequence values $x(n)$ may be found by simply picking off the coefficients of z^{-n}.

Example

Consider the z-transform

$$X(z) = \log\left(\frac{Z + c}{Z}\right) \qquad |Z| > |c|$$

Solution:

The power series expansion of this function is

$$X(z) = \log\left(\frac{Z + c}{Z}\right) = \log(1 + c\,Z^{-1}) \qquad |Z| > |c|$$

$$= \sum_{n=1}^{\infty} \frac{1}{n} (-1)^{n+1}\, c^n\, Z^{-n} \qquad n > 0$$

Therefore, the sequence x(n) having this z-transform is

$$x(n) = \begin{cases} \frac{1}{n}(-1)^{n+1} c^n & n > 0 \\ 0 & otherwise \end{cases}$$

Contour Integration

Contour integration is another approach that may be used to find the inverse z-transform of X(z). This procedure relies on Cauchy's integral theorem, which states that if *C* is a closed contour that encircles the origin in a counterclockwise direction,

$$\frac{1}{2\pi j} \oint_C z^{-k} dz = \begin{cases} 1 & k = 1 \\ 0 & k \neq 1 \end{cases}$$

With

$$X(z) = \sum_{n=-\infty}^{\infty} x(n) z^{-n}$$

Cauchy's integral theorem may be used to show that the coefficients *x*(*n*) may be found from *X*(*z*) as follows:

$$x(n) = \frac{1}{2\pi j} \oint_C X(z) z^{n-1} dz$$

C is a closed contour within the region of convergence of *X*(*z*) that encircles the origin counterclockwise. Contour integrals of this form may often be evaluated with the help of Cauchy's residue theorem,

$$x(n) = \frac{1}{2\pi j} \oint_C X(z) z^{n-1} dz$$
$$= \sum (residues\ of X(z) z^{n-1}\ at\ the\ poles\ inside\ C)$$

If *X*(*z*) is a rational function of z with a first-order pole at $z = \alpha_k$,

$$x(n) = \frac{1}{2\pi j} \oint_C X(z) z^{n-1} dz$$
$$= \sum (residues\ of X(z) z^{n-1}\ at\ the\ poles\ inside\ C)$$

$$Res[X(z)z^{n-1}\ at\ z\ =\ \alpha_k] = [(1 - z^{-1}\ \alpha_k) X(z) z^{n-1}]$$

Contour integration is particularly useful if only a few x(n) values are needed.

Example

Find the inverse of each of the following *z*-transforms:

i. $X(z) = 1 + 2(Z^2 + Z^{-2}) \qquad 0 < |z| < \infty$

ii. $X(z) = \frac{1}{1-0.5Z^{-1}} + \frac{1}{1-0.2Z^{-1}} \qquad 0.5 < |z|$

iii. $X(z) = \frac{1}{1+2Z^{-1}+Z^{-2}} \qquad |z| > 2$

iv. $X(z) = \frac{1}{(1-Z^{-1})(1-Z^{-2})} \qquad |z| > 1$

Solution:

i. Because *X*(*z*) is a finite-order polynomial, *x*(*n*) is a finite-length sequence. Therefore, *x*(*n*) is the coefficient that multiplies z^{-1} in *X*(*z*). Thus, *x*(0) = 4 and *x*(2) = *x*(-2) = 3.

ii. This *z*-transform is a sum of two first-order rational z functions. Because the region of convergence of *X*(*z*) is the exterior of a circle, *x*(*n*) is a right-sided sequence. Using the *z*-transform pair for a right-sided exponential, we may invert *X*(*z*) easily as follows:

$$\boldsymbol{x(n) = (0.5)^n\ u(n) + 3\left(\frac{1}{3}\right)^n\ u(n)}$$

iii. Here, we have a rational function of z with a quadratic denominator in z. Before we can find the inverse z-transform, we need to factor the denominator and perform a partial fraction expansion:

$$X(z) = \frac{1}{1 + 2Z^{-1} + Z^{-2}} = \frac{1}{(1 + 2Z^{-1})(1 + Z^{-1})}$$

$$X(z) = \frac{1}{(1 + 2Z^{-1})} - \frac{1}{(1 + Z^{-1})}$$

Because $x(n)$ is right-sided, the inverse z-transform is

$$\boldsymbol{x(n) = 2(-2)^n\, u(n) - (-1)^n\, u(n)}$$

v. One way to invert this z-transform is to perform a partial fraction expansion. With

$$X(z) = \frac{1}{(1 - Z^{-1})(1 - Z^{-2})}$$

The solution is given as follows:

$$\boldsymbol{x(n) = 0.25[1 + (-1)^n + 2(n + 1)]u(n)}$$

Example

Find the inverse z-transform of the second-order system.

$$X(z) = \frac{1+0.25Z^{-1}}{(1-0.5Z^{-1})^2} \qquad |z| > 0.5$$

Here we have a second-order pole at $z = ½$. The partial fraction expansion for $X(z)$ is

$$X(z) = \frac{K_1}{(1 - \frac{1}{2}z^{-1})} + \frac{K_2}{\left(1 - \frac{1}{2}z^{-1}\right)^2}$$

The constants K_1 and K_2 are -0.5 and 1.5, respectively, so

$$X(z) = \frac{-0.5}{(1 - \frac{1}{2}z^{-1})} + \frac{1.5}{\left(1 - \frac{1}{2}z^{-1}\right)^2}$$

and

$$\boldsymbol{x(n) = -(0.5)^{n+1}\, u(n) + 3(n+1)(0.5)^{n+1}\, u(n)}$$

Example

Find the inverse z-transform of $X(z) = \sin z$.

Solution

To find the inverse z-transform of $X(z) = \sin z$, we expand $X(z)$ in a Taylor series about $z = 0$ as follows:

Because

$$X(z) = \sum_{n=-\infty}^{\infty} x(n)z^{-n}$$

So

$$X(z) = X(z)|_{z=0} + z\frac{dX(z)}{dz}|_{z=0} + \frac{z^2}{2!}\frac{d^2X(z)}{dz^2}|_{z=0} + \cdots + \frac{z^n}{n!}\frac{d^nX(z)}{dz^n}|_{z=0} + \cdots$$

$$= z - \frac{z^3}{3!} + \frac{z^5}{5!} - \cdots = \sum_{n=0}^{\infty}(-1)^n\frac{z^{2n+1}}{(2n+1)!}$$

Because

$$X(z) = \sum_{n=-\infty}^{\infty} x(n)z^{-n}$$

We may associate the coefficients in the Taylor series expansion with the sequence values *x*(*n*). Thus, we have

$$x(n) = \sum_{n=0}^{\infty} (-1)^n \frac{z^{2n+1}}{(2n+1)!}$$

$$= (-1)^n \frac{1}{(2|n|+1)!} \qquad n = -1, -3, -5, \ldots$$

Example

Evaluate the following integral:

$$\frac{1}{2\pi j} \oint_C \frac{1 + 2Z^{-1} - Z^{-2}}{(1 - 0.5Z^{-1})(1 - \frac{2}{3}Z^{-1})} z^3 dz$$

Where the contour of integration *C* is the unit circle.

Solution:

Recall that for a sequence *x*(*n*) that has a *z*-transform *X*(*z*). The sequence may be recovered using contour integration as follows:

$$x(n) = \frac{1}{2\pi j} \oint_C X(z) z^{n-1} dz$$

Therefore, the integral that is to be evaluated corresponds to the value of the sequence *x*(*n*) at *n* = 4 that has a *z*-transform

$$X(z) = \frac{1 + 2Z^{-1} - Z^{-2}}{(1 - 0.5Z^{-1})(1 - \frac{2}{3}Z^{-1})}$$

Thus, we may find *x*(*n*) using a partial fraction expansion of *X*(*z*) and then evaluate the sequence at *n* = 4. With this approach, however, we are finding the x(n) values for all n. Alternatively, we could perform a long division and divide the X(z) numerator by the denominator. The coefficient multiplying z^{-4} would then be the value of *x*(*n*) at *n* = 4 and the integral value. However, because we are only interested in the sequence value at *n* = 4, the easiest approach is to evaluate the integral directly using the Cauchy integral theorem. The integral value is equal to the sum of the residues of the X(z)z3 poles inside the unit circle. Because

$$X(z)z^3 = \frac{1 + 2Z^{-1} - Z^{-2}}{(1 - 0.5Z^{-1})(1 - \frac{2}{3}Z^{-1})} z^3$$

has poles at z =1/2 and z =2/3,

$$REs[X(z)z^3]|_{z=0.5} = -\frac{3}{16}$$

And

$$REs[X(z)z^3]|_{z=2/3} = \frac{112}{81}$$

Therefore, we have

$$\frac{1}{2\pi j}\oint_C X(z)z^{n-1}dz = -\frac{3}{16} + \frac{112}{81} = 1.195$$

Example

find out the z-transform for the following sequences:

a. $x(n) = 15\,u(n)$
b. $x(n) = 10\,\sin(0.25\pi n)\,u(n)$
c. $x(n) = (0.5)^n\,u(n)$
d. $x(n) = (0.5)^n\,\sin(0.25\pi n)u(n)$

e. $x(n) = e^{-0.1n}\cos(0.25\pi n)\,u(n)$

Solution:

(a) From Table 9.1, we get

$$X(z) = Z(x(n)) = Z(15\,u(n)) = \frac{15}{z-1}$$

(b) From Table 9.1, we obtain

$$X(z) = Z(x(n)) = Z(10\sin(0.25\pi n)\,u(n))$$
$$= \frac{10\sin(0.25\pi)\,z}{z^2 - 2z\cos(0.25\pi n) + 1} = \frac{7.07\,z}{z^2 - 1.414z + 1}$$

(c) From Table 9.1, we get

$$X(z) = Z(x(n)) = Z((0.5)^n\,u(n)) = \frac{z}{z-0.5}$$

(d) From Table 9.1, we get

$$X(z) = Z(x(n)) = Z((0.5)^n\sin(0.25\pi n)u(n))$$
$$= \frac{0.5\times\sin(0.25\pi)\,z}{z^2 - 2\times 0.5\times z\cos(0.25\pi n) + (0.5)^2}$$
$$= \frac{0.3536\,z}{z^2 - 1.4142z + 0.25}$$

(e) From Table 9.1, we get

$$X(z) = Z(x(n)) = Z(e^{-0.1n}\cos(0.25\pi n)\,u(n))$$
$$= \frac{z\left(z - e^{-0.1}\cos(0.25\pi)\right)}{z^2 - 2\,z\,e^{-0.1}\cos(0.25\pi) + (e^{-0.1})^2}$$
$$= \frac{z(z-0.6397)}{z^2 - 1.279\,z + 0.8187}$$

Example

Find the inverse of the following z-transform,

$$\boldsymbol{X}(\mathrm{z}) = \frac{1}{(1-\mathrm{z}^{-1})(1-0.5\mathrm{z}^{-1})}$$

Eliminating the negative power of z^2 by multiplying the number or determining by

$$X(z) = \frac{\mathrm{z}^2}{\mathrm{z}^2(1-\mathrm{z}^{-1})(1-0.5\mathrm{z}^{-1})}$$

Dividing both sides by z leads to

$$\frac{X(z)}{z} = \frac{z}{(z-1)(z-0.5)}$$

$$= \frac{A}{(z-1)} + \frac{B}{(z-0.5)}$$

$$A = \lim_{z=1} \frac{z}{(z-0.5)} = 2$$

$$B = \lim_{z=0.5} \frac{z}{(z-1)} = 1$$

$$x(n) = Z^{-1}(X(\mathrm{z})) = Z^{-1}\left(\frac{z}{(z-1)}\right) + Z^{-1}\left(\frac{z}{(z-0.5)}\right)$$
$$= u(n) + (0.5)^n\, u(n)$$

Example

Find the inverse of the following z-transform

i.

$$X(z) = 1 + \frac{z}{(z-1)} + \frac{z}{(z-0.8)}$$

ii.

$$X(z) = \frac{2z}{(z-1)^2} + \frac{5z}{(z-0.8)^2}$$

iii.

$$X(z) = \frac{z^{-6}}{z+1} + \frac{z^{-5}}{z-0.7} + z^{-4}$$

Solution:

Dividing both sides by $\boldsymbol{z}$, we have

i.

$$\boldsymbol{X(z) = 1 + \frac{z}{(z-1)} + \frac{z}{(z-0.8)}}$$

$$\boldsymbol{x(n) = Z^{-1}\big(X(z)\big) = Z^{-1}(1) + Z^{-1}\left(\frac{z}{(z-1)}\right) + Z^{-1}\left(\frac{z}{(z-0.8)}\right)}$$
$$\boldsymbol{= \delta(n) + u(n) + (0.8)^n\, u(n)}$$

ii.

$$\boldsymbol{X(z) = \frac{2z}{(z-1)^2} + \frac{5z}{(z-0.8)^2}}$$

$$\boldsymbol{x(n) = Z^{-1}\big(X(z)\big) = Z^{-1}\left(\frac{2z}{(z-1)^2}\right) + Z^{-1}\left(\frac{5z}{(z-0.8)^2}\right)}$$
$$\boldsymbol{= 2\, n\, u(n) + 5\, n\, (0.8)^n\, u(n)}$$

$$\boldsymbol{= 2\, r(n) + 5\ (0.8)^n\, r(n)}$$

iii.

$$\boldsymbol{X(z) = \frac{z^{-6}}{z+1} + \frac{z^{-5}}{z-0.7} + z^{-4}}$$

$$\boldsymbol{x(n) = Z^{-1}(X(z)) = Z^{-1}\left(\frac{z^{-6}}{z+1}\right) + Z^{-1}\left(\frac{z^{-5}}{z-0.7}\right) + Z^{-1}(z^{-4})}$$

$$\boldsymbol{x(n) = Z^{-1}\left(z^{-7}\frac{z}{z+1}\right) + Z^{-1}\left(z^{-6}\frac{z}{z-0.7}\right) + Z^{-1}(z^{-4})}$$

$$\boldsymbol{x(n) = u(n-7) + (0.7)^n u(n-6) + \delta(n-4)}$$

Example

Determine the convolution of the following two sequences using z-transform,

$$x_1(n) = 3\delta(n) + 2\delta(n-1)$$
$$x_2(n) = 2\delta(n) - \delta(n-1)$$

Solution:

Taking the z-transform of the two sequences, we have

$$X_1(z) = Z\big(3\delta(n) + 2\delta(n-1)\big) = 3 + 2z^{-1}$$
$$X_2(z) = Z\big(2\delta(n) - \delta(n-1)\big) = 2 - z^{-1}$$

Using the z-transform property for thc convolution of two sequences, we have

$$X(z) = X_1(z)\, X_2(z) = (3 + 2z^{-1})(2 - z^{-1})$$

Therefore,

$$X(z) = 6 + z^{-1} - 2z^{-2}$$

Taking the inverse z-transform of both sides (and using the shift theorem), we have

$$x(n) = Z^{-1}\big(X(z)\big) = 6Z^{-1}(1) + Z^{-1}(z^{-1}) - 2Z^{-1}(z^{-2})$$
$$= 6\delta(n) + \delta(n-1) - 2\delta(n-2)$$

Example

Given a transfer function depicting a DSP system

$$H(z) = \frac{z-1}{z+0.5}$$

Determine

(a) The impulse response $h(n)$.
(b) The step response $s(n)$.
(c) The system response $y(n)$, if the input is given as $x(n) = (0.5)^n\, u(n)$.

Solution

Part (a): In this case $x(n) = \delta(n)$, thus $X(z) = 1$. As

$$H(z) = \frac{Y(z)}{X(z)}$$

Therefore, in this case, the z-transform of the output is equal to the transfer function:

$$H(z) = Y(z)$$

By taking the inverse z-transform of the transfer function, we can find out the unit impulse response $h(n)$, of the system. The transfer function is written as

$$\frac{H(z)}{z} = \frac{z-1}{z(z+0.5)}$$

This can further be written in the form of partial fractions as

$$\frac{H(z)}{z} = \frac{A}{z} + \frac{B}{(z+0.5)}$$

where

$$A = \left.\frac{z-1}{(z+0.5)}\right|_{z=0} = \frac{0-1}{(0+0.5)} = -2$$

$$B = \frac{z-1}{z}\bigg|_{z=0.5} = \frac{0.5-1}{0.5} = -1$$

Thus we have

$$\frac{H(z)}{z} = \frac{-2}{z} + \frac{-1}{(z+0.5)}$$

Or

$$H(z) = -2 - \frac{z}{(z-0.5)}$$

Taking the inverse z-transform of both sides (and using Table 9.1), we get

$$h(n) = -2\,\delta(n) - (0.5)^n\,u(n)$$

Which is the required impulse response of the system.

Part (b): In this case $x(n) = u(n)$, thus $X(z) = \frac{z}{z-1}$. As

$$H(z) = \frac{Y(z)}{X(z)}$$

Therefore, in this case,

$$Y(z) = H(z)\,X(z) = \frac{(z-1)}{(z+0.5)}\frac{z}{(z-1)}$$

It was written as

$$\frac{Y(z)}{z} = \frac{z-1}{(z+0.5)(z-1)} = \frac{1}{(z+0.5)}$$

Taking inverse z-transform of both sides, we get

$$y(n) = (-0.5)^n\,u(n)$$

Which is the required step response of the system.

Part (c): In this case $x(n) = (0.25)^n\, u(n),\ X(z) = \frac{z}{z-0.25}$

$$H(z) = \frac{Y(z)}{X(z)}$$

Therefore, in this case,

$$Y(z) = H(z)X(z) = \frac{(z-1)}{(z+0.5)} \cdot \frac{z}{(z-0.25)}$$

It was written as

$$\frac{Y(z)}{z} = \frac{z-1}{(z+0.5)(z-0.25)}$$

further can be written in the form of partial fractions as

$$\frac{Y(z)}{z} = \frac{A}{(z+0.5)} + \frac{B}{(z-0.25)}$$

where

$$A = \left.\frac{z-1}{(z-0.25)}\right|_{z=-0.5} = \frac{-0.5-1}{(-0.5-0.25)} = 2$$

$$B = \left.\frac{z-1}{(z+0.5)}\right|_{z=0.25} = \frac{0.25-1}{0.25+0.5} = -1$$

Thus we have

$$\frac{Y(z)}{z} = \frac{2}{(z+0.5)} + \frac{-1}{(z-0.25)}$$

Or

$$Y(z) = \frac{2z}{(z+0.5)} - \frac{z}{(z-0.25)}$$

Taking inverse z-transform of both sides, we get

$$y(n) = 2(-0.5)^n\, u(n) - (0.25)^n\, u(n)$$

9.3 Evaluation of LTI System Response Using Z-Transform

Figure 9.8 shows an LTI System Where h(n) is the impulse response of the system, H(z) is the transfer function, the input signal x(n), and the output response is y(n), the transfer function in z-domain given as

$$H(z) = H\left(e^{j\omega}\right) = \frac{Y(z)}{X(z)} \quad (10.1)$$

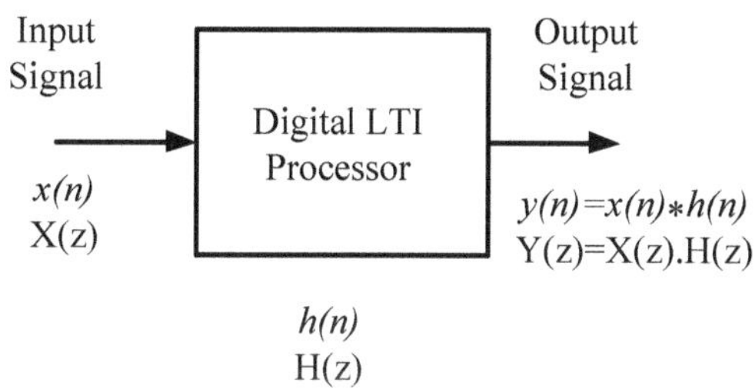

Figure 9.8 LTI processor system

9.4 Digital System Implementation from its Function

Since the z-transform is a linear transformation, the system implementation is similar to that in the time domain. The most convenient form for system synthesis is the z-transform of the general difference equation given by

$$Y(z) = \sum_{k=1}^{M} a_k z^{-k} Y(z) + \sum_{k=q}^{b} b_k z^{-k} X(z) (10.2)$$

1-Gain: Figure 9.9 shows a Gain block.where k is the value of gain.

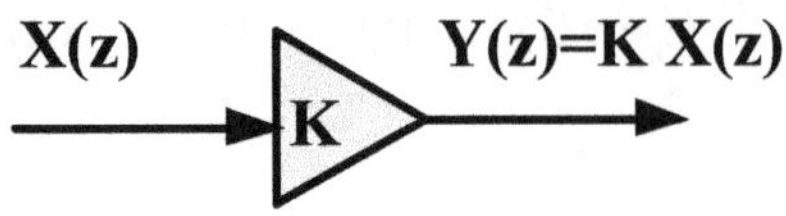

Figure 9.9 Gain block

2-Delay: Figure 9.10 shows a delay block.

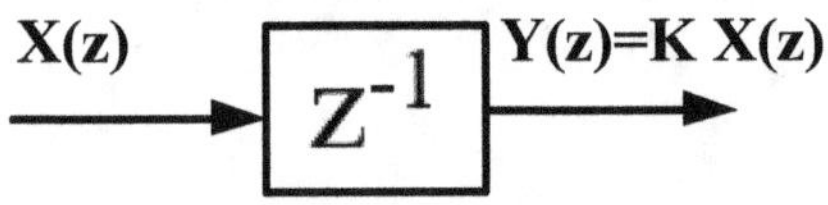

Figure 9.10 Delay block

3-Addition Figure 9.11 shows an addition block used to add two or more signals.

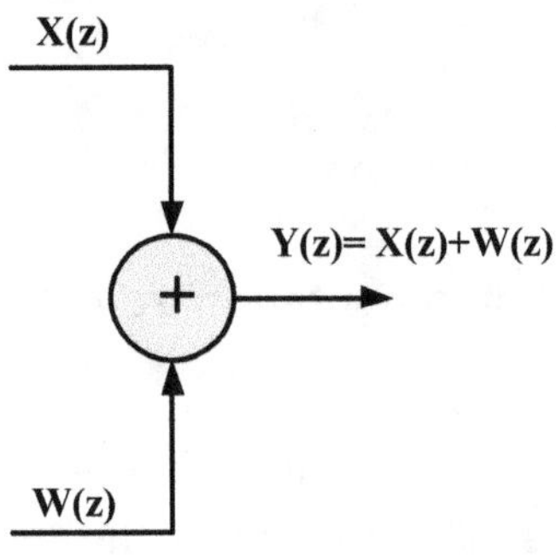

Figure 9.11 Addition block

Example

Find the impulse response and the transfer function of the following system, as shown in Figure 9.12

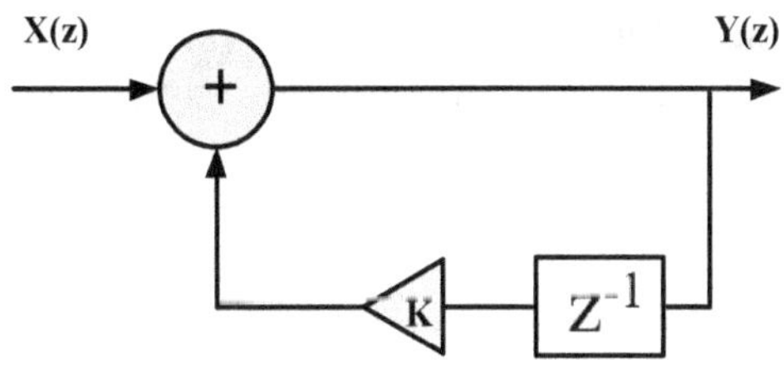

Figure 9.12 System configuration.

Solution

$$y(n) = x(n) + K\, y(n-1)$$

$$Y(z) = X(z) + K\, Y(z) z^{-1}$$

For impulse response $x(n) = \delta(n)$

X(z)=1

$$Y(z) = 1 + K\, Y(z) z^{-1}$$

$$Y(z) = \frac{1}{1 - K\, z^{-1}}$$

9.5 Types of Filters

There are four types of the filter according to the pass frequencies

- ***Low-Pass Filters:***

  Low-pass filters are designed to pass low frequencies, from zero to a certain cut-off frequency, and to block high frequencies.

- ***High-Pass Filters:***

  High-pass filters pass high frequencies and block low frequencies from a certain cut-off frequency.

- ***Band-Pass Filters:***

Band-pass filters are designed to pass a certain frequency range, which does not include zero and block other frequencies.

- ***Band-Stop Filters:***

  Band-stop filters block a certain frequency range that does not include zero and pass other frequencies.

Figure 9.13 shows the types of these filters.

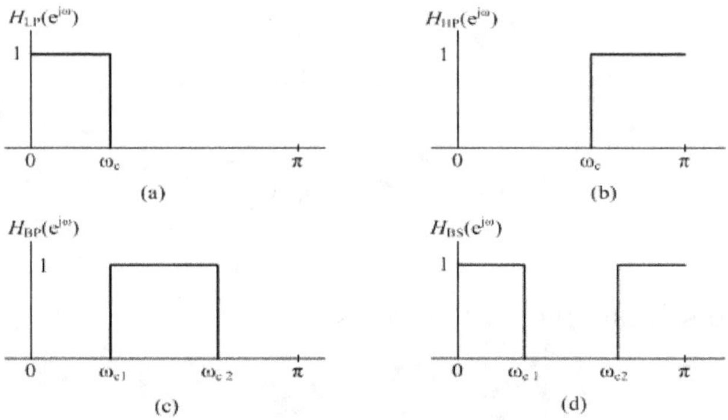

Figure 9.13 Types of filters

Example 9.14 plots the magnitude frequency response for the low pass filter given by the transfer function

$$H(z) = \frac{z}{(z - 0.25\ )}$$

Figure 9.14 shows an LPF Response.

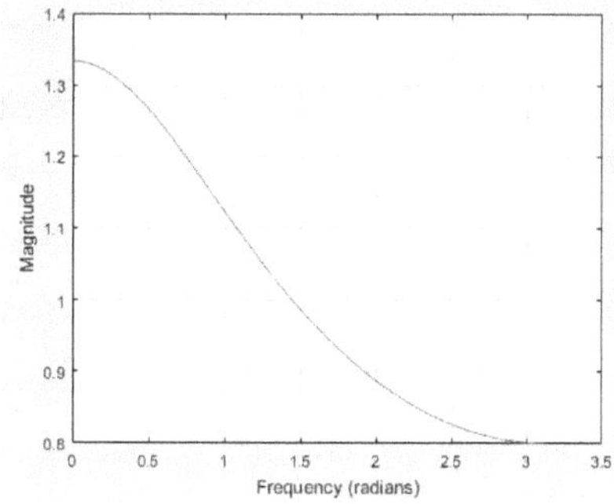

Figure 9.14 LPF response.

Example

Plots the magnitude frequency response for the high pass filter given by the transfer function

$$H(z) = 1 - 1.5\,z^{-1}$$

Solution

Figure 9.15 shows an HPF Response

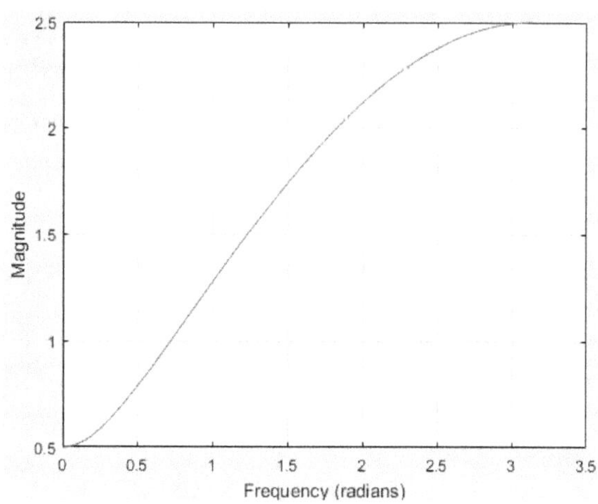

Figure 9.15 HPF response

Example

Plots the magnitude frequency response for the band-pass filter given by the transfer function

$$H(z) = \frac{0.5z^2 - 0.5}{z^2 - 0.25\,z + 0.25}$$

Solution

Figure 9.16 shows a Response of BPF

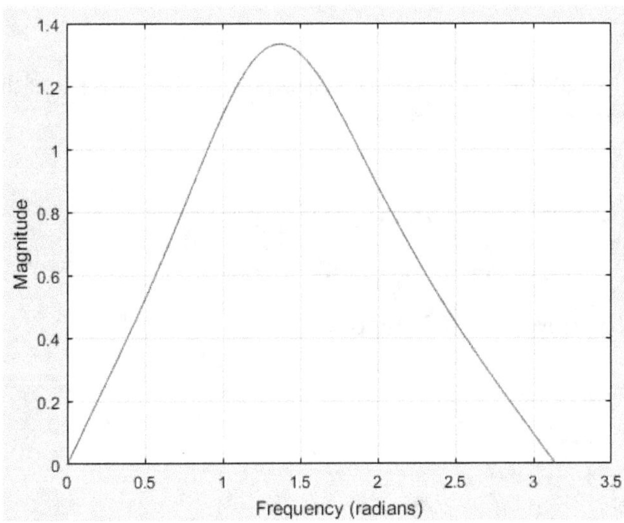

Figure 9.16 BPF response

Example

The magnitude frequency response for the stop pass filter given by the transfer function as

$$H(z) = \frac{1 - 0.9\,z^{-1} + 0.81\,z^{-2}}{1 - 0.6\,z^{-1} + 0.36\,z^{-2}}$$

Solution

Figure 9.17 shows an SPF Response.

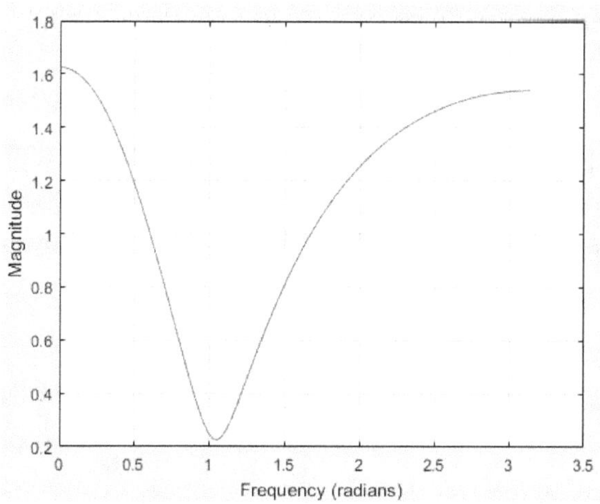

Figure 9.17 SPF response

Comparison of IIR and FIR digital filters

Difference between IIR and FIR

	IIR	**FIR**
efficiency	More than FIR.	Less.
Design	Complex in design than FIR.	Easy to design.
Phase Linearity	Difficult to control, non-linear.	linear phase always possible
Stability	It can be unstable because of Feedback.	Always stable(no Feedback)
Order	less	more
History	derived from analog filters	no analog history

9.6 Difference Equation and Transfer Function

Let The general difference equation given by

$$y(n) = b_0 x(n) + b_0 x(n-1) + \cdots + b_M x(n-M) - a_1 y(n-1) + \cdots - a_N y(n-N)$$

The supposition that all initial conditions of this system are zero, and the X(z) and Y(z) representing the z-transforms of the sequences x(n) and y(n), respectively, taking the z-transform of the equation above yields

$$Y(z) = b_0 X(z) + b_0 X(z) z^{-1} + \cdots + b_M X(z) z^{-M} - a_1 Y(z) z^{-1} + \cdots - a_N Y(z) z^{-N}$$

The z-transfer function H(z) is defined as the ratio of the z-transform of the output Y(z) to the z-transform of the input X(z).To get the transfer function Rearranging the equation to get the following

$$H(z) = \frac{Y(z)}{X(z)} = \frac{b_0 + b_1 z^{-1} + \cdots + b_M z^{-M}}{1 + a_1 z^{-1} + \cdots + a_N z^{-N}} = \frac{A(z)}{B(z)}$$

Where H(z) is defined as the transfer function with its numerator and denominator polynomials defined as

$$A(z) = b_0 + b_1 z^{-1} + \cdots + b_M z^{-M}$$

$$B(z) = 1 + a_1 z^{-1} + \cdots + a_N z^{-N}$$

The z-transfer function represents the digital filter in the z-domain, as shown in figure below:

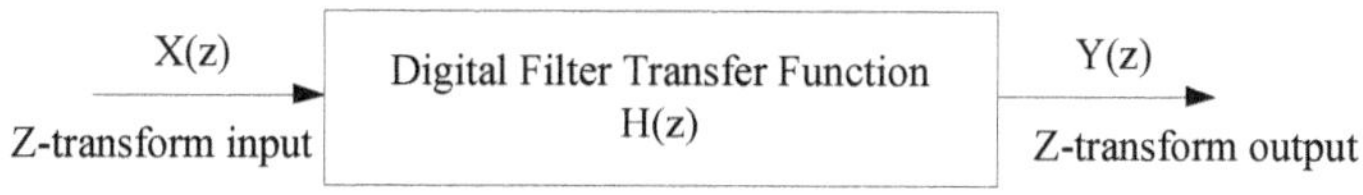

Figure 9.18 z-transfer function and represent the digital filter

Example

The following difference equation describes a DSP system.

$$y(n) = x(n) + x(n-1) - 1.2x(n-2) - 2y(n-1) - 0.8y(n-2)$$

Find the transfer function H(z), the Numerator polynomial A(z), and the denominator polynomial equation B(z).

Solution:

Moving the last two terms to the left side of the difference equation and factoring Y(z)

On the left side, X(z) on the right side, we obtain

$$y(n) = x(n) + x(n-1) - 1.2x(n-2) - 2y(n-1) - 0.8y(n-2)$$

$$Y(z) = X(z) + X(z)Z^{-1} - 1.2X(z)Z^{-2} - 2Y(z)Z^{-1} - 0.8Y(z)Z^{-2}$$

$$Y(z) + 2Y(z)Z^{-1} + 0.8Y(z)Z^{-2} = X(z) + X(z)Z^{-1} - 1.2X(z)Z^{-2}$$

$$[1 + 2Z^{-1} + 0.8\,Z^{-2}]Y(z) = [1 + Z^{-1} - 1.2Z^{-2}]\,X(z)$$

$$\frac{Y(z)}{X(z)} = \frac{1 + Z^{-1} - 1.2Z^{-2}}{1 + 2Z^{-1} + 0.8\,Z^{-2}}$$

$$\frac{Y(z)}{X(z)} = \frac{Z^2 + Z - 1.2}{Z^2 + 2Z + 0.8}$$

Example

The following difference equation describes a DSP system

$$y(n) = x(n) - 0.35x(n-1) + 1.2x(n-2))$$

Find the transfer function H(z), the Numerator polynomial A(z), and the denominator polynomial equation B(z).

Solution

Taking the z-transform on both sides of the previous difference equation, we achieve

$$y(n) = x(n) - 0.35x(n-1) + 1.2x(n-2))$$

$$Y(z) = X(z) - 0.35X(z)Z^{-1} + 1.2X(z)Z^{-2}$$

$$Y(z) = [1 - 0.35Z^{-1} + 1.2Z^{-2}]\,X(z)$$

$$\frac{Y(z)}{X(z)} = [1 - 0.35Z^{-1} + 1.2Z^{-2}]$$

Example

Given the following transfer function

$$H(z) = \frac{z^{-1} - 0.6\,z^{-2}}{1 + 1.2z^{-1} + 0.55z^{-2}}$$

Convert it into its pole-zero form.

Solution

We first multiply the numerator and denominator by z^2 to obtain the transfer function whose numerator and the denominator polynomials have the positive powers of z, as follows

$$H(z) = \frac{z^2(z^{-1} - 0.6\,z^{-2})}{z^2(1 + 1.2z^{-1} + 0.55z^{-2})}$$

$$H(z) = \frac{z - 0.6}{z^2 + 1.2z + 0.55}$$

The zero of H(z): $z - 0.6 = 0 \rightarrow z = 0.6$

The poles of H(z): $z^2 + 1.2z + 0.55 = 0 \rightarrow z = -0.6 \pm j0.4359$

The figure has seen the poles inside the unit circle, so the system is stable.

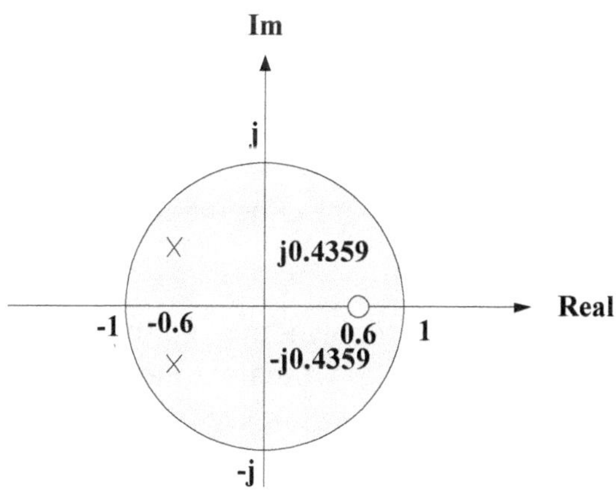

Figure 9.19 Z-plane representation

Example

Given the following transfer function

$$H(z) = \frac{z(1 - z^{-2})}{(1 + 1.3z^{-1} + 0.36z^{-2})}$$

Convert it into its pole-zero form.

Solution

We first multiply the numerator and denominator by z^2 to obtain the transfer function whose numerator and the denominator polynomials have the positive powers of z, as follows

$$H(z) = \frac{(1 - z^{-2})\, z^2}{(1 + 1.3z^{-1} + 0.36z^{-2})\, z^2} = \frac{z^2 - 1}{z^2 + 1.3z + 0.36}$$

Putting the numerator polynomial equal to zero and then finding the roots gives us the zeros of the transfer function,

$$z^2 - 1 = 0$$

$$(z - 1)(z + 1) = 0$$

Therefore, we get $z_1 = 1$ and $z_2 = -1$ as the roots.

Now, setting the denominator polynomial equal to zero and finding the roots gives us the poles of the transfer function,

$$z^2 + 1.3z + 0.36 = 0$$

$$z = \frac{-1.3 \pm \sqrt{(1.3)^2 - 4(1)(0.36)}}{2(1)} = \frac{-1.3 \pm \sqrt{1.69 - 1.44}}{2}$$
$$= \frac{-1.3 \pm \sqrt{0.25}}{2} = \frac{-1.3 \pm 0.5}{2} = -0.4, -0.9$$

Therefore, the poles are $p_1 = -0.4$ and $p_{21} = -0.9$. The transfer function can now be written in the pole-zero form as

$$\boldsymbol{H(z) = \frac{(z-1)(z+1)}{(z+0.4)(z+0.9)}}$$

9.7 Sampling Theorem

The sampling theorem states that a band-limited signal which has no frequency components higher than f_H Hz can be recovered completely from a set of samples taken at the rate of $f_s \square (\geq 2f_H)$ samples per second.

The above sampling theorem is also called the uniform sampling theorem for *baseband* or *low-pass signals*.

The minimum sampling rate is $2f_H$ samples per second are called the Nyquist rate; it is reciprocal $1/2f_H$, measured in seconds, is called the Nyquist interval.

In sampling, a strict distinction between baseband and passband signals is made as follows:

For *baseband* signals, $B \geq f_L \ \ Hz$

For *passband* signals, $B < f_L \ \ Hz$

Where B is the bandwidth of the signal $g(t)$, and

f_L is the lowest frequency of the signal $g(t)$.

9.7.1 Sampling Theory

The sample concept of sampling given as switch as shown in Figure 9.20, and the *sampled signal using switching sampling is given in Figure 9.21*

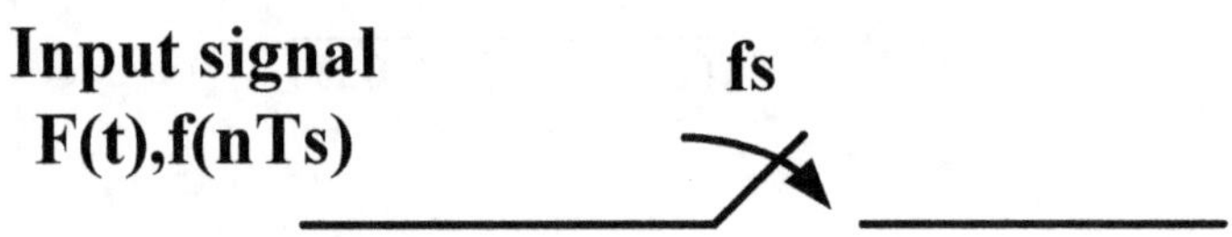

Figure 9.20 basic concept of sampling

fs=sampling frequency rate in sample/sec or Hz.

$$T_s = \frac{1}{f_s} \qquad (12.18)$$

Ts sampling period time (sec).

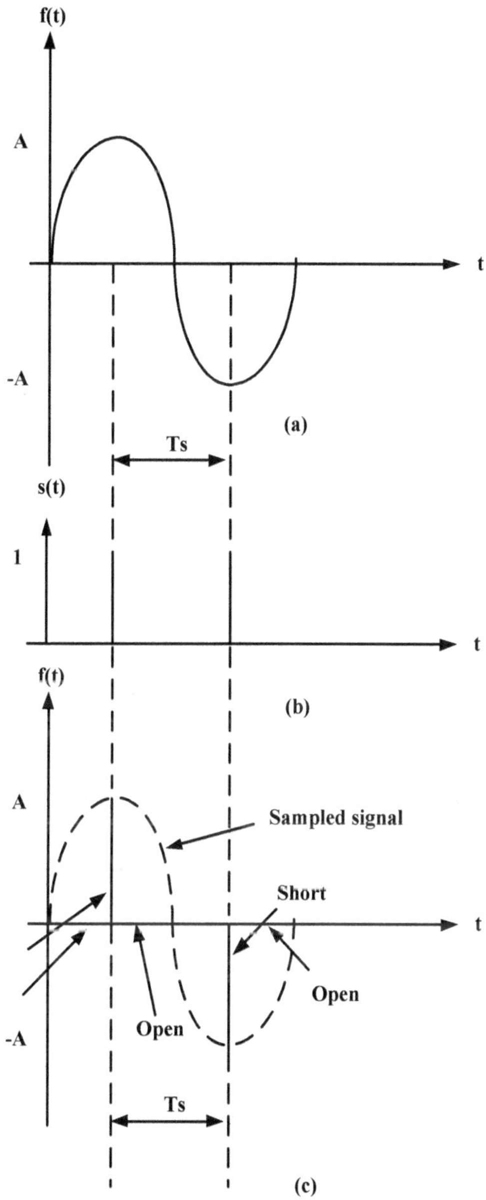

Figure 9.21 sampled signal using switching sampling

9.7.2 Nyquist Condition

$$f_s \geq 2f_m \qquad (12.19)$$

Nyquist rate frequency in sample/sec or Hz.

$$f_{s(\min)} = 2f_m \qquad (12.20)$$

Example

Calculate the sampling rate frequency for the signal $f(t) = 2\sin(2\pi \times 10^3 t)$

Solution

$$\omega_m = 2\pi \times 10^3 \frac{rad}{sec} = 2\pi f_m$$

$$f_m = 1kHz \rightarrow T_m = \frac{1}{f_m} = 1\,msec$$

$$f_s = 2\,f_m \rightarrow f_s = 2 \times 1 = 2\,msec$$

$$T_s = \frac{1}{f_s} = 0.5\ msec$$

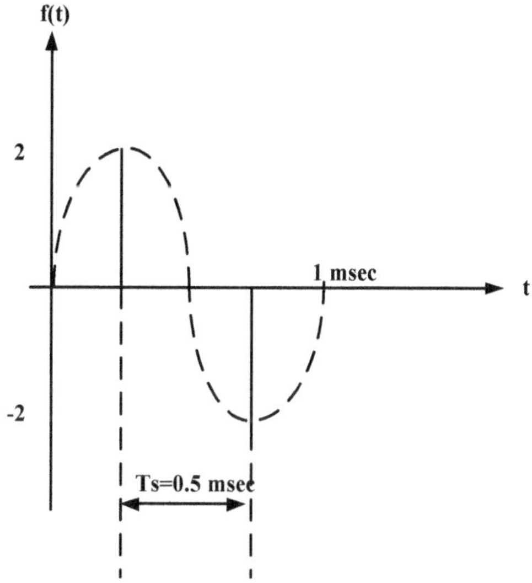

Figure 9.22 Sample signal using switching sampling of Example 12.1

Multiple choice questions

1. Determine the convolution sum of two sequences
2. x(n) = {3, 2, 1, 2} and h(n) = {1, 2, 1, 2}

   a) y(n) = {3,8,8,12,9,4,4}

   b) y(n) = {3,8,3,12,9,4,4}

   c) y(n) = {3,8,8,12,9,1,4}

   d) y(n) = {3,8,8,1,9,4,4}

Answer: a

3. Sampling theorem:

   a) fm<fs

   b) fs>fm

   c) fs>=2fm

   d) fs=2fm

Answer: c

3. Application of Convolution:

a) FIR Filtering

b) Addition

c) Manipulation

d) None of these

Answer: a

4. Condition for the aliasing problem:

a) fs<fm

b) fs<2fm

c) fs=fm

d) all of these

Answer: d

5. DFT stands as:

a. Discrete Fourier transform

b. digital function transform

c. digital frequency transform

d. none

Answer: a

7. The FFT algorithms:

a. eliminate the redundant calculation and enable analyzing a signal's spectral properties.

b. enable the redundant calculation and redundancy to analyze the spectral properties of a signal.

c. a & b

d. none.

Answer: a

8. The relation between DFT and Fourier series coefficients of a periodic sequence is

a. X(K) = Ck/N

b. X(K)= Ck

c. X(K) = NCk

d. X(K)=1/Ck

Answer: c

10. Drawbacks of DSP are

a. Digital processing needs pre and post-processing devices

b. high-cost

c. No memory storage

d. none of above

Answer: a

11. The minimum sampling frequency for *xa(t) is real with Xa(f) non-zero only for*

18 KHz < |f| < 22 KHz is

a. *8.8 KHz*

b. *9 KHz*

c. *11 KHz*

d. *17.6 KHz*

Answer: a

12. Find the response of an FIR filter with impulse response h(n)={1,2,4} to the input sequence

x(n)={1,2}.

a. y(n)={1,4,8,8}

b. y(n)={1,4,6,6}

c. y(n)={1,2,8,8}

d. none of above

Answer: a

13. IIR filters

a) Use feedback

b) Are sometimes called recursive filters

c) Can oscillate if not properly designed

d) all of the above

Answer: d

14. The output of two digital filters can be added. Or, the same effect can be achieved by

a) adding their coefficients

b) subtracting their coefficients

c) convolving their coefficients

d) averaging their coefficients, and then using a Blackman window

Answer: a

15. A DSP convolves each discrete sample with four coefficients, and they are all equal to 0.25. This must be a

a) low-pass filter

b) high-pass filter

c) band-pass filter

d) band-stop filter

Answer: a

16. Two digital filters can be operated in a cascade. Or, the same effect can be achieved by

a) adding their coefficients

b) subtracting their coefficients

c) convolving their coefficients

d) averaging their coefficients, and then using a Blackman window

Answer: c

17. FIR filters have ...., and IIR filters have ....

a. Zeros, poles & zeros

b. poles & zeros, Zeros

c. Zeros, zeros

d. none of above

Answer: a

18. Drawbacks FIR filters are:

a. more computation than an IIR with a similar effect.

b. prevent phase distortion

c. less computation

d. all of above

Answer: a

19. Calculate DFT of x (n) = {1, 0, 1, 0}.

a. x (k) = {2, 0, 2, 0}

b. x (k) = {1, 0, 1, 0}

c. x (k) – {2, 0, 1, 0}

d. none

Answer: a

20. Determine the convolution sum of two sequences x(n) = {3, 2, 1, 2} and h(n) = {1, 2, 1, 2}

a) y(n) = {3,8,8,12,9,4,4}

b)y(n) = {3,8,3,12,9,4,4}

c) y(n) = {3,8,8,12,9,1,4}

d) y(n) = {3,8,8,1,9,4,4}

Answer: a

21. The FFT algorithms:

a. eliminate the redundant calculation and enable the analysis of the spectral properties of a signal.

b. enable the redundant calculation and redundancy to analyze the spectral properties of a signal.

c. a & b

d. none

Answer: a

22. The minimum sampling frequency for *$x_a(t)$ is real with $X_a(f)$ non-zero only for*

9 KHz < $|f|$ <12 KHz is

a. 4.5 KHz

b. 6 KHz

c. 9 KHz

d. 12 KHz

Answer: b

23. The minimum sampling frequency for *$x_a(t)$ is complex with $X_a(f)$ non-zero only for*

30 KHz <$|f|$ < 35 KHz is

a. 6 KHz

b. 5 KHz

c. 15 KHz

d. 17.5 KHz

Answer: b

24. If a continuous-time filter with an impulse response $h_a(t)$ is sampled with a sampling frequency of f_S, what happens to the cut-off frequency w_C of the discrete-time filter as f_S is increased?

a. w_C increases

b. w_C decreases

c. w_C remains constant

d. w_C depends upon f_S

Answer: b

25. Fast convolution techniques:

a. overlap save

b. overlap-add

c. a & b

d. none of above

Answer: c

26. Find the response of an FIR filter with impulse response h(n)= {1,2,4} to the input sequence x(n)={1,2}.

a. y(n)={1,4,8,8}

b. y(n)={1,4,6,6}

c. y(n)={1,2,8,8}

d. none of above

Answer: a

27. A Blackman window can eliminate ripple in FIR filters. The tradeoff is

a) larger transition bandwidth

b) smaller transition bandwidth

c) a non-linear phase response

d) possible instability

Answer: a

28. The letter A below indicates the filter

a) stopband

b) transition band

c) passband

d) ripple

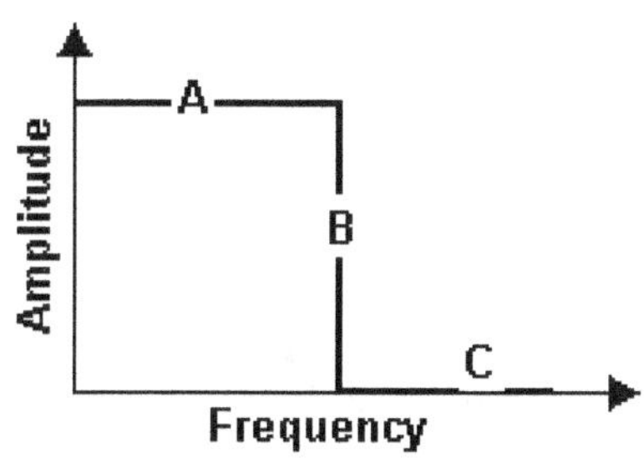

Answer: c

29. Coefficient symmetry is important in FIR filters because it provides

    a) a smaller transition bandwidth

    b) less passband ripple

    c) less stopband ripple

    d) a linear phase response

Answer: d

30. Which of the following is used to alter FIR filter coefficients, so they smoothly approach zero at both ends?

    a) rectangular window

    b) Blackman window

    c) Laplace window

    d) Hilbert window

Answer: b

31. FIR filters often require more computation because you must do .... for each term in the impulse response.

    a. a multiply-add

    b. add

    c. multiply

    d. all of above

Answer: d

32. Suppose that $x_a(t)$ is bandlimited to 8 kHz (that is, $X_a(f) = 0$ for $|f| > 8000$), then what is

the Nyquist rate for $x_a(t)$?

a. 16 KHz

b. 4 KHz

c. 8 KHz

d. 12 KHz

Answer: a

Chapter 10

Digital Communications

Digital communication systems are becoming increasingly attractive because of the ever-growing demand for data communication and digital transmission.

Advantages of Digital Communication

As the signals are digitized, there are many advantages of digital communication over analog communication, such as:

- The effect of distortion, noise, and interference is much less in digital signals as they are less affected.
- Digital circuits are more reliable.
- Digital circuits are easy to design and cheaper than analog circuits.
- The hardware implementation in digital circuits is more flexible than analog.
- The occurrence of cross-talk is very rare in digital communication.
- The signal is unaltered as the pulse needs a high disturbance to alter its properties, which is very difficult.
- Signal processing functions such as encryption and compression are employed in digital circuits to maintain the secrecy of the information.
- The probability of error occurrence is reduced by employing error-detecting and correcting codes.

Basic Digital Communication Nomenclature

The following are some of the basic digital signal nomenclatures that appear in digital communication literature:

Information source: This device produces information to be communicated using the DCS. The information source can be analog or discrete.

Textual message: This is a sequence of characters. For digital transmission, as in Figure 10. 1-a, the message will be a sequence of digits or symbols from a finite symbol set or alphabet.

Character: A character is a member of an alphabet or set of symbols. Characters may be mapped into a sequence of binary digits. See Figure 10. 1-b. *Binary digit (a bit):* This is the fundamental information unit for all digital systems. The term bit also is used as a unit of information content.

Bit stream: This is a sequence of binary digits (ones and zeros). Figure 10. 1-c shows the " HOW " message with the 7-bit ASCII character code.

Symbol (digital message): A symbol is a group of "k" bits considered as a unit. This unit, known as a *message symbol* m_i(i=1,2,....,M)from a finite symbol set or alphabet (see Figure 10. 1-d). The size of the alphabet, M, is $M = 2^k$, where k is the number of bits in the symbol.

Digital waveform: A voltage or current waveform representing a digital symbol. The waveform characteristics (amplitude, width, and position for pulses or amplitude, frequency, and phase for sinusoids) allow its identification as one of the symbols in the finite symbol alphabet. Figure 10. 1-e shows an example of a bandpass digital waveform.

Data rate: This quantity in bits per second (bits/s) is given by

$$R = \frac{K}{T} \log_2 M \qquad \text{bits/s}$$

Where $M = 2^k$-symbol alphabet, T is the k-bit symbol duration, and M is the M – ray signaling, each symbol in an $M{-}ray$ the alphabet can be related to a unique sequence of k-bit.

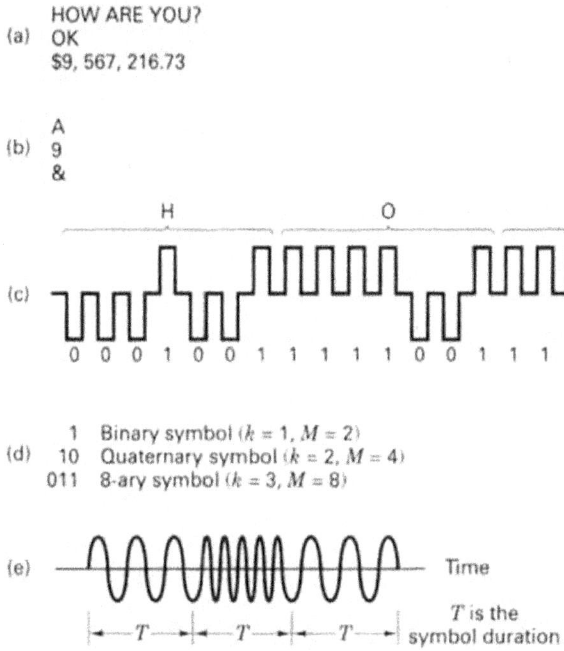

Figure 10.1 Nomenclature examples. (a) Textual messages. (b) Characters. (c) Bitstream (7-bit ASCII). (d) Symbols $m_i(i=1,2,....,M)$, $M = 2^k$. (e) Bandpass digital waveform $s_i(i=1,2,....,M)$.

Comparison between digital signals and analog signals:

	Analog	**Digital**
Signal	An Analog signal is a continuous signal which represents physical measurements.	Digital signals are discrete time signals generated by digital modulation.
Waves	Denoted by sine waves	Denoted by square waves
Representation	Uses a continuous range of values to represent information	Uses discrete or discontinuous values to represent information
Example	The human voice in the air, analog electronic devices.	Computers, CDs, DVDs, and other digital electronic devices.

Technology	Analog technology records waveforms as they are.	Samples analog waveforms into a limited set of numbers and records them.
Response to Noise	More likely to get affected, reducing accuracy	Less affected since noise responses are analog in nature
Flexibility	Analog hardware is not flexible.	Digital hardware is flexible in implementation.
Uses	It can be used in analog devices only. Best suited for audio and video transmission.	Best suited for Computing and digital electronics.
Bandwidth	Analog signal processing can be done in real-time and consumes less Bandwidth.	There is no guarantee that digital signal processing can be done in real-time and consumes more Bandwidth to carry out the same information.
Memory	Stored in the form of the wave signal	Stored in the form of binary bit
Power	Analog instrument draws large power	Digital instrument draws only negligible power

Performance of Digital Communication System

The performance of digital communication can be discussed through E*ffectiveness* and *Reliability.*

1. **Effectiveness (validity)**

To explain the Effectiveness, we must illustrate two important things, the *Transmission Rate* and the *Bandwidth*.

Transmission Rate: The information is conveyed (transmitted) by a change in the values of the signal in time. Since the frequency of a signal is a direct measure of the rate of change in values of the signal, the more the frequency of a signal, the more the achievable data rate or information transfer rate.

Information Transmission Rate: measured in *bits per second* R_b (bps)

Symbol Transmission Rate: measured in *symbols per second* or *Bauds* R_B (Bauds)

$$R_b = R_B log_2 M , M = 2^k , k \text{ \# bits}$$

Example 1: If we transmit a message consisting of seven symbols in one second {Red, Orange, Yellow, Green, Cyan, Blue, Purple}, {000, 001, 010, 011, 100, 101, 110} calculate the information transmission rate for this transmit process?

Sol: Each symbol represent by 3 bits, so M=2^3, R_B =7 Bauds InformationtransmissionrateR_b=$R_B log_2 M$=7×3=21bps

Bandwidth Efficiency: It is known as *spectral* or *spectrum efficiency*, which refers to the information rate that can be transmitted over a given bandwidth in a specific communication system. The bandwidth efficiency (η_b) of a digital communication system is measured in *bit/s/Hz*. Also, the bandwidth efficiency (η_B) was measured in *Bauds/Hz*. For example, the transmission technique that uses one kilohertz of Bandwidth to transmit 1,000 bits per second has a modulation efficiency of 1 (bit/s)/Hz.

$$\eta_b = R_b / B \text{ Measured in bps/Hz}$$
$$\eta_B = R_B / B \text{ Measured in Bauds/Hz}$$

Example

The TDMA version of the North American Digital Cellular (NADC) system (π/4 DQPSK based system) achieves a 48 Kbits-per-second data rate over a 30 kHz bandwidth or 1.6 bits per second per Hz, calculate the bandwidth efficiency and compare it with the value in table 10.1?

Table 10.1 The theoretical bandwidth efficiency limits for the main modulation types

Modulation format	*Theoretical bandwidth efficiency limits*
MSK	1 bit/second/Hz
BPSK	1 bit/second/Hz
QPSK	2 bit/second/Hz
8PSK	3 bit/second/Hz
16 QAM	4 bit/second/Hz
32 QAM	5 bit/second/Hz
64 QAM	6 bit/second/Hz
256 QAM	8 bits/second/Hz

Solution

$$\eta_b = \frac{R_b}{B} = \frac{48000bps}{30000Hz} = 1.6\ bps/Hz$$

We can see that the theoretical efficiency would be two bits per second per Hz; in practice, it is 1.6 bits per second per Hz.

2. Reliability

The reliability of the digital system is defined as the probability of making the correct decision at the system's output when a digit enters the system. The reliability of any communication system depends on the signal-to-noise ratio (SNR) calculation or the Bit/Symbol Error Rate (BER, SER) (Probability of Error).

Signal-to-Noise Ratio (SNR or S/N): this is a measure used in communication systems to compare the level of the desired signal to the level of background noise. It is the ratio of signal power to noise power, often expressed in decibels.

$$SNR = \frac{P_{signal}}{P_{noise}}$$

Bit/Symbol Error Rate (BER, SER) or Probability of Error is the number of bit errors per unit of time. The BER is the number of bit errors divided by the total number of transferred bits during a studied time interval. In digital transmission, bit errors are the number of received bits of a data stream over a communication channel that has been altered due to noise, interference, distortion, or bit synchronization errors.

BER $P_{eb} = Error\ bits / Total\ bits$

SER $P_{eB} = Error\ symbols / Total\ symbols$

Communication System Models

The communication systems are classified into several types according to signals and their transmission methods during this systems.

1. Analog communication systems, Figure 10. 2 shows us an example of such a system.

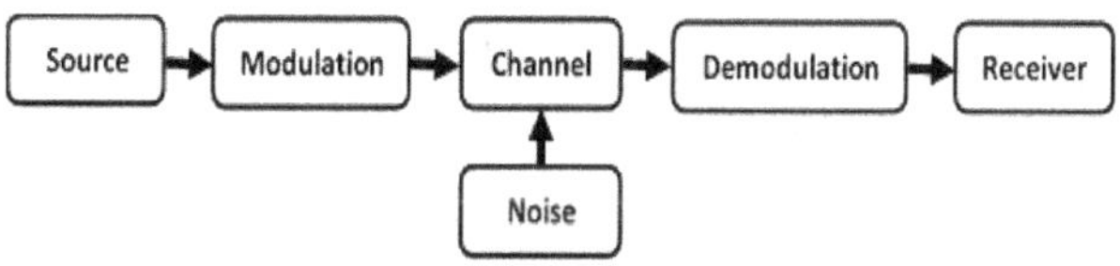

Figure 10.2 Analog passband communication systems

2. Digital Communication Systems, Figures 10.3 & 10.4 show us examples of such systems.

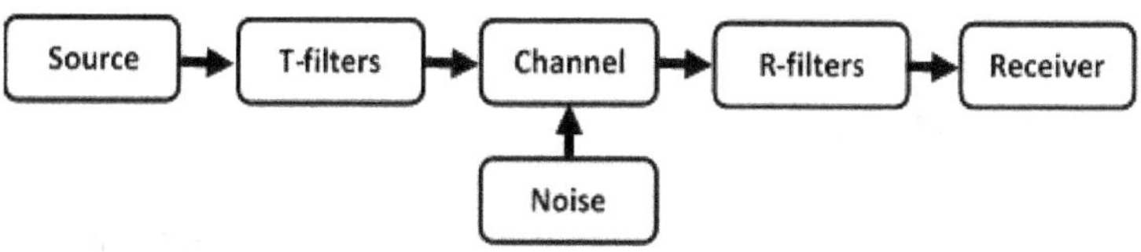

Figure 10.3 Digital baseband communication systems

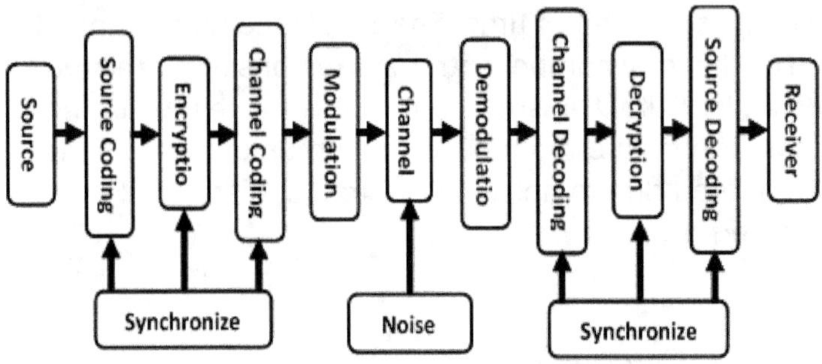

Figure 10.4 Digital passband communication systems

3. Digital Transmission of Analog Signals (Digital TV), as shown in Figure 10.5.

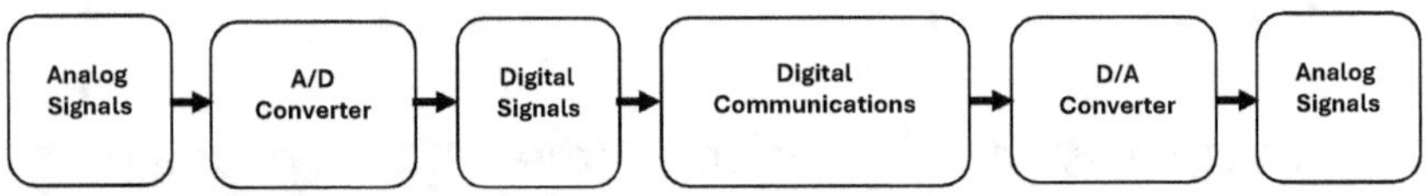

Figure 10.5 Digital transmission of analog signals

4. Analog Transmission of Digital Signals (For instance: Modem), as shown in Figure 10.6.

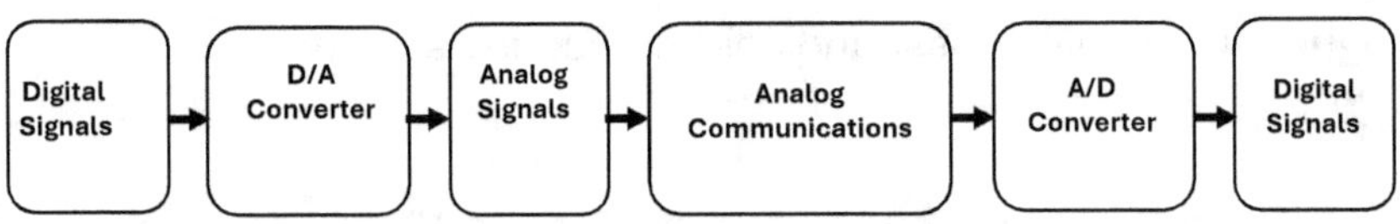

Figure 10.6 Analog transmission of digital signals

Transmission Methods in Communication Systems

There are three transmission methods in Communication systems (simplex, half duplex, and full duplex).
Simplex Communications: (only one way) it is a communication channel that sends information in one direction only, or it is a communications channel that operates in one direction at a time. as shown in Figure 10.7.

Figure 10.7 Simplex communications

Half Duplex Communications: (may be two-way, but one-way at any time) In a half-duplex system, there are still two clearly defined paths/channels, and each party can communicate with the other but not simultaneously; the communication is in one direction at a time. Figure 10.8 shows the half-duplex system.

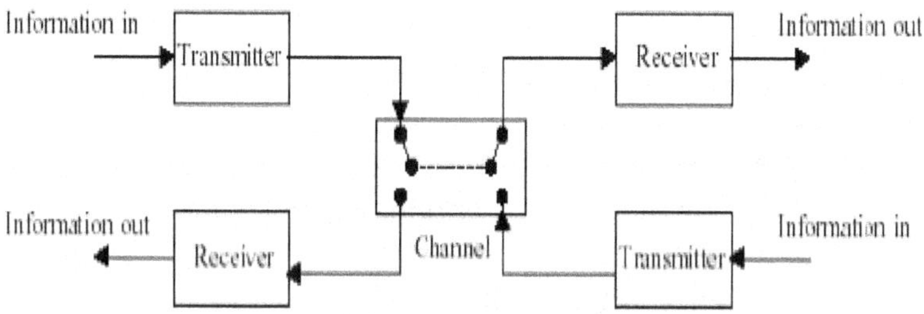

Figure 10.8 Half-duplex communications

Full Duplex Communications: (two-way communications simultaneously). In Figure 10.9, a full-duplex system, both parties can communicate with each other simultaneously.

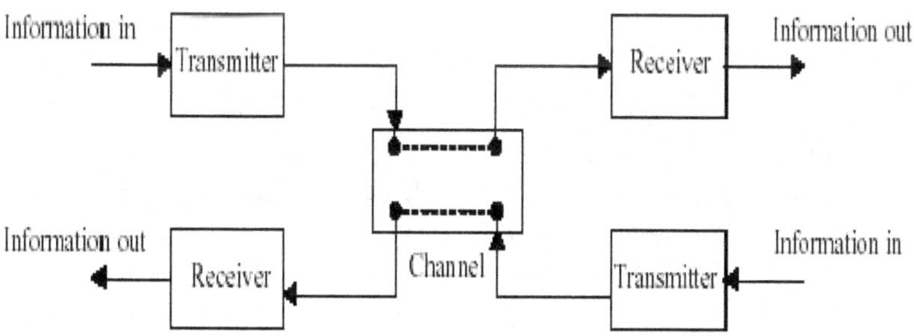

Figure 10.9 Full duplex communications

Multiplexing techniques

In communication, a medium can carry only one signal at any time under the simplest conditions. For multiple signals to share one medium, the medium must somehow be divided, giving each signal a portion of the total Bandwidth. This is where the concept of multiplexing comes in!

Multiplexing techniques can be divided into four basic categories:

1. Frequency Division Multiplexing (FDM)
2. Time Division Multiplexing (TDM)
3. Code Division Multiplexing (CDM)
4. Wavelength Division Multiplexing (WDM) (Optical)

- In FDM, the available Bandwidth is divided into several smaller independent logical channels, each with a small bandwidth. It assigns “frequency ranges” to each “user” or “signal” on a medium.
- In TDM, sharing is accomplished by dividing available “transmission time” on a medium/channel among users. Each channel user is allotted a small time interval to transmit a message.

TDM can be implemented in two ways: Synchronous TDM and AsynchronousTDM.

Synchronous TDM: The multiplexer allocates the same time slot to each device at all times, whether or not a device has anything to transmit. Time slot 1, for example, is assigned to devise one alone and cannot be any other device, as shown in Figures 10.10 and 10.11.

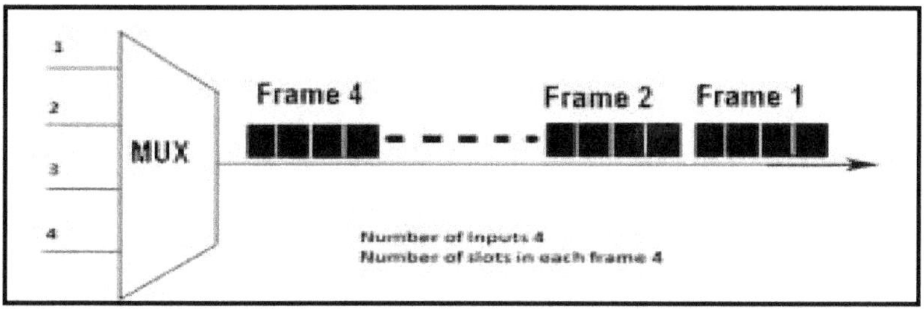

Figure 10.10 Synchronous TDM

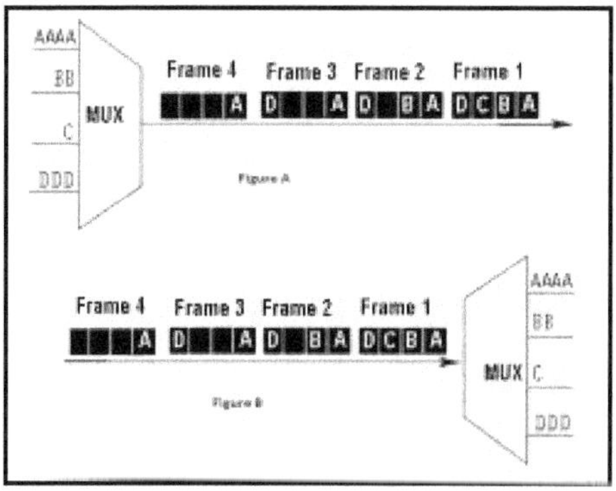

Figure 10.11 How the Synchronous TDM work

In Synchronous TDM, a frame consists of one complete cycle of time slots. Thus the number of slots in a frame equals the number of inputs. Figure 10.12 (A & B) is an example of how the Synchronous TDM work.

Asynchronous TDM: In asynchronous TDM, each slot in a frame is not dedicated to fixing the device. Each slot contains an index of the device to be sent a message. Thus, the number of slots in a frame is not necessarily equal to the number of input devices. Figure 10.12 shows us the Asynchronous TDM. In asynchronous TDM, a frame contains a fixed number of time slots.

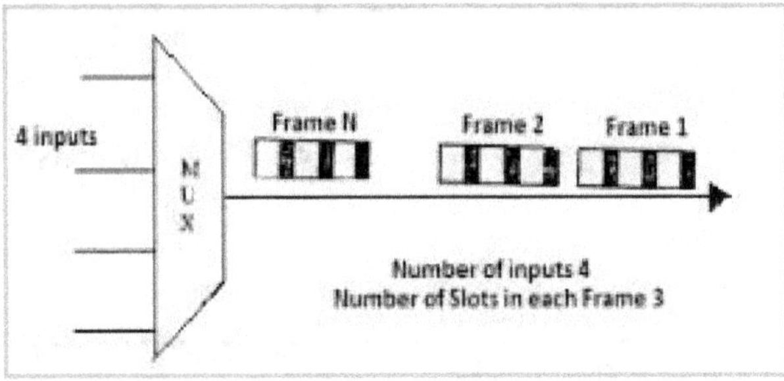

Figure 10.12 Asynchronous TDM

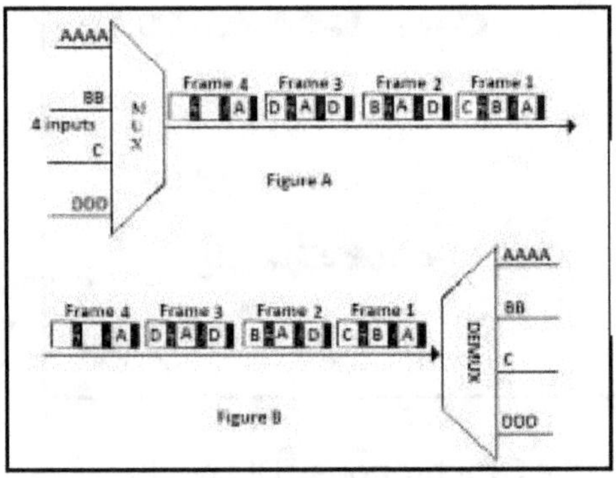

Figure 10.13 How asynchronous TDM work

Each slot has an index of which device to receive. Figure 10.13 (A & B) shows how asynchronous TDM works.

CDM is a form of multiplexing where the transmitter encodes the signal using a pseudo-random sequence. CDM involves the original digital signal with a spreading code. This spreading spreads the signal's spectrum greatly and reduces the power over anyone part of the spectrum.

WDM is a technology that multiplexes several optical carriers signals onto a single optical fiber using different wavelengths (i.e., colors) of laser light. This technique enables bidirectional communications over one strand of fiber and the multiplication of capacity.

Difference Between FDM and TDM	
-FDM divides the channel into multiple, but smaller frequency ranges to accommodate more users.	- TDM divides a channel by allocating a time period for each channel.
- TDM provides much better flexibility compared to FDM.	
- FDM proves much better latency compared to TDM. Latency is the time it takes for the data to reach its destination. As TDM allocates time periods, only one channel can transmit at a given time, and some data would often be delayed, though it's often only in milliseconds. Since channels in FDM can transmit at any time, their latencies would be much lower compared to TDM.	
- TDM and FDM can be used in tandem. FDM and TDM are often used in tandem, to create even more channels in each frequency range.	

Modulation Techniques in Telecommunication Systems

In telecommunication systems, modulation is the process of varying one or more properties of the carrier signal, with a modulating signal typically containing information to be transmitted. Modulation techniques are divided into two parts according to the type of signal dealing with it; *analog modulation* and *digital modulation*.

Continuous Modulation	***Analog Modulation***	***Linear***	**AM** (amplitude modulation) **DSBSC** (double-sideband suppressed-carrier) **SSB** (single-sideband modulation) **VSB** (vestigial sideband modulation)
		Non-linear	**FM** (frequency modulation) **PM** (phase modulation)
	Digital Modulation	**ASK** (amplitude-shift keying) **FSK** (frequency-shift keying) **PSK** (phase-shift keying) **DPSK** (differential phase-shift keying) **QAM** (quadrature amplitude modulation) **MSK** (minimum shift keying), etc.	
Pulse Modulation	***Pulse Analog Modulation***	**PAM** (pulse-amplitude modulation) **PDM** (pulse-depth modulation) **PPM** (pulse-position modulation)	
	Pulse Digital Modulation	**PCM** (pulse-code modulation) **DM** (ΔM) (Delta modulation) **DPCM** (Differential PCM) **ADPCM** (Adaptive DPCM) **∑.Δ** (Delta-sigma modulation), etc.	

Pulse Modulation

Pulse modulation schemes allow the narrowband analog signal to be transferred as a digital signal with a fixed bit rate, which can be transferred over an underlying digital transmission system. To transform an analog waveform into a form that is compatible with digital communication, the following steps are taken:

1. Sampling.
2. Quantization.
3. Encoding.

Figure 10.14 shows us the components of an A/D converter in the digital transmission of analog signals:

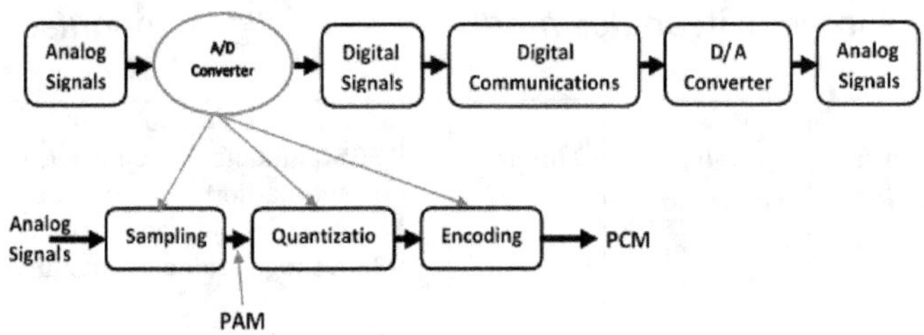

Figure 10.14 Components of A/D converter in the digital transmission of analog signals

Concept of Sampling

Sampling is defined as "The process of measuring the instantaneous values of the continuous-time signal in a discrete form.", or it is the reduction of a continuous-time signal to a discrete-time signal.

The sample is a piece of data taken from the whole data, which is continuous in the time domain, or it is a value or set of values at a point in time and/or space.

When a source generates an analog signal, and if that has to be digitized, having 1s and 0s, i.e., High or Low, the signal has to be discretized in time. This discretization of the analog signal is called sampling. Figure 10.15 indicates a continuous-time signal

$x(t)$ and a sampled signal $x_s(t)$. When $x(t)$ is multiplied by a periodic impulse train, the sampled signal $x_s(t)$ is obtained. *Sampling Rate:* The gap between the samples should be fixed to discretize the signals. That gap can be termed a sampling period T_S.

Sampling frequency $= \frac{1}{T_s} = f_S$

where T_S is the sampling time, and f_S is the sampling frequency or rate.

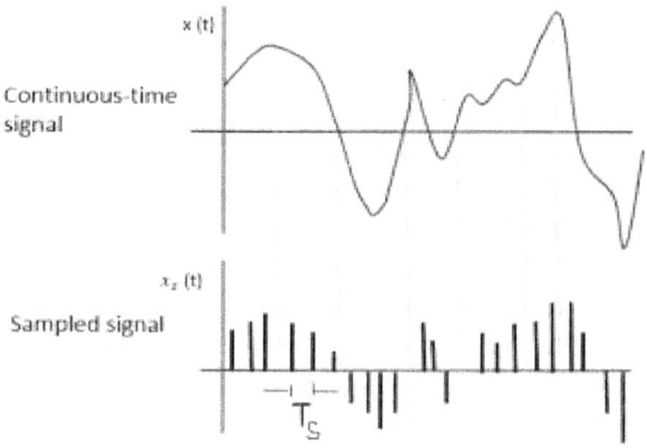

Figure 10.15 The sampling rate denotes the number of samples taken per second or for a finite set of values.

Nyquist Rate: Suppose a signal is band-limited with no frequency components higher than f_m Hertz. That means f_m is the highest frequency. For such a signal, the sampling rate should be twice the highest frequency for effective reproduction of the original signal.

$f_S = 2f_m$ (This rate of sampling is called as Nyquist rate)

where f_m the highest frequency

The sampling theorem states that "a signal can be exactly reproduced if it is sampled at the rate f_S which is greater than twice the maximum frequency f_m."

To understand the sampling theorem, let us consider a band-limited signal, i.e., a signal whose value is non-zero between some $-f_m$ and f_m Hertz. Such a signal is represented as

$$x = 0 \text{ for } |f| > f_m.$$

For the signal $x(t)$, the band-limited signal in the frequency domain can be represented as shown in Figure 10.16.

If the signal $x(t)$ is sampled <u>above</u> the Nyquist rate, the original signal can be recovered, and if it is sampled <u>below</u> the Nyquist rate, the signal cannot be recovered.

Figure 10.17 shows the Fourier transform of a signal $x_s(t)$. Here, theinformationisreproducedwithoutanyloss. Thereisno mixing up, and hence recovery is possible.

Figure 10.18 shows us what happens if the sampling rate is equal to twice the highest frequency (f_S = 2W). The information is replaced without any loss. Hence, this is also a good sampling rate.

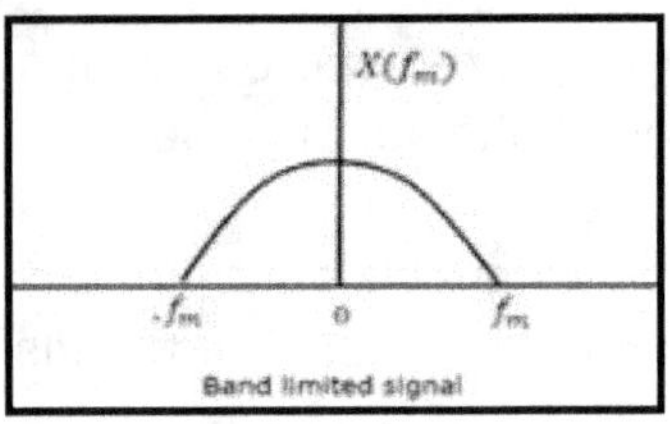

Figure 10.17 The band-limited signal

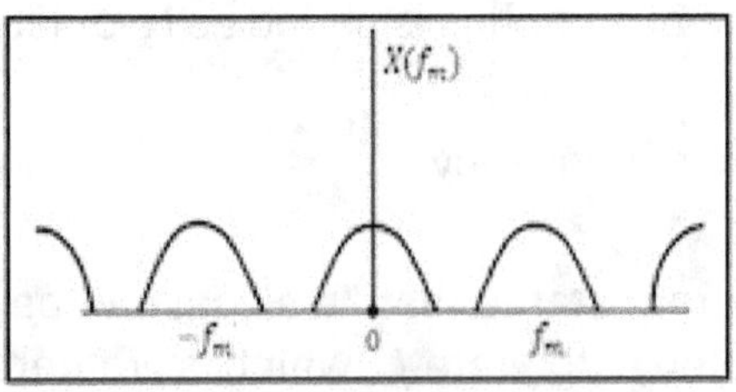

Figure 10.17 Fourier transform of a signal

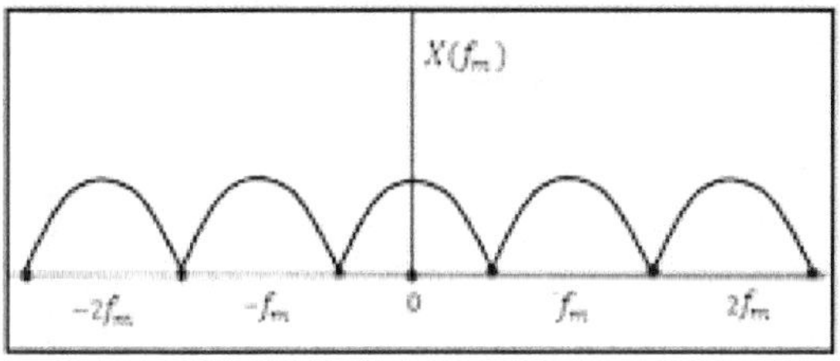

Figure 10.18 Sampling rate is equal to twice the highest frequency.

Now, let us look at the condition, f_S<$2W$ the resultant pattern will look like Figure 10.19.

We can observe the above pattern in Figure 10.19 that the overlapping of information is done, leading to mixing up and losing information. This unwanted phenomenon of overlapping is called *Aliasing*.

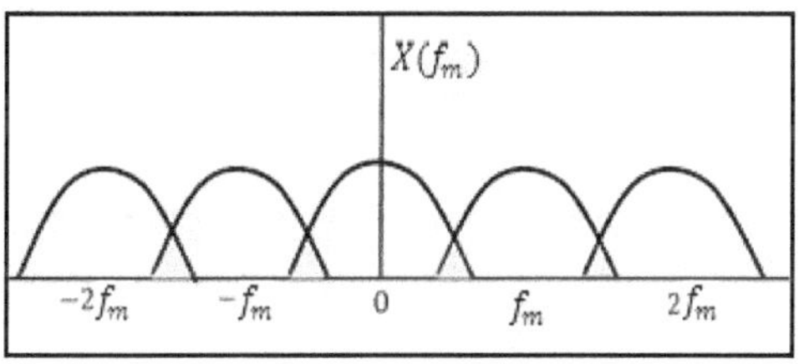

Figure 10.19 Overlapping of information

Concept of Quantization

The sampling method chooses a few points on the analog signal, which are then joined to round off the value to a near-stabilized value. Such a process is called *quantization*.

Quantization represents the amplitude's sampled values by a finite set of levels, which means converting a continuous amplitude sample into a discrete-time signal. Figure 10.20 shows how an analog signal gets quantized. One line (like a stair) represents the quantized signal, while another represents the analog signal.

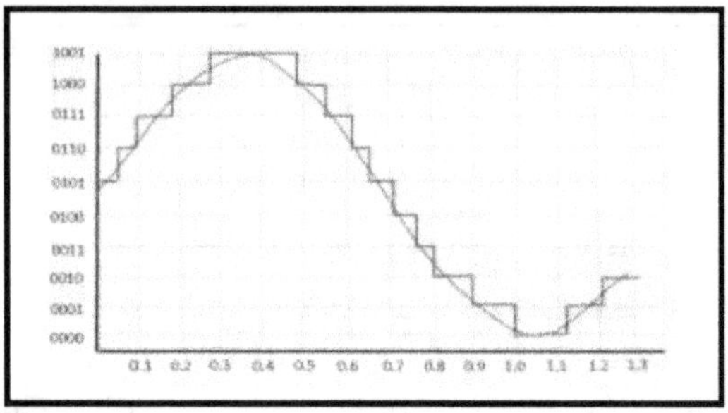

Figure 10.20 Quantization of analog signal

Types of Quantization

There are two types of Quantization - *Uniform Quantization* and *Nonuniform Quantization*. The type of quantization in which the quantization levels are uniformly spaced is termed *Uniform Quantization*. The type of quantization in which the quantization levels are unequal and mostly the relation between them is logarithmic is termed a *Non-uniform Quantization*. Figure 10.21 shows the uniform quantization and non-uniform quantization.

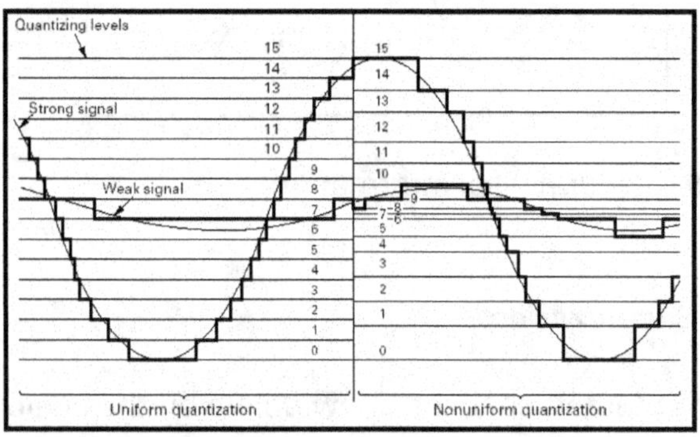

Figure 10.21 Uniform and non-uniform quantization

There are two types of uniform quantization. They are the *Mid-Rise* type and *Mid-Tread* type.

The Mid-Rise type is called because the origin lies in the middle of a rising part of the stair-case like a graph. The quantization levels in this type are even in number (Figure 10.22). The mid-tread type is so called because the origin lies in the middle of a thread of the stair-case like a graph. The quantization levels in this type are odd in number. Both the mid-rise and mid-tread types of uniform quantizers are symmetric about the origin.

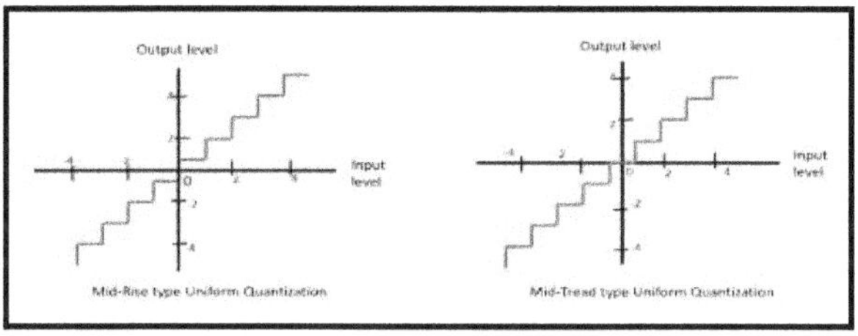

Figure 10.22 Quantization levels

Pulse Code Modulation (PCM) Waveform (Line coding)

Data, as well as signals that represent data, can either be digital or analog. Line coding is the process of converting *digital data to digital signals*. This technique converts a sequence of bits to a digital signal. Several types of PCM waveforms are described and illustrated below;

Unipolar NRZ (Non-Return-to-Zero) signaling: This type's advantage is the need for one power supply to generate the signals. From Figure 10. 25, in this type of signaling, we can represent bit 1 as a high level (+5V) or at a low level (-V) and 0 bit in the base level. T_b is the time of sending one bit, and T_s is the duration of one symbol.

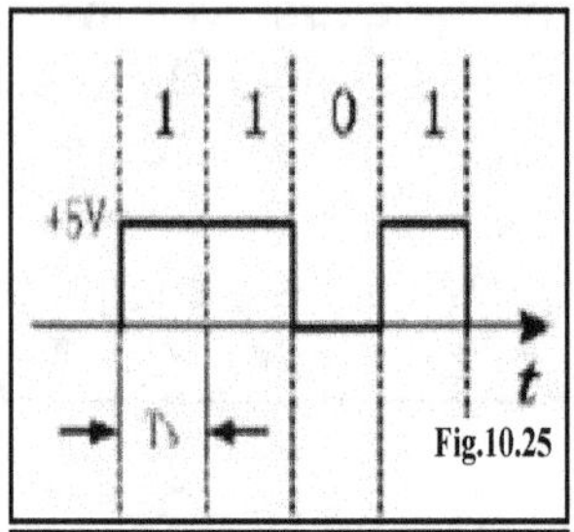

Fig.10.25

Unipolar RZ (Return-to-Zero) Signaling: in this type of signaling, the signal return to zero after pulse duration (T) of T_b time $(0<T<T_b)$. Figure 10. 26. Show the unipolar RZ. T: pulse duration and T_b is a time of sending one bit.

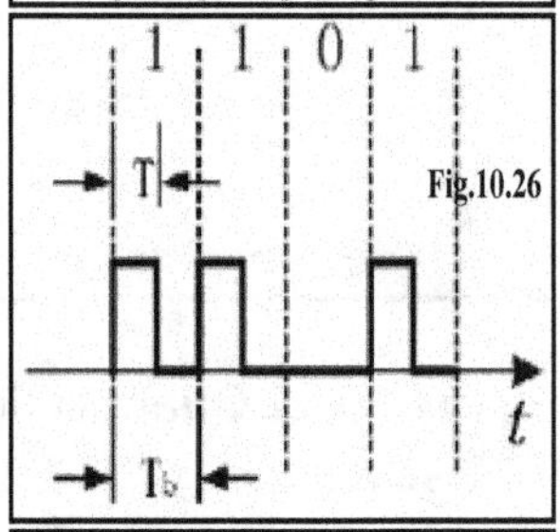

Fig.10.26

Polar NRZ Signaling: This type's disadvantage is the need for two power supplies to generate the signals. In Figure 10.27. The representation of bit 1 comes from a high level (+5V) and bit 0 from a low level (-5V), so this type needs two signal generators.

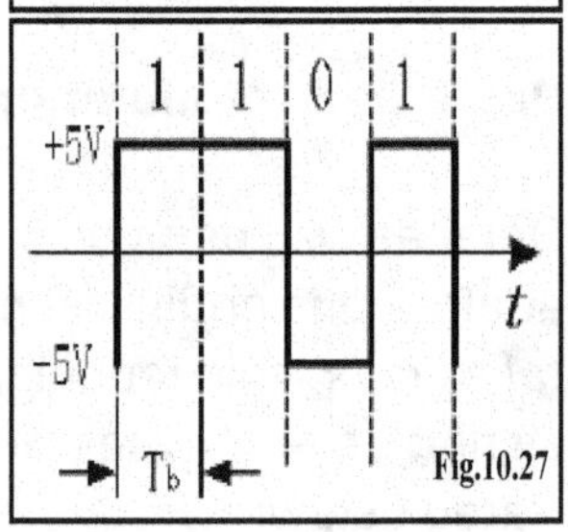

Fig.10.27

Polar RZ Signaling: in this type of signaling, the signal return to zero after pulse duration (T) of T_b time $(0<T<T_b)$. Bit 1 represents a high level (+5V), and bit 0 represents in low level (-5V). Figure 10.28 shows the bipolar RZ.

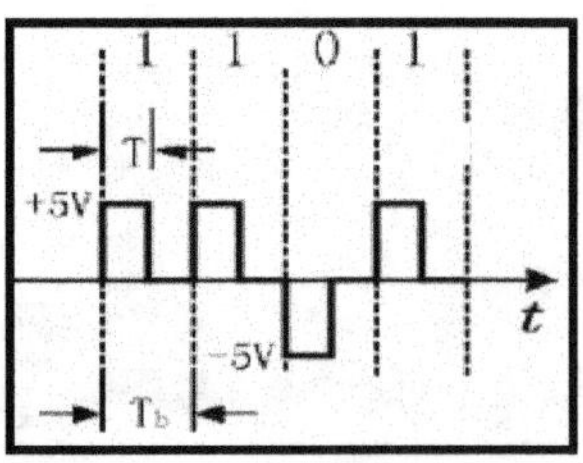

Bipolar RZ-AMI (Alternate Mark Inversion): The Bipolar RZ-AMI is a signaling scheme used in telephone systems. The ones are represented by equal-amplitude alternating pulses (+5V) and (-5V). The zeros are represented by the absence of pulses(base level). This type of signaling needs two power supplies and has three voltage levels (+, 0,-) (Pseudo-ternary signaling). Figure 10.23 shows the RZ-AMI representation.

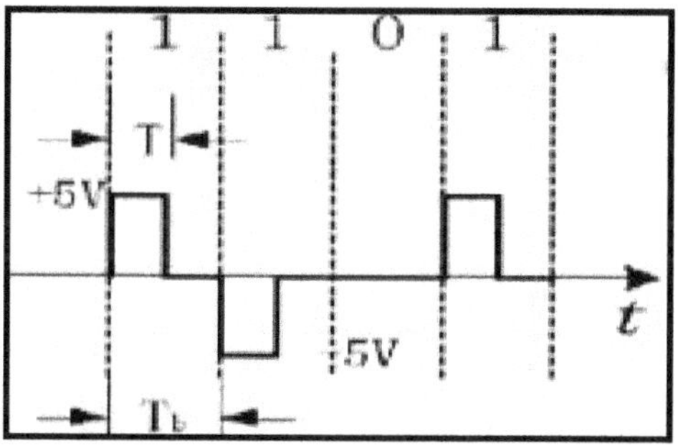

Figure 10.23 RZ-AMI representation

Bipolar NRZ (Non-Return-to-Zero) signaling: In this type of signaling, there are three groups; NRZ-L (L refers to level), NRZ-M (M refers to mark), and NRZ- S (S refers to space). NRZ-L is used in digital logic circuits. A binary one is represented by one voltage level, and another voltage level represents a binary zero. There is a change in level whenever the data change from a one to a zero or from a zero to a one. With NRZ-M, the one, or mark, is represented by a change in level, and the zero, or space, represents no change in level. NRZ-S is the complemented of NRZ-M: A one is represented by no change in level, and a change in level represents a zero. Figure 10.24 shows all the types of NRZ-L, NRZ-M, and NRZ-S.

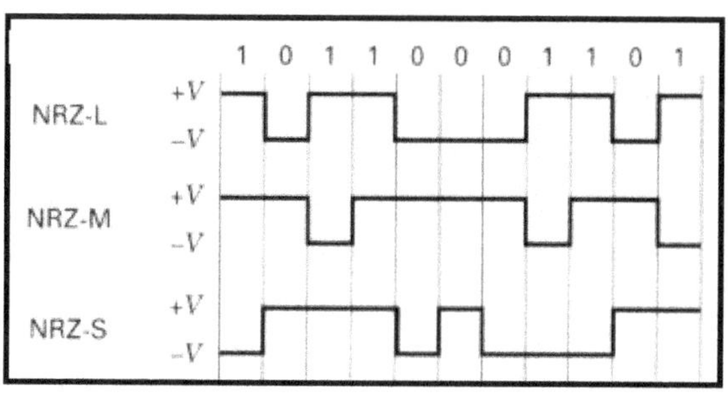

Figure 10.24 NRZ-L, NRZ-M, and NRZ-S

Manchester Code: Manchester coding (also known as phase encoding, or PE) is a PCM waveform in which the encoding of each data bit is either low then high or high then low, of equal time, as shown in Figure 10. 31.

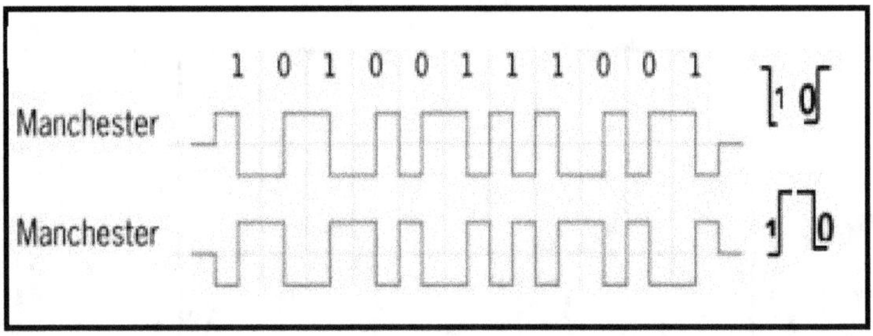

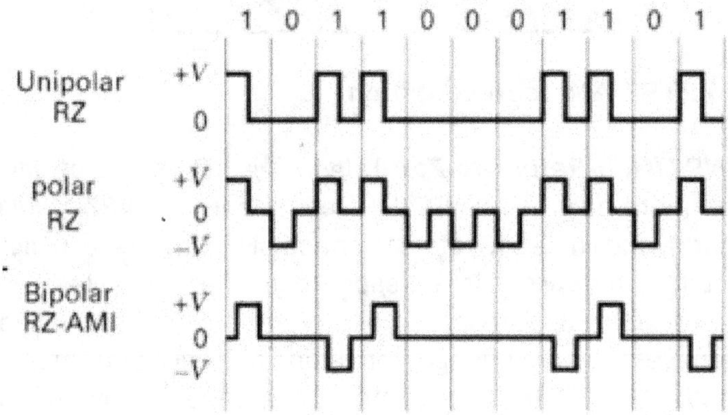

Example: Draw the waveforms of PCM in terms of Unipolar RZ, Polar RZ, and Bipolar RZ-AMI for the message 1 0 1 1 0 0 0 1 1 0 1 with high-level voltage (+V) and a low-level voltage (-V).

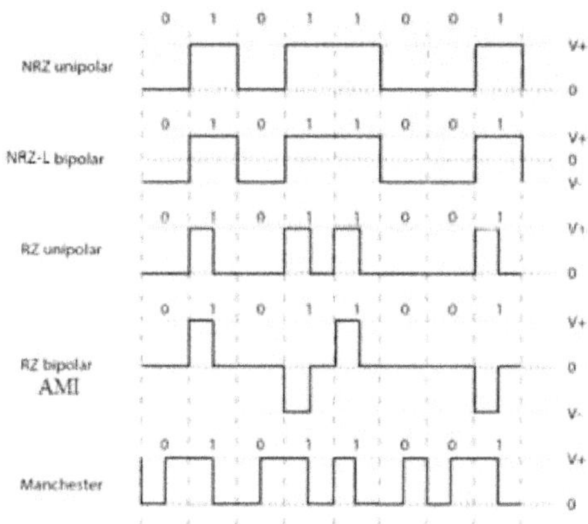

Example: For the message 0 1 0 1 1 0 0 1, draw the waveforms of PCM in terms of Unipolar NRZ, Bipolar NRZ-L, Unipolar RZ, Bipolar RZ-AMI, and Manchester, with high-level (+V) and low-level (-V) voltages.

Digital Modulation Techniques

Digital Modulation provides more information capacity, high data security, quicker system availability, and great-quality communication. Depending upon the need, many types of digital modulation techniques and their combinations exist.

Binary: ASK, FSK, PSK, DPSK, and QPSK.
Mary: ASK, FSK, PSK, and DPSK.

Binary Amplitude-Shift Keying (BASK)

BASK is a form of amplitude modulation representing digital data as variations in the amplitude of a carrier wave. In an ASK system, the binary symbol one is represented by transmitting a fixed-amplitude carrier wave and fixed frequency for a bit duration of T_b seconds. If the signal value is one, then the carrier signal will be transmitted; otherwise, a signal value 0 will be transmitted.

$$s_1(t) = \sqrt{\frac{2E_b}{T_b}}\cos(2\pi f_c t\ )\ for\ "1"$$

$$s_2(t) = 0 for\ "0" \qquad 0 \leq t \leq T$$

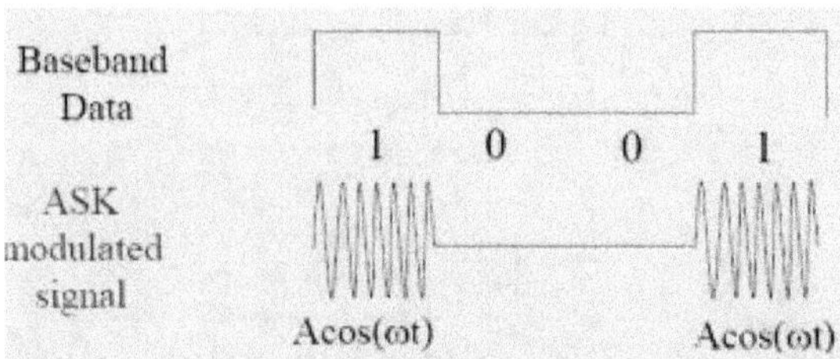

Where $s(t)$ is the modulated signal, E_b is transmitted signal energy per bit, T_b is bit duration. The average transmitted signal energy is (the two binary symbols must be equiprobable) $E_{av} = \frac{E_b}{2}$

Binary Frequency Shifted-Keying (BFSK)

FSK is a frequency modulation scheme in which digital information is transmitted through discrete frequency changes of a carrier signal. BFSK uses discrete frequencies to transmit binary (0s and 1s) information. With this scheme, the "1" is called the mark frequency, and the "0" is called the space frequency.

$$s_i(t) = \sqrt{\frac{2E_b}{T_b}}\cos(2\pi f_i t\ ) \qquad 0 \leq t \leq T_b$$

$$s_1(t) = \sqrt{\frac{2E_b}{T_b}}\cos(2\pi f_1 t\ )\ for\ "1"\ corresponding\ to\ i = 1$$

$$s_2(t) = \sqrt{\frac{2E_b}{T_b}}\cos(2\pi f_2 t\ )\ for\ "0"\ corresponding\ to\ i = 2$$

$$f_i = \frac{n_c+}{T_b}$$

For some fixed integer n_c, $i = 1,2$

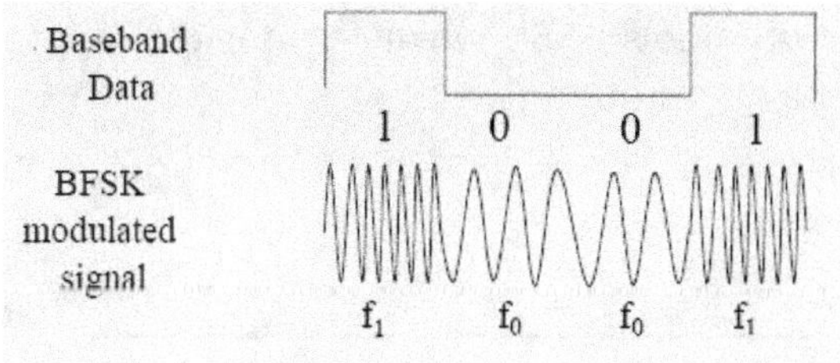

Sunde's BFSK: When the frequencies f_1 and f_2 are chosen so that they differ by an amount equal to the reciprocal of the bit duration T_b.

Generation of BFSK: Figure 10. 32 shows the generation of BFSK modulation.

<u>**Example:**</u> Draw Sunde's BFSK waveform for the binary input sequence 0011011001 for a bit duration T_b = 1 s.

<u>Sol:</u> If i = 1, n_c =1 f_1= 2/ T_b for $s_1 t$ "0" and i =2, n_c=1 f_2 = 3/T_b for $s_2 t$"1".

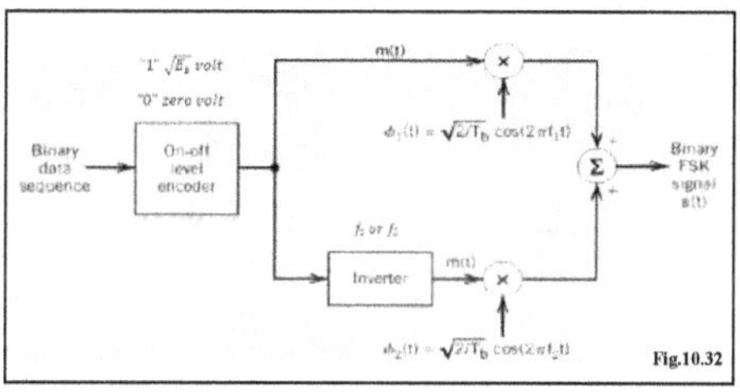

Fig.10.32

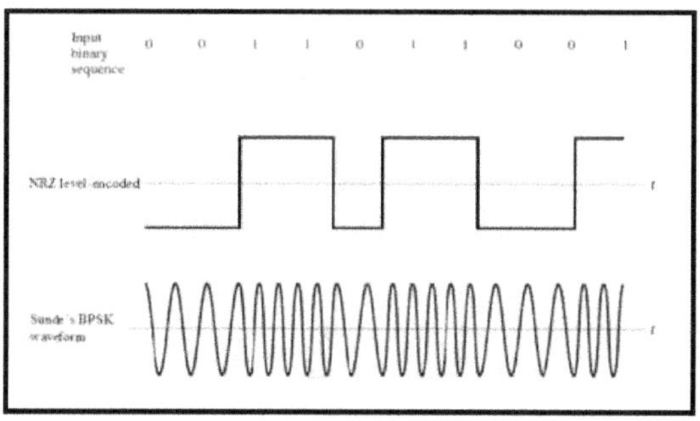

Example: Draw Sunde's BFSK waveform for the input binary sequence 101 for a bit duration T_b= 1s.

Sol: If i= 1, n_c=2 f_1=3/$T_b$for s_1 "1" and i= 2, n_c= 2 f_2 =4/$T_b$fors_2"0".

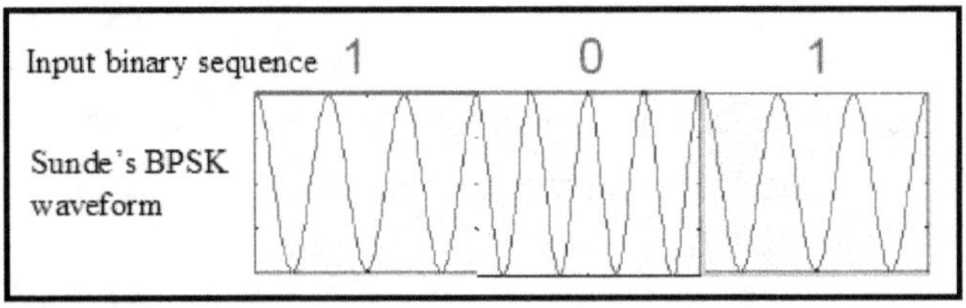

Binary Phase-Shift Keying (BPSK)

BPSK (also sometimes called PRK, phase reversal keying, or 2PSK) is the simplest form of phase shift keying (PSK). It uses two phases separated by 180° and can also be termed 2-PSK. It does not matter exactly where the constellation points are positioned; Figure 10. shows them on the real axis, at 0°and 180°.

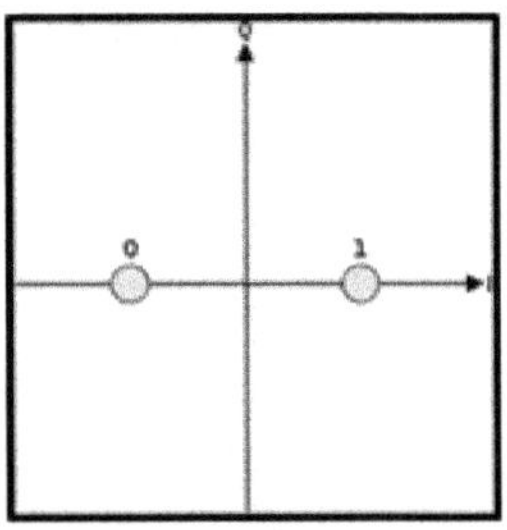

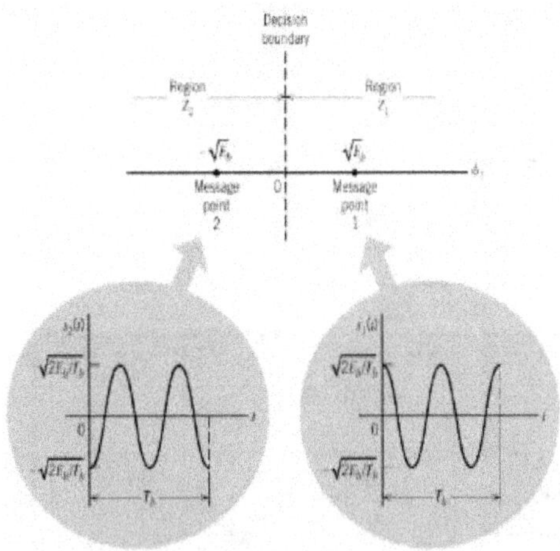

$$s_1(t) = \sqrt{\frac{2E_b}{T_b}}\cos(2\pi f t)\ for\ "1",\ s_2\ (t) = \sqrt{\frac{2E_b}{T_b}}\cos(2\pi f t + \pi) = -\sqrt{\frac{2E_b}{T_b}}\cos(2\pi f_c t)\ for\ "0", 0 \le t \le T$$

Generation of BPSK: Figure 10.25 shows the generation of BPSK modulation.

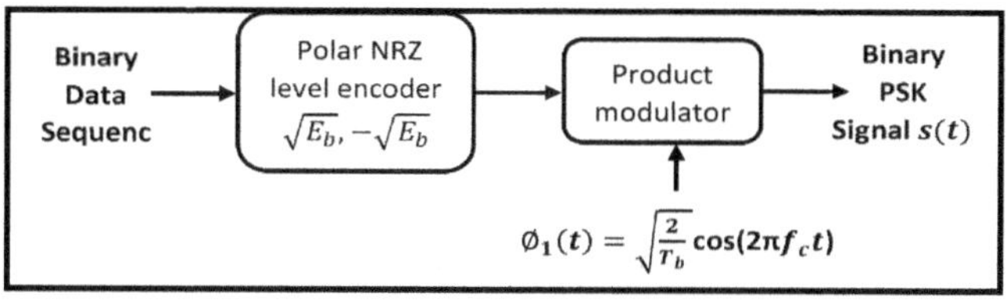

Figure 10.25 Generation of BPSK modulation.

Detection of BPSK: compare to the phase of the carrier

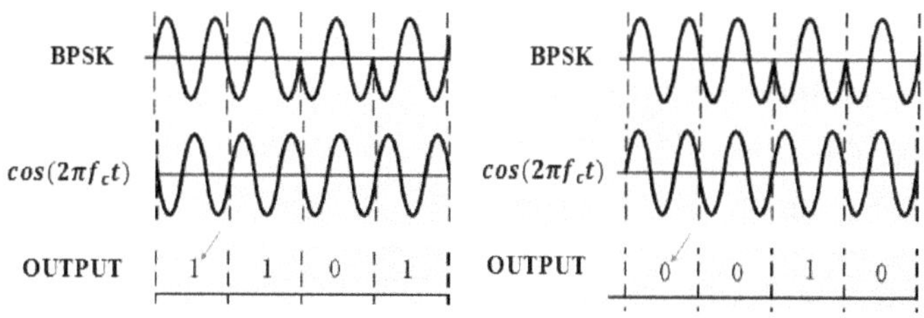

Examples: BPSK

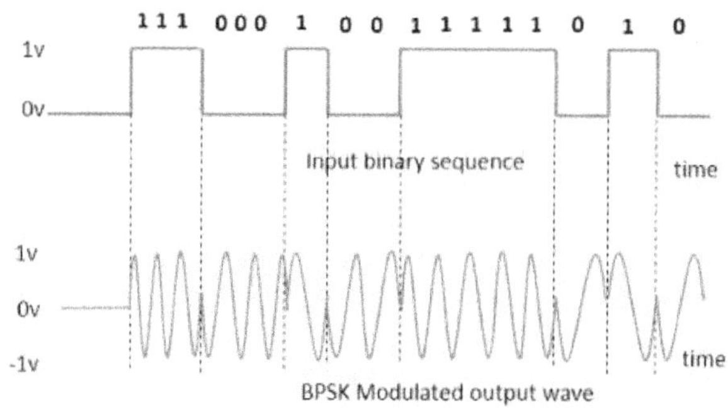

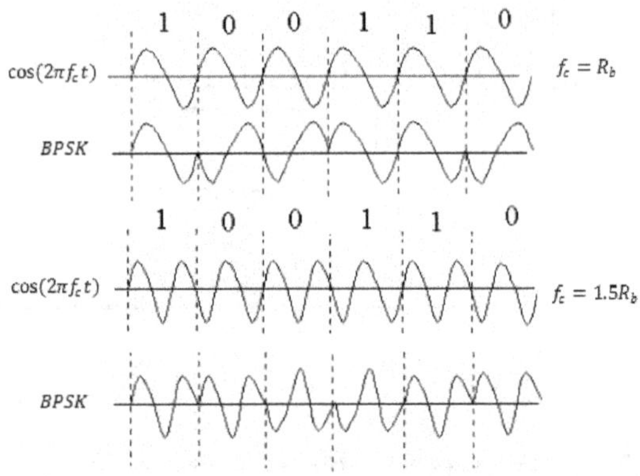

Differential Phase Shift Keying (DPSK)

The phase of the modulated signal is shifted relative to the previous signal element. No reference signal is considered here. The signal phase follows the high or low state of the previous element. This DPSK technique doesn't need a reference oscillator. Figure 10. 34 represents the model waveform of DPSK. It is seen in Figure 10. 34 that if the data bit is Low, i.e., 0, then the signal phase is not reversed but continues as it was. If the data is High, i.e., 1, then the signal phase is reversed, as with NRZI, inverted on 1 (a form of differential encoding).

Generation of DPSK: BDPSK= Differential Code's BPSK. To draw the waveform of BPSK to be compared to its previous phase.

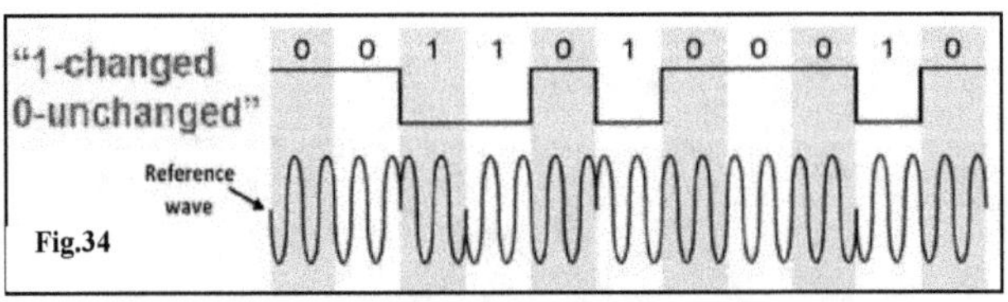

Fig.34

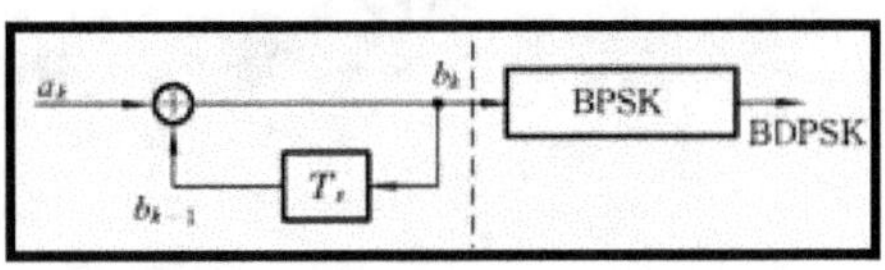

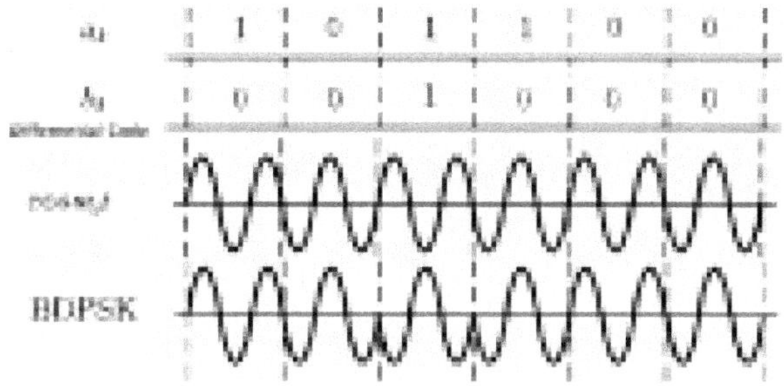

Example: Draw the BDPSK waveform for the binary input sequence 101100.

Solution

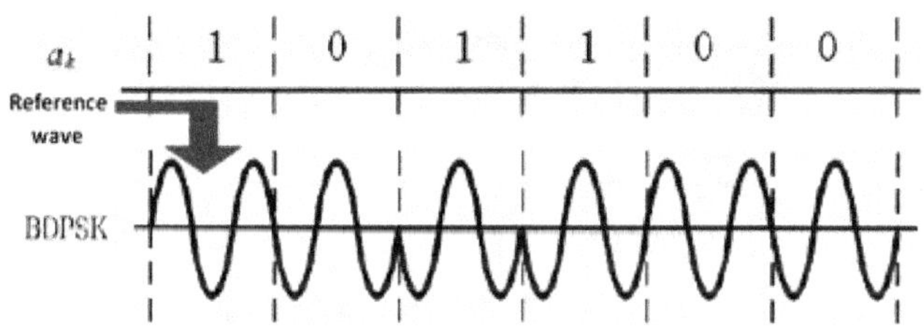

Quadrature Phase Shift Keying (QPSK)

QPSK is a variation of BPSK and a Double Side Band Suppressed Carrier (DSBSC) modulation scheme, which simultaneously sends two bits of digital information called digits. Instead of converting digital bits into a series of digital streams, it converts them into bit pairs. This decreases the data bit rate to half, allowing the other users space. The purpose of QPSK is to increase *bandwidth efficiency.*

$$s_i(t) = \sqrt{\frac{2E}{T}} \cos\left(2\pi f_c t + (2i-1)\frac{\pi}{4}\right),\ i = 1,2,3,4$$

The signal is $s_i(t)$: $s_1(t) = \sqrt{\frac{2E}{T}} \cos\left(2\pi f_c t + \frac{\pi}{4}\right)$ **"10"**

$$s_2(t) = \sqrt{\frac{2E}{T}} \cos\left(2\pi f_c t + \frac{3\pi}{4}\right) \quad \textbf{"00"}$$

$$s_3(t) = \sqrt{\frac{2E}{T}} \cos\left(2\pi f_c t + \frac{5\pi}{4}\right) \quad \textbf{"01"}$$

$$s_4(t) = \sqrt{\frac{2E}{T}} \cos\left(2\pi f_c t + \frac{7\pi}{4}\right) \quad \textbf{"11"}$$

where E is the signal energy per symbol, and T is the symbol duration.

The constellation diagram for QPSK with Gray coding is shown in Figure 10.26. Each adjacent symbol only differs by 1 bit.

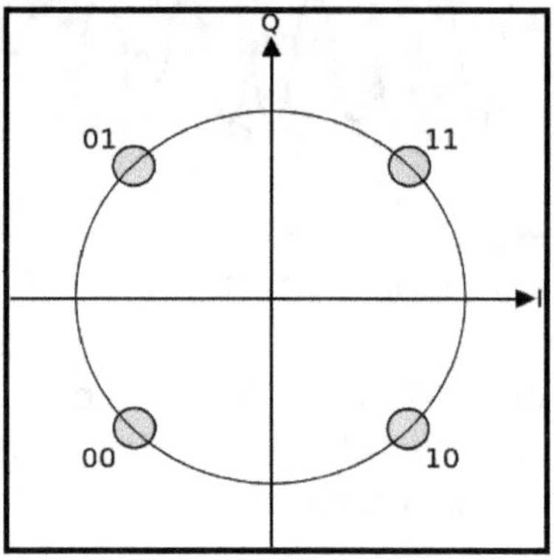

Figure 10.26 Constellation diagram for QPSK

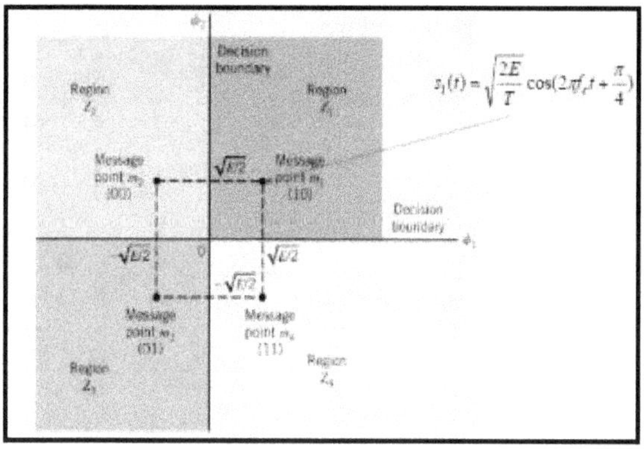

Generation of QPSK:

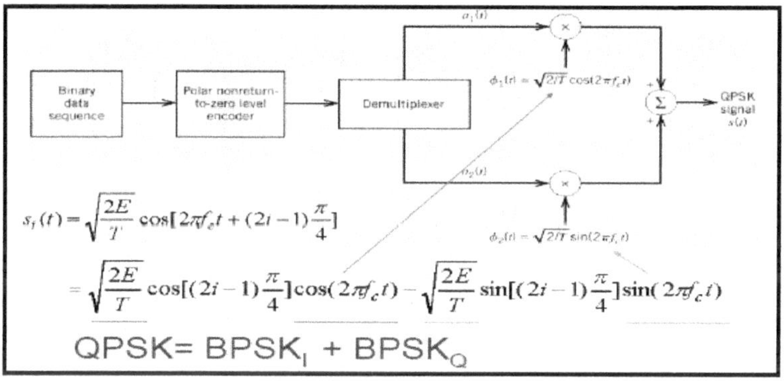

The *waveform of QPSK:* to draw the waveform of QPSK, we can use the table below:

Decimal	MSB-LSB for binary number	Triangle function	Value of angle
0	00	-cos -sin	135°
1	01	-cos sin	225°
2	10	cos -sin	45°
3	11	cos sin	315°

The general form of QPSK waveform

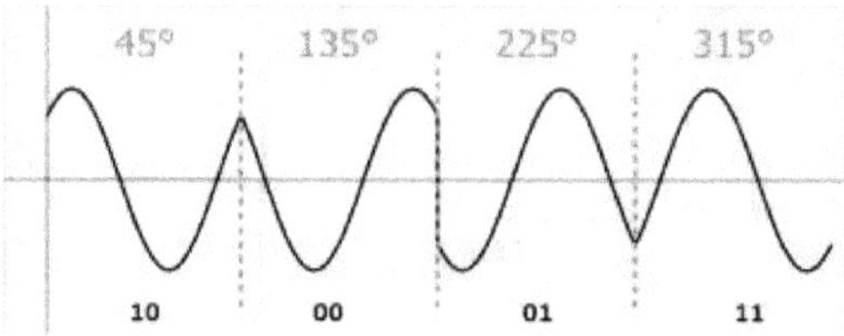

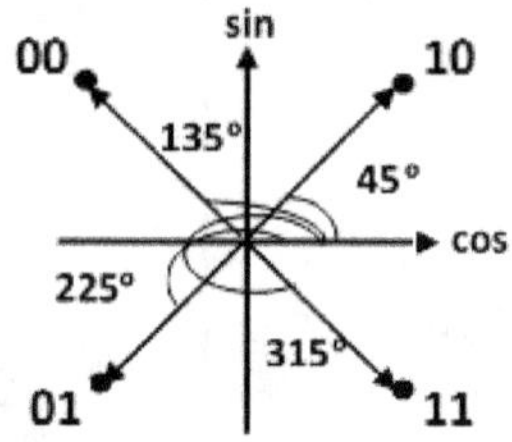

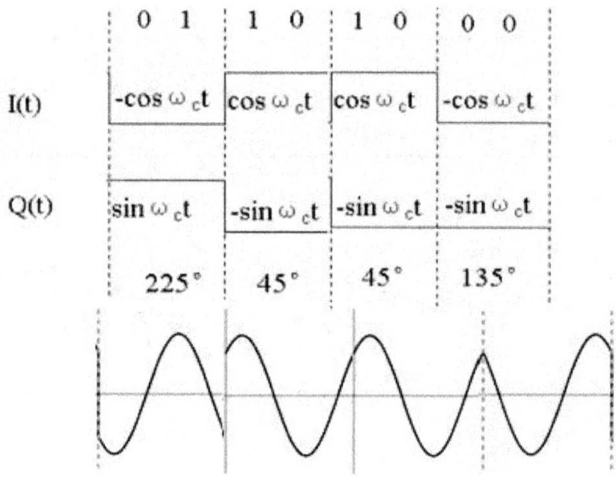

<u>Example:</u> Draw the QPSK waveform for the binary input sequence 011011001001.

<u>Sol:</u>

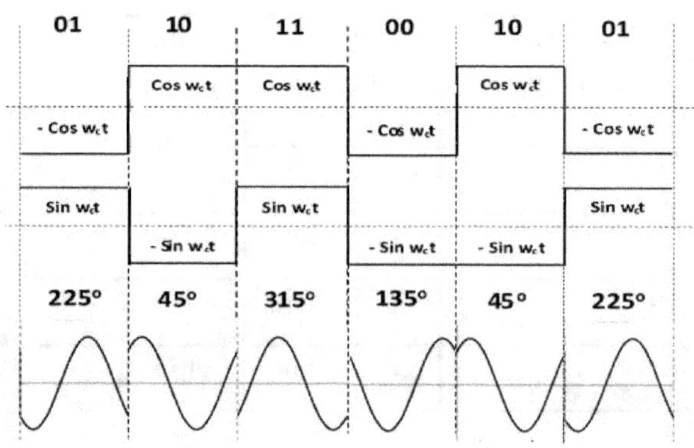

Types of QPSK:

1. ***Offset QPSK (OQPSK):*** Offset quadrature phase-shift keying (OQPSK) is a variant of phase-shift keying modulation using four different phase values to transmit. It is sometimes called staggered quadrature phase-shift keying (SQPSK). Taking four phase values (2 bits) at a time to construct a QPSK symbol can allow the signal phase to jump by as much as 180°at a time.

   The phase transitions are limited to 90^{0}, and the transitions on the (I and Q) channels are staggered. This yields much lower amplitude fluctuations than non-offset QPSK and is sometimes preferred in practice. Figure 10. 36 shows the constellation of OQPSK. The signal doesn't cross zero because only one bit of the symbol is changed simultaneously.

2. ***π/4–QPSK:*** This variant of QPSK uses two identical constellations, which are rotated by 45° (π/4 radians, hence the name) concerning one another. Usually, the even or odd symbols are used to select points from one of the constellations, and the other symbols select points from the other constellation. This also reduces the phase shifts from a maximum of 180°, but only to a maximum of 135°, so the amplitude fluctuations of π/4–QPSK are between OQPSK and non-offset QPSK. Inπ/4 QPSK, the set of constellation points is toggled with each symbol, so transitions through zero cannot occur. This scheme produces the lowest envelope variations.

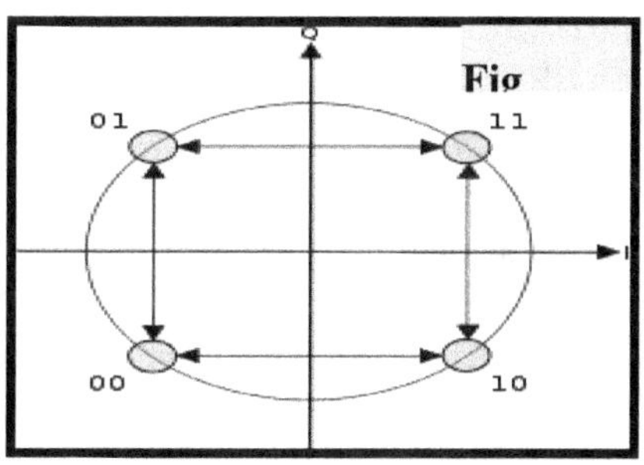

Multiple choice questions

1. In the spread spectrum technique
   A [ ]) a modulated signal is modulated again
   B [ ]) a modulated signal is modulated twice again
   C [ ]) the power of a modulated signal is increased
   D [v]) the noise component of a modulated signal is decreased

2. QAM stands for
   A [ ]) Quasi-Amplitude Modulation
   B [ ]) Quadrature Angle Modulation
   C [v]) Quadrature Amplitude Modulation
   D [ ]) None of the above

3. The Bandwidth required for amplitude modulation is
   A [ ]) half the frequency of the modulating signal
   B [ ]) equal to the frequency of the modulating signal
   C [v]) twice the frequency of the modulating signal
   D [ ]) four times the frequency of the modulating signal

4. BPSK stands for
   A [v]) Binary Phase Shifting Key
   B [ ]) Broad Phase Shifting Key
   C [ ]) Bit Phase Shifting Key
   D [ ]) Binary Pulse Shifting Key

5. In a system using QASK and 4 bits to form a symbol, the required Bandwidth is
   A [ ]) same as BPSK
   B [ ]) half of that in BPSK
   C [v]) one-fourth of that in BPSK
   D [ ]) none of the above

6. Bandwidth for practical PCM system is (where n is no. of channels, N length of PCM codes)
   A [v]) 2n N fm Hz
   B [ ]) 4n fm Hz
   C [ ]) 2n fm

D [ ]) (N + 1)fm

7. Circuit switching, message switching, and Packet Switching networks belong to which of the following systems?
   A [v]) Computer communication system
   B [ ]) Electromechanical communication system
   C [ ]) PBX system
   D [ ]) EPBX system

8. The Shannon-Hartley law
   A [v]) refers to noise
   B [ ]) defines Bandwidth
   C [ ]) describes the signaling rate
   D [ ]) refers to distortion

9. In communication systems, the frequency range used for voice is
   A [ ]) 15 Hz to 150 Hz
   B [ ]) 30 Hz to 1000 Hz
   C [ ]) 100 Hz to 1500 Hz
   D [v]) 300 Hz to 3800 Hz

10. In delta modulation, the output signal to quantization noise ratio S0/Nq is (N is the number of bits)
   A [ ]) $(3/\pi^2)N^2$
   B [v]) $(3/\pi^2)N^3$
   C [ ]) $(3/\pi)N^3$
   D [ ]) NONE OF THESE

11. Industrial noise has a frequency range of
   A [ ]) 1 to 100 Hz
   B [ ]) 200 to 5000 Hz
   C [ ]) above 200 MHz
   D [v]) 1 to 600 MHz

12. Line of sight transmission is used in
   A [ ]) radio frequencies below 1 MHz
   B [ ]) VHF only

C [ ]) audio signals only
D [v]) VHF and UHF

Chapter 11

Power Electronics

Power electronics is a branch of electrical engineering that deals with controlling and converting electrical power using electronic devices. Power electronics has applications in various industries, including energy generation and distribution, transportation, and industrial automation.

The main objective of power electronics is to efficiently convert electrical energy from one form to another, using various electronic devices such as diodes, transistors, thyristors, and integrated circuits. Power electronics systems can convert electrical power from AC to DC, DC to AC, DC to DC, and AC to AC, depending on the application requirements.

Some of the key applications of power electronics include:

Electric motor control: Power electronics are used to control the speed and torque of electric motors in various industrial and transportation applications.

Renewable energy systems: Power electronics efficiently convert the electrical power generated by renewable energy sources such as solar panels and wind turbines into usable electrical power.

Power supplies: Power electronics convert AC power from the electrical grid to the DC power required by electronic devices such as computers and smartphones.

Electric vehicle charging: Power electronics are used to efficiently charge the batteries of electric vehicles.

High-voltage direct current (HVDC) transmission: Power electronics efficiently transmit electrical power over long distances using HVDC transmission systems.

Power electronics systems must be designed to handle high voltages and currents and to operate efficiently over a wide range of load conditions. They must also be designed to handle various types of interference, such as electromagnetic interference (EMI) and radio frequency interference (RFI).

Power electronics is a rapidly evolving field, with new devices and techniques being developed to improve the efficiency and performance of power electronics systems. As the demand for energy-efficient and renewable energy systems continues to grow, the importance of power electronics in modern technology will continue to increase.

This chapter covers the most important applications in power electronics circuits, the most important of which are uncontrolled and controlled rectifier circuits. Also, the dc chopper circuit will explain.

11.1 Rectifiers (AC-DC Converters)

One of the first and most widely used power electronic device applications has been rectification. Rectification refers to converting an ac voltage or current source to dc voltage and current. Rectifiers refer to power electronic converters where the electrical power flows from the ac side to the dc side.

11.1.1 Rectifier Types

- Single Phase Rectifier
- Three-Phase Rectifier

Classification of single-phase rectifiers

1) Half-wave rectifiers
   i- Uncontrolled rectifier
   ii- Controlled rectifier

2) Full-wave rectifiers
   i- Uncontrolled rectifier
   ii- Controlled rectifier

In the rectifier analysis, the following simplifying assumptions will be made.

- The internal impedance of the ac source is zero.
- Power electronic devices used in the rectifier are ideal switches.

Uncontrolled Rectifiers: Provide a fixed dc as the output voltage for a given ac supply where diodes are used only.

Controlled Rectifiers: Provide an adjustable dc output voltage by controlling the phase at which the devices are turned on, where thyristors and diodes are used.

The full wave rectifiers can divide into:

A. *Half-controlled*: Allows electrical power flow from ac to dc (i.e., rectification only).

B. *Fully - controlled*: Allow power flow in both directions (i.e., rectification and inversion).

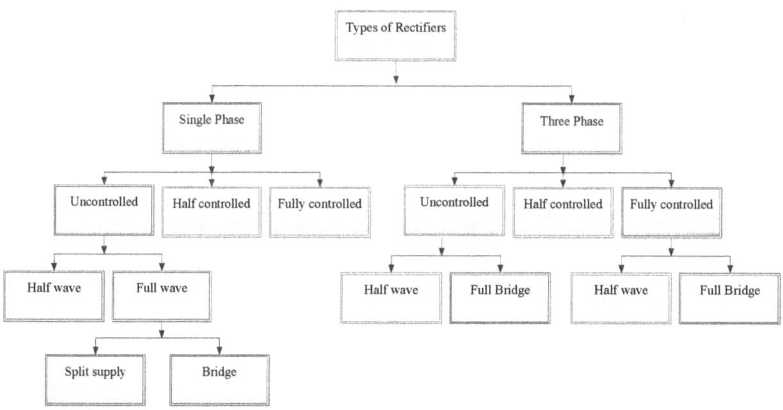

Figure 11.1 Classification of rectifiers.

11.1.2 Performance Parameters

The performance of a rectifier can be evaluated in terms of the following parameters.

1. Output dc Power

Output dc power = Average output voltage x Average output current

$$P_{dc} = V_{dc} I_{dc}$$

2. Total Power

Total power = RMS output voltage x RMS output current

$$P_{total} = V_{rms} I_{rms}$$

3. The efficiency of a Rectifier,

$$\text{Rectification efficiency } (\mathcal{J}) = \frac{P_{dc}}{P_{ac}} = \frac{I_{av} \cdot V_{av}}{I_{rms} \cdot V_{rms}}$$

Let f be the instantaneous value of any voltage or current associated with a rectifier circuit, then the following terms, characterizing the properties of f, can be defined.

Average (dc) value of f (F_{av}): Assuming f to be periodic over the period T

$$F_{av} \ or \ F_{mean} = \frac{1}{T}\int_0^T f(t)dt$$

4. Output ac Component

The output voltage can have two components: (1) the dc value and (2) the ac components or ripple. The RMS value of the ac component of the output voltage is

$$V_{ac} = \sqrt{V_{rms}^2 - V_{dc}^2}$$

5. Form Factor, *FF*

RMS (effective) value of f (F_{rms}): For periodic current over the period T

$$F_{rms} = \sqrt{\frac{1}{T}\int_0^T f^2(t)dt}$$

The form factor of f (FF): It measures the shape of the output voltage. It is defined as

$$FF = \frac{F_{rms}}{F_{av}}$$

6. Ripple Factor (*RF)*

It is a measure of the ripple content.

$$RF = \frac{V_{ac}}{V_{dc}}$$

or,

$$RF = \sqrt{(\frac{V_{rms}}{V_{dc}})^2 - 1} = \sqrt{FF^2 - 1}$$

7. Transformer Utilization Factor (*TUF)*

A transformer is often used to introduce galvanic isolation between the rectifier input and the AC mains and to adjust the rectifier AC input voltage to a level suitable for the required application. One of the parameters used to define the characteristics of the transformer is the Transformer Utilization Factor (TUF)

$$TUF = \frac{P_{dc}}{V_S\ I_S}$$

where

V_S: RMS voltage of the transformer secondary, and
I_S: RMS current of the transformer secondary.

8. Displacement Factor (*DF):*

$$DF = Cos\ \Phi$$

Where Φ is the phase angle between the fundamental of the input current and voltage.

9. Harmonic Factor (*HF)*

$$HF = \sqrt{\frac{I_S^{\,2} - I_1^{\,2}}{I_1^{\,2}}}$$

where I_1 is the fundamental RMS component of the input current.

10. Power Factor(*PF)*

$$PF = \frac{V_S\, I_1 \,\text{Cos}\,\Phi}{V_S\, I_S} = \frac{I_1}{I_S}\,\text{Cos}\,\Phi$$

$$= \text{Distortion Factor . Displacement Factor}$$

11.1.3 Uncontrolled Rectifiers

The rectifiers convert ac to dc using power diodes and are called uncontrolled rectifiers because they give constant out-voltage and fixed value if the input voltage (ac voltage) is constant. The diode is an appropriate component of uncontrolled rectifiers circuits because of one-way conducted properties and classifications of circuits are based on:

i. The number of phases: one phase and three phases.
ii. Waveform shape: half wave or full wave.

In this chapter, we will look at single-phase circuits if the diode has ideal properties, i.e., its resistance to the current is zero if the diode is forward-biased, and its resistance is very high if it is reverse-biased.

Half Wave Uncontrolled Rectifiers

i) Single Phase Half Wave Rectifier with Resistive Load

This circuit is one of the simplest circuits of the rectifier. During the positive half of the input voltage wave, the rectifier is in the forward bias state, allowing the current to pass through to the load resistance. In the case of an ideal diode used, the value of the voltage drops on both ends of the rectifier is zero. Consequently, the voltage on both ends of the load resistance is the same as the positive half of the input voltage, as shown in Figure 11.2.

There are two advantages of connecting using the transformer. First, it allows raising and reducing the source voltage as needed. Second, it achieves electrical insulation between the alternating current source and the rectifier to prevent sudden electric shocks in the secondary coil circuit.

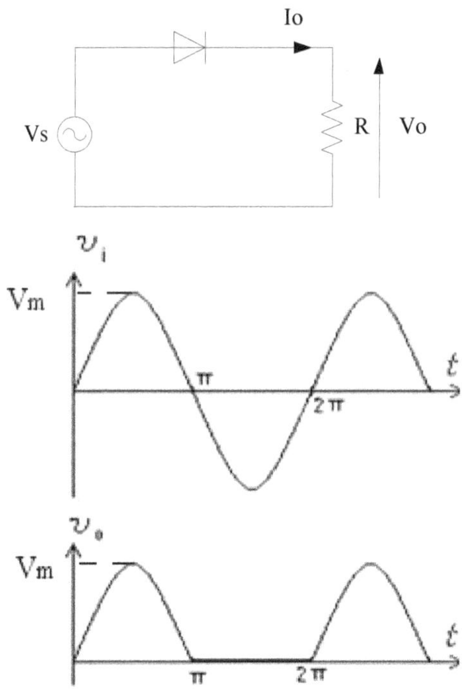

Figure 11.2 Half-wave uncontrolled rectifier.

The continuous output voltage can be calculated from the following formula:

$$V_{out} = V_{dc} = \frac{Vm}{\pi} = \frac{\sqrt{2}}{\pi} \times V_{rms}$$

The output current is calculated from the following formula:

$$I_{out} = I_{dc} = \frac{Vdc}{R} = \frac{Vm}{\pi R}$$

where

V_m represents the maximum value of the source voltage, and
V_{rms} effective source voltage.

To calculate the average value of the output voltage of the one-half wave is the value measured by the continuous voltmeter and can be calculated from the:

$$V_{av} = V_m / \pi$$

Where V_{av} is the average value of the output voltage.

Example

A single phase single wave has a purely resistive load of 8 Ω and a source voltage of 100 V at 60Hz. Calculate the output voltage and current.

Solution

$$V_o = V_{dc} = V_m/\pi$$

$$=(\sqrt{2} \times 100)/\pi = 45 \text{ V}$$

$$I_o = I_{dc} = V_{dc}/R = 45/8$$

$$= 5.63 \text{ A}$$

Example

Find the V_{av} of the combined voltage half wave if the maximum value of the source voltage is 50 V.

Solution

$V_{av} = V_m / \pi$

$= 50 / 3.14$

$= 15.9\ V$

Formula Derivative:

$$V_{av} \ or \ V_{dc} \ or \ V_{lmean} = \frac{1}{T}\int_0^T f(t)dt$$

$$= \frac{1}{2\pi}\int_0^\pi V_{s\,max} sin\theta d\theta$$

$$= \frac{V_{s\,max}}{2\pi}\int_0^\pi sin\theta d\theta$$

$$= \frac{V_{s\,max}}{2\pi}[cos\pi - cos0]$$

$$v_{s\,rms} = \frac{V_{s\,max}}{\sqrt{2}}$$

$$\therefore \ V_{lmean} = \frac{V_{s\,max}}{\pi}$$

or $I_{lmean} = \frac{1}{2\pi}\int_0^\pi I_{s\,max} sin\theta d\theta \ \ \therefore \ I_{lmean} = \frac{V_{lmean}}{R}$

$$v_{o\,rms} = \sqrt{\frac{1}{T}\int_0^T V^2(t)dt}$$

$$v_{0\,rms} = \sqrt{\frac{1}{2\pi}\int_0^\pi V_{smax}{}^2 sin^2\theta d\theta}$$

$$= V_{smax}\sqrt{\frac{1}{2\pi}\int_0^{\pi}\frac{1}{2}(1 - cos2\theta)d\theta}$$

$$= V_{smax}\sqrt{\frac{1}{4\pi}\int_0^{\pi}(1 - cos2\theta)d\theta}$$

$$=V_{smax}\sqrt{\frac{1}{4\pi}\int_0^{\pi} 1d\theta - \frac{1}{2}\int_0^{\pi} cos2\theta\, d\theta}$$

$$= V_{smax}\sqrt{\frac{1}{4\pi}[\,\theta - \frac{sin2\theta}{2}\,]_0^{\pi}}$$

$$= V_{smax}\sqrt{\frac{1}{4\pi}[\,\pi - \frac{sin2\pi}{2} - 0 + \frac{sin2 \times 0}{2}]}$$

$$= V_{smax}\sqrt{\frac{1}{4\pi}[\pi]}$$

or $i_{0\,rms} = \sqrt{\frac{1}{2\pi}\int_0^{\pi} I_{smax}{}^2 sin^2\theta d\theta}$ $\therefore\ i_{0\,rms} = \frac{v_{o\,rms}}{R}$

$$P_{dc} = \frac{(\frac{Vm}{\pi})^2}{R} = \frac{V_m{}^2}{\pi^2 R}$$

$$P_{total} = \frac{V_{rms}{}^2}{R} = \frac{(V_m / 2)^2}{R} = \frac{1}{4}\frac{V_m{}^2}{R}$$

1. $$\frac{P_{dc}}{P_{total}} = \frac{1}{\pi^2 R}4R = 0.405\ (40.5\%)$$

2. $$FF = \frac{V_{rms}}{V_{dc}} = \frac{\frac{Vm}{2}}{Vm/\pi} = \frac{\pi}{2} = 1.57\ (157\%)$$

$$RF = \sqrt{FF^2 - 1}$$

3. $$= \sqrt{1.57^2 - 1}$$

$$= 1.21\,(121\%)$$

4. $$TUF = \frac{P_{dc}}{V_S . I_S} = \frac{\frac{V_m^{\ 2}}{\pi^2 R}}{(\frac{V_m}{\sqrt{2}}) * (\frac{V_m}{2R})} = \frac{2\sqrt{2}}{\pi^2} = 0.287\#$$

Example

For the single phase half wave uncontrolled rectifier circuit shown in Figure 11.3, $R = 5\Omega$, $V_S = 150\ sinwt$. Calculate $V_{Lmean}, I_{Lmean}, v_{s_{r.m.s}}, v_{o_{r.m.s}}, i_{o_{r.m.s}}$ and RF.

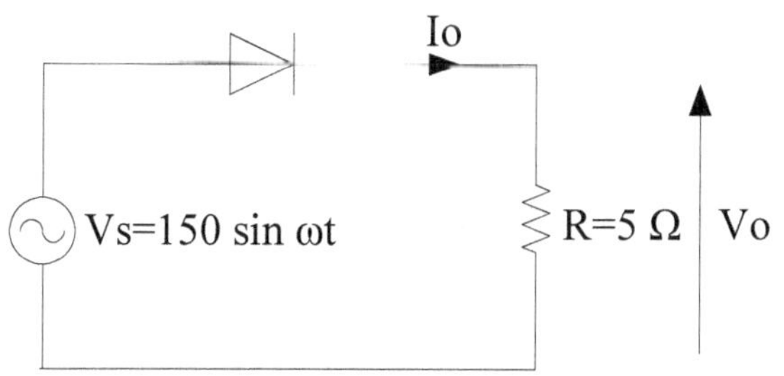

Figure 11.3 Single phase half wave uncontrolled rectifier circuit diagram.

Solution

$$V_{lmean} = \frac{V_{s\,max}}{\pi} = \frac{150}{3.14} = 47.7\ V$$

$$I_{lmean} = \frac{V_{lmean}}{R} = \frac{47.7}{5} = 9.5\ A$$

$$v_{s\ rms} = \frac{V_{s\ max}}{\sqrt{2}} = \frac{150}{\sqrt{2}} = 106.06\ V$$

$$v_{o\ rms} = \frac{V_{s\ max}}{2} = \frac{150}{2} = 75\ V$$

$$i_{o\ rms} = \frac{v_{o\ rms}}{R} = \frac{75}{5} = 15\ A$$

$$RF = \sqrt{\frac{{V_{o\ rms}}^2 - {V_{lmean}}^2}{{V_{lmean}}^2}} = \sqrt{\frac{(75)^2 - (47.7)^2}{(47.7)^2}}$$
$$= 1.213$$

ii) Single Phase Half Wave Rectifier with Inductive Load

Due to the inductive load, the conduction period of the diode will extend beyond 180° until the current becomes zero. Consider the case of the R-L load, as shown in Figure 11.4. The voltage source is a sine wave, the positive half period when $0 < \omega t < \pi$, and the negative half period when $\pi < \omega t < 2\pi$. When the voltage source becomes positive, the diode starts conducting, and the source keeps the diode in conduction until ωt reaches π radians. At that instant, defined by $\omega t = \pi$ radians, the current through the circuit is not zero, and some energy is stored in the inductor.

The voltage across an inductor is positive when the current increases and it becomes negative when the current through it tends to fall. When the voltage across the inductor is negative, it is in such a direction as to forward bias the diode. The polarity of the voltage across the inductor is shown in the waveforms in Figure 11.3. When the source voltage changes from a positive to a negative value, the voltage across the diode changes its direction, and there is current through the load at the instant $\omega t = \pi$ radians, and the diode continues to conduct till the energy stored in the inductor becomes zero. After that, the current tends to flow in the reverse direction and the diode blocks conduction. The entire applied voltage appears across the diode as a reverse bias voltage. An expression for the current through the diode can be obtained by solving the differential equation representing the circuit.

(a)

(b)

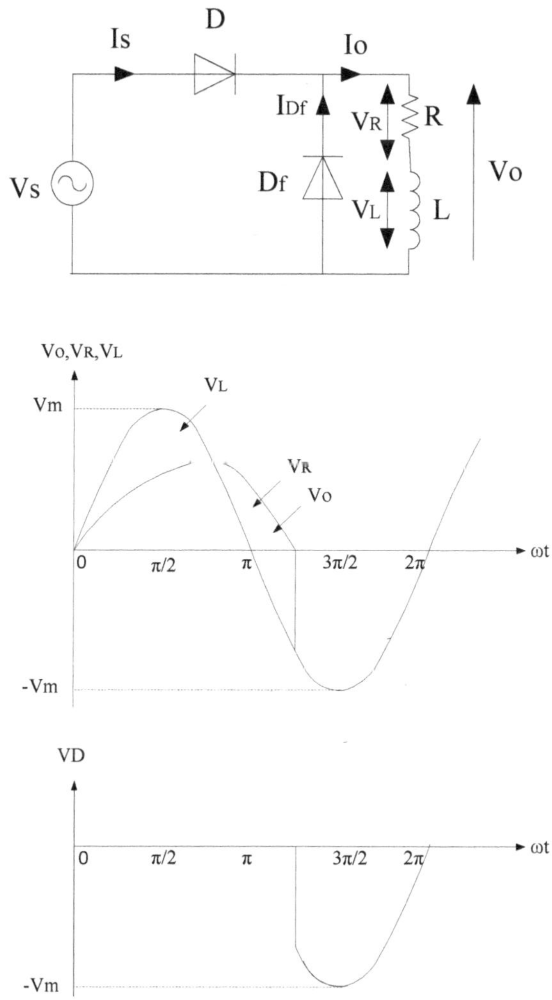

Figure 11.4 Single phase half wave rectifier with an inductive load.

(a) Circuit diagram.

(b) Waveforms.

During diode conduction,

$$L\frac{di}{dt} + i\,R = V_m \sin \omega t$$

A solution to this differential equation is :

$$i = \frac{V_m}{Z}\left[Sin(\omega t - \Phi) + Sin\ \Phi\ e^{\frac{-\omega t}{\tan \Phi}}\right]; \qquad 0 \le \omega t \le \beta$$

$$Z = \sqrt{R^2 + \omega^2 L^2}\,; \tan \Phi = \frac{\omega L}{R}$$

At $\omega t = \beta, i = 0$

$$\therefore 0 = Sin(\beta - \Phi) + \sin \Phi\ e^{\frac{-\beta}{\tan \Phi}}$$

This is a transcendental equation and can be solved by iterative techniques. The extinction angle can be determined for a given load impedance angle Φ.

The average output voltage is

$$V_{dc} = \frac{V_m}{2\pi}\int_0^{\beta} \sin \omega t\ d\omega t$$

$$= \frac{V_m}{2\pi}(1 - \cos\beta)$$

The average output current is

$$I_{dc} = \frac{V_m}{2\pi\ R}(1 - \cos\beta)$$

Example

For the single phase half wave uncontrolled rectifier circuit shown in Figure 11.5, $R = 10\Omega$, L=9 mH, $V_s = 150 sinwt$. Calculate and trace $V_{Lmean}, I_{Lmean}, v_{s_{r.m.s}}$ and FF.

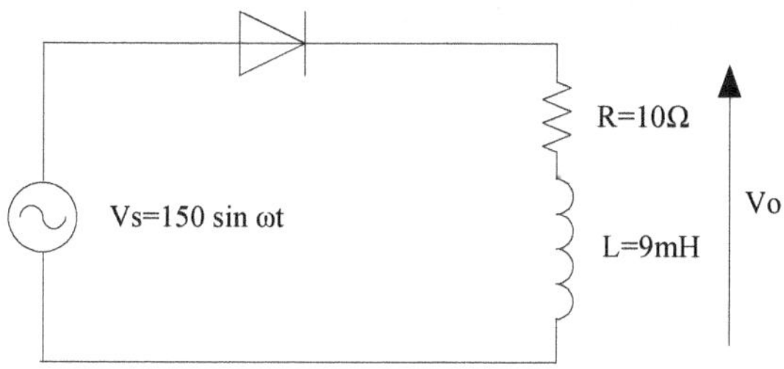

Figure 11.5 Circuit diagram.

Solution

$$\varphi = \tan^{-1}\frac{\omega L}{R} = \tan^{-1}\frac{2 \times 3.14 \times 50 \times 9 \times 10^{-3}}{10} = 15.7^{0}$$

$$V_{lmean} = \frac{1}{2\pi}\int_{0}^{\pi+\varphi} V_{s\,max} sin\theta d\theta$$

$$V_{lmean} = \frac{V_{s\,max}}{2\pi}[1 + \cos\varphi]$$

$$=\frac{150}{2\pi}[1 + \cos 15.7°] = 46.8\,V$$

$$so\ I_{lmean} = \frac{V_{lmean}}{R} = \frac{46.8}{10} = 4.68\,A$$

$$V_{s\,rms} = \frac{V_{s\,max}}{\sqrt{2}} = \frac{150}{\sqrt{2}} = 106.06\,V$$

$$v_{0\,rms} = V_{smax}\sqrt{\frac{1}{4\pi}[\pi + \varphi - \frac{sin2(\pi + \varphi)}{2}}$$

$$= 150\sqrt{\frac{1}{4 \times 3.14}[\,3.14 + 15.7^0 \times \frac{\pi}{180} - \frac{sin2(180 + 15.7)}{2}}$$

$$= 150\sqrt{0.08\,[\,3.14 + 0.27 - \frac{0.52}{2}]} \approx 75.2\,V$$

$$Form\,factor\,(FF) = \frac{v_{o\,rms}}{V_{lmean}} = \frac{75.2}{46.8} = 1.6$$

Full-Wave Rectifiers

Although the rectifier half-wave has some applications, the use of a rectifier full-wave is more prevalent in the sources of power for DC, and the difference between full-wave and the half-wave rectifier is that the rectifier full-wave allows one-way current through the load in each half cycle. While a single half-wave allows the current to pass through the positive half of the wave only, and as a result, we get the positive pulse number from the output of the full-wave rectifier equal to twice the number of positive pulses we get from a single half-wave output over the same period. Figure 11.6 shows a full wave rectifier.

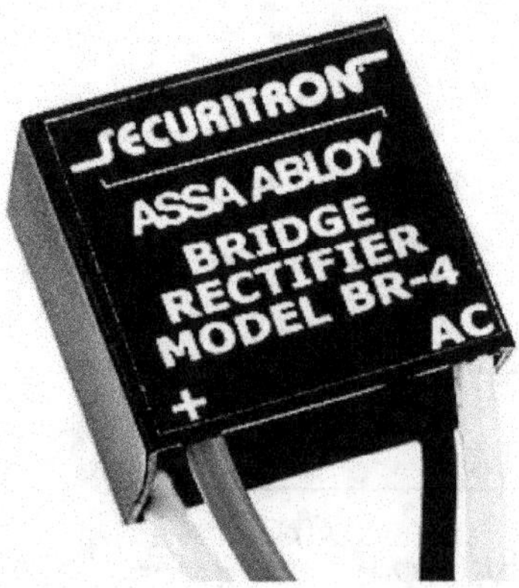

Figure 11.6 Full wave rectifier. (https://www.grainger.com/product).

Full-Wave Center Tapped Rectifier With Resistive Load

Figure 11.7 shows a complete wave rectifier circuit using an adapter with an intermediate junction point. In this circuit, when the sinusoidal wave is in the positive half, the rectifier D_1 is in the front bias position, so the rectifier (D_2) is in a reverse bias state.

In the case of a Sinusoidal wave in the negative half, the rectifier (D_1) is in reverse bias, and the D_2 is in the front bias state. As a result, it conducts the current; thus, half of the positive and negative wave appears on the output. Figures 11.8 a and b show the stages in which the full wave is obtained.

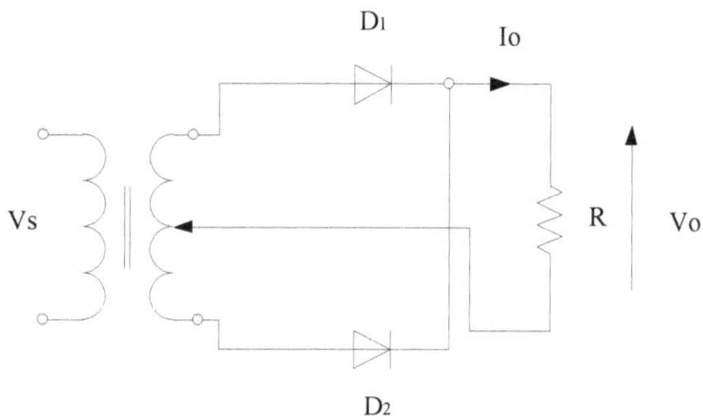

(a)

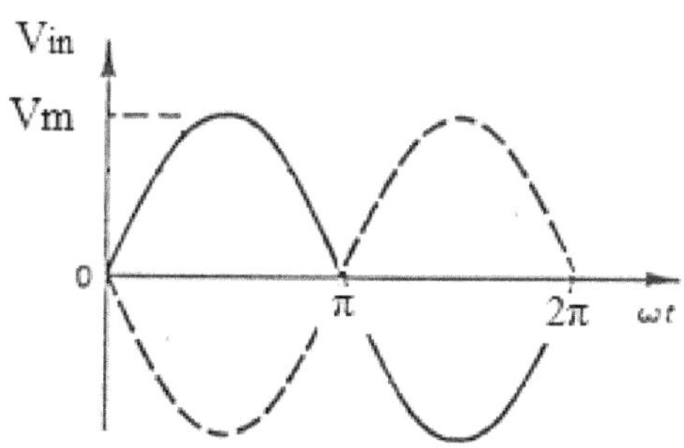

(b)

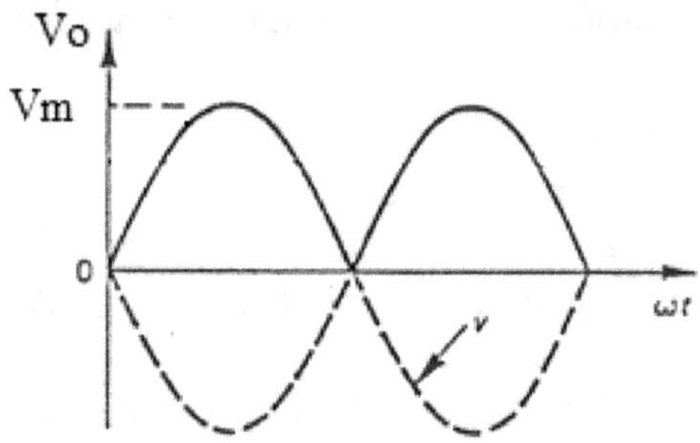

Figure 11.7 Full wave center tapped rectifier. (a) circuit diagram. (b) waveforms.

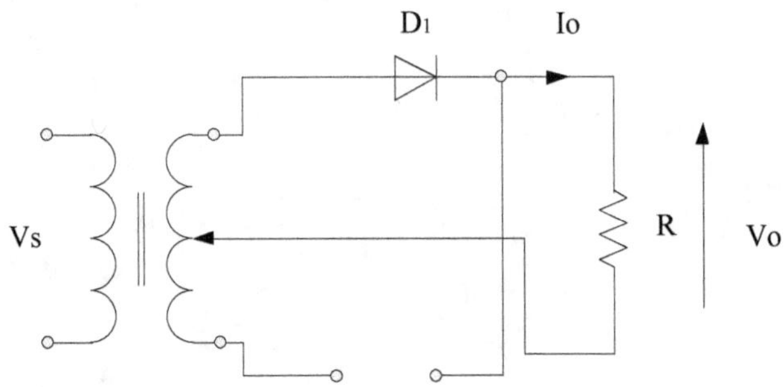

(a)

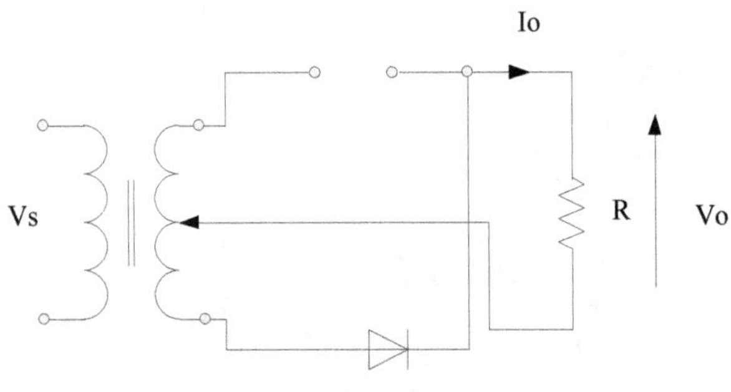

(b)

Figure 11.8 The stages in which the full wave is obtained.

(a) Diode 1 ON.
(b) Diode 2 ON.

The average value of the voltage, in this case, shall be calculated from the following formula:

$$V_{av} = 2V_m / \pi$$

In this case, the average value of the voltage (Vav) (full wave) Is twice that obtained in a half-wave state.

Example

Find the V_{av} of the load voltage a full wave if the maximum value of the output voltage is 15 V, and create the load current if the load resistance is 3 Ω.

Solution

$$V_{av} = 2Vm / \pi$$
$$= 2 \times 15 / 3.14$$
$$= 9.55 \text{ v}$$

$$I_{dc} = V_{av} / R$$
$$= 9.55 / 3$$
$$= 3.18 \text{ A}$$

Example

The rectifier shown in Figure 11.6 has a purely resistive load of R . Calculate:

i. The rectifier efficiency.
ii. Form factor.
iii. Ripple factor.
iv. Peak inverse voltage (PIV) for diode D_1

Solution

i.

$$\eta = \frac{V_{dc} \cdot I_{dc}}{V_{rms} \cdot I_{rms}}$$

$$= \frac{\frac{2\,V_m}{\pi} \cdot \frac{2\,V_m}{\pi\,.\,R}}{\frac{V_m}{\sqrt{2}} \cdot \frac{V_m}{\sqrt{2}\,.\,R}} = 81.05\%$$

ii.

$$F.F = \frac{V_{rms}}{V_{dc}}$$

$$= \frac{\frac{V_m}{\sqrt{2}}}{\frac{2\,V_m}{\pi}} = 111\%$$

iii.

$$R.F = \frac{V_{ac}}{V_{dc}} = \sqrt{F.F^2 - 1}$$

$$= \sqrt{1.11^2 - 1} = 48\%$$

iv.

PIV=2.Vm

Single–Phase Full Wave (Bridge) Rectifier

i) Single – Phase Full Wave (Bridge) Rectifier with a resistive load

This rectifier needs four diodes to form the circuit, as shown in Figure 11.9.

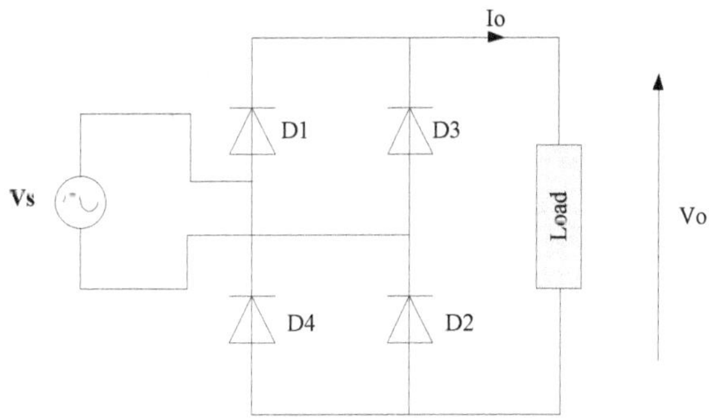

Figure 11.9 Single–phase full wave (bridge) rectifier.

The diodes (D_1, D_2) are ON during the positive part of the input voltage, and the diodes (D_3, D_4) are reverse-biased. The current follow through the diode (D_1, D_2) and during the negative half of the input voltage wave through D_3, and D_4, where are ON and Diodes (D_1, D_2) are in reverse bias. The current follow through the diode in (D_3, D_4).

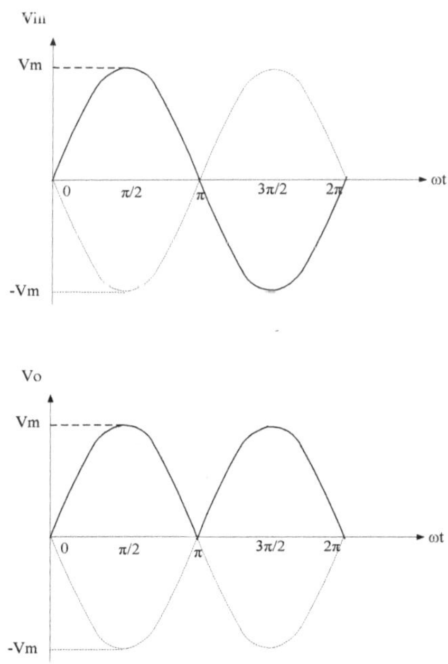

Figure 11.10 Input and output voltage of single-phase bridge rectifier.

The voltage and current calculation equations are the same as used in the full-wave rectifier with the center tap transformer. However, it should be noted that the diode in the rectifier of the bridge rectifier is subjected to a reverse voltage equal to half of the voltage exerted by the diode in the rectifier with the center tap transformer, which reduces the diode's rating. Figure 11.11, a, and b shows the stages of the rectifier of a full wave using four diodes.

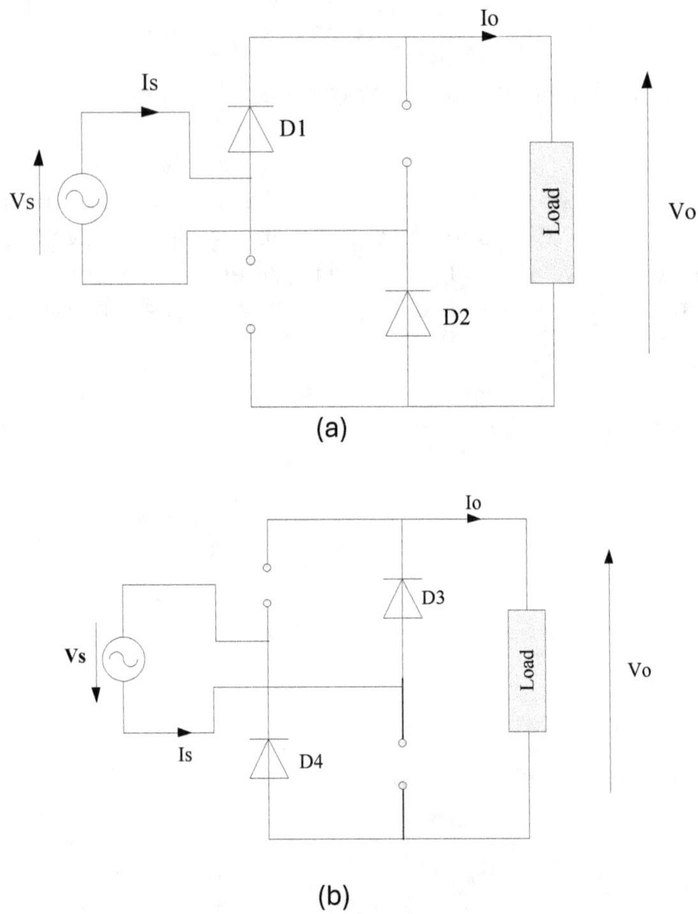

Figure 11.11 The stages of operation using a single-phase bridge rectifier. (a) D_1 and D_2 are ON., (b)D_3 and D_4 are ON.

Formula Derivative:

$$V_{lmean} = \frac{1}{2\pi}\int_0^{\pi} V_{s\,max} sin\theta d\theta \times 2$$

$$= \frac{V_{s\,max}}{\pi}\int_0^{\pi} sin\theta d\theta$$

$$= \frac{V_{s\,max}}{\pi}[-cos\theta]_0^{\pi}$$

$$= -\frac{V_{s\,max}}{\pi}[cos\pi - cos0]$$

$$\therefore \ V_{lmean} = \frac{2\,V_{s\,max}}{\pi}$$

$$\therefore \ I_{lmean} = \frac{V_{lmean}}{R}$$

or $I_{lmean}=\frac{1}{2\pi}\int_0^{\pi} I_{s\,max} sin\theta d\theta \times 2$

$$v_{s\,rms} = \frac{V_{s\,max}}{\sqrt{2}}$$

$$v_{0\,rms} = \sqrt{\frac{1}{2\pi}\int_0^{\pi} V_{smax}{}^2 sin^2\theta d\theta \times 2}$$

$$= V_{smax}\sqrt{\frac{1}{\pi}\int_0^{\pi}\frac{1}{2}(1 - cos2\theta)d\theta}$$

$$= V_{smax}\sqrt{\frac{1}{2\pi}\int_0^{\pi}(1 - cos2\theta)d\theta}$$

$$= V_{smax}\sqrt{\frac{1}{2\pi}\int_0^{\pi} 1d\theta - \frac{1}{2}\int_0^{\pi} cos2\theta\, d\theta}$$

$$= V_{smax}\sqrt{\frac{1}{2\pi}[\pi - \frac{sin2\pi}{2} - 0 + \frac{sin2\times 0}{2}]}$$

$$= V_{smax}\sqrt{\frac{1}{2\pi}[\pi]}$$

$$v_{0\,rms} = \frac{V_{smax}}{\sqrt{2}}$$

or $$i_{0\,rms} = \sqrt{\frac{1}{2\pi}\int_0^{\pi} I_{smax}{}^2 sin^2\theta d\theta \times 2} \therefore\ i_{0\,rms} = \frac{v_{o\,rms}}{R}$$

Example

For the full wave uncontrolled rectifier (bridge) circuit shown in Figure below, $R = 10\Omega$, Calculate $V_{Lmean}, I_{Lmean}, v_{s_{r.m.s}}. v_{o_{r.m.s}}$, and $i_{o_{r.m.s}}$

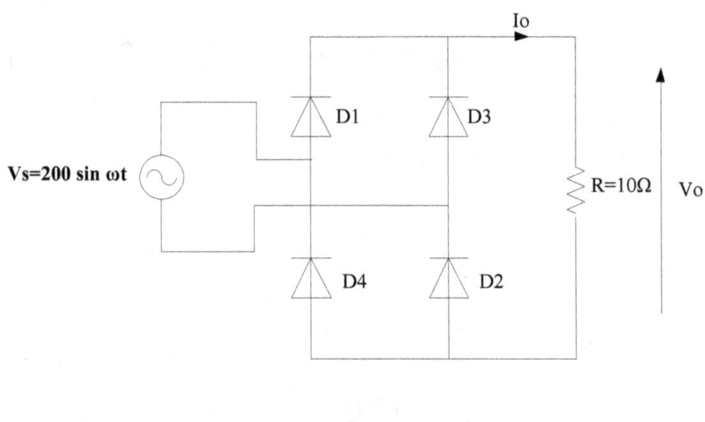

Solution

$$V_{L_{mean}} = \frac{1}{2\pi}\int_0^{\pi} V_{max} sin\,\theta d\theta \times 2 = \frac{2V_{max}}{\pi} = 127.3\ V$$

$$I_{L_{mean}} = \frac{V_{L_{mean}}}{R} = \frac{127.3}{10} = 12.73\ A$$

$$v_{s_{rms}} = \frac{V_{smax}}{\sqrt{2}} = \frac{200}{\sqrt{2}} = 141.42\ V$$

$$v_{0\ rms} = \sqrt{\frac{1}{2\pi}\int_0^{\pi} V_{smax}{}^2 sin^2\theta d\theta \times 2} = \frac{V_{max}}{\sqrt{2}} = \frac{200}{\sqrt{2}} = 141.42\ V$$

$$i_{0\ rms} = i_{s\ rms} = \frac{v_{o\ rms}}{R} = \frac{141.42}{10} = 14.14\ A$$

ii) Single – Phase Full Wave (Bridge) Rectifier with Inductive Load:

For an RL series-connected load in Figure 11.12, the analysis method is like that for the half-wave rectifier with the freewheeling diode discussed previously. After a transient that occurs during start-up, the load current $i_{o(t)}$ reaches a periodic steady-state condition. For the bridge circuit, the current is transferred from one pair of diodes to the other pair when the source changes polarity. The voltage across the RL load is a full-wave rectified sinusoid, as it was for the resistive load. The full-wave rectified Sinusoidal voltage across the load can be expressed as a Fourier series consisting of a dc term and even harmonics.

$$V_{dc} = \frac{2\,V_m}{\pi}; \quad V_{rms} = \frac{V_m}{\sqrt{2}}$$

$$RF = \sqrt{(\frac{V_{rms}}{V_{dc}})^2 - 1} = \sqrt{(\frac{\pi}{2\sqrt{2}}) - 1} = 0.483$$

This is significantly less than the half-wave rectifier, with a highly inductive load, which is the usual practical application, virtually constant load current flows. The bridge diode currents are then square blocks of current with magnitude $I_{dc.}$, in this application:

The average current in each diode $= I_{D,av} = \frac{I_{dc}}{2}$

RMS current in each diode $= I_{D,rms} = \frac{I_{dc}}{\sqrt{2}}$

Diode current, $FF = \frac{I_{D,rms}}{I_{D,av}} = \sqrt{2}$

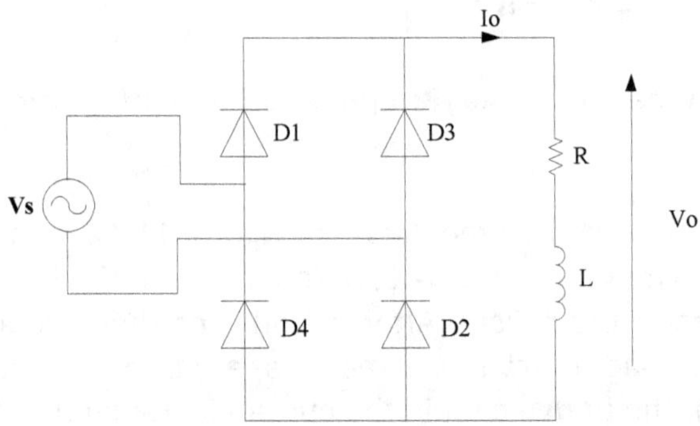

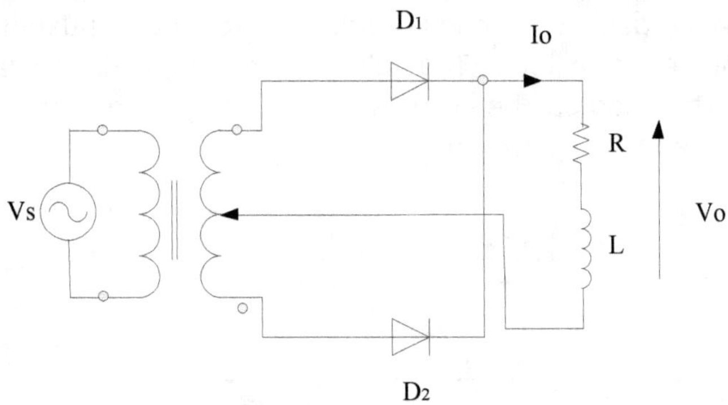

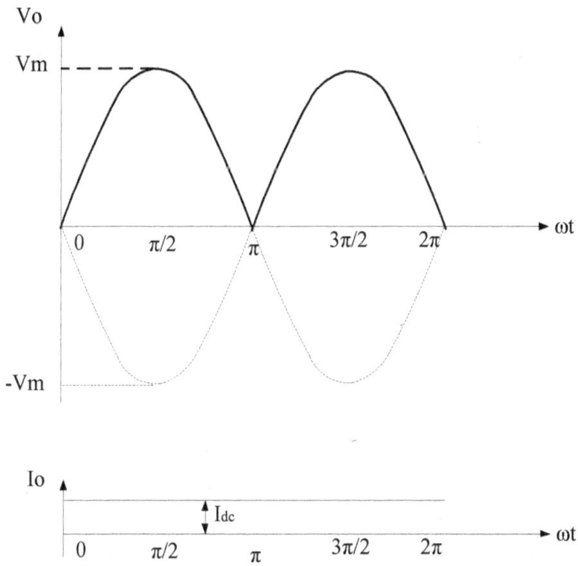

Figure 11.12 Single–phase full wave (bridge and center tap) rectifier with an inductive load and its related waveforms.

Example

Repeat above example with a heavily inductive load and find RF and efficiency.

Solution

$$V_{L_{mean}} = \frac{2V_{S_{max}}}{\pi} = \frac{2 \times 200}{\pi} = 127.3\ V$$

$$I_{L_{mean}} = \frac{V_{L_{mean}}}{R} = \frac{127.3}{10} = 12.73\ A$$

$$v_{S_{rms}} = v_{o_{rms}} = \frac{V_{smax}}{\sqrt{2}} = \frac{200}{\sqrt{2}} = 141.42\ V$$

$$i_{s\ rms} = i_{o\ rms} = \sqrt{\frac{1}{2\pi}\int_0^{\pi} I_{lmean}{}^2\theta d\theta \times 2} = I_{lmean} = 12.73\ A$$

$$Form\ factor\ (FF) = \frac{v_{o\ rms}}{V_{lmean}} = \frac{141.42}{127.3} = 1.11$$

$$Ripple\ Factor\ (RF) = \sqrt{FF^2 - 1} = \sqrt{(1.11)^2 - 1} = 0.48$$

$$efficiency\ (\eta) = \frac{P_{av}}{P_{ac}} = \frac{V_{lmean} \times I_{lmean}}{v_{o\ rms} \times i_{o\ rms}} = \frac{127.3}{141.42} = 90\%$$

11.1.4 Rectifiers with Filter Circuits

In the previous rectifier circuits, the output is a rectifier voltage, the direction of the variable value in the form of pulses, and to reduce the value of the ripples in the voltage, and we should be using some types of filters that apply to the output of the rectifier circuits., Either capacitor (C) or inductance and capacitance (LC).

For a given load, a larger capacitor will reduce ripple but cost more, creating higher peak currents in the transformer secondary and the supply feeding it. In extreme cases where many rectifiers are loaded onto a power distribution circuit, it may prove difficult for the power distribution authority to maintain a correctly shaped sinusoidal voltage curve.

For a given tolerable ripple, the required capacitor size is proportional to the load current and inversely proportional to the supply frequency and the number of output peaks of the rectifier per input cycle. The load current and the supply frequency are generally outside the control of the designer of the rectifier system, but the number of peaks per input cycle can be affected by choice of rectifier design.

A half-wave rectifier will only give one peak per cycle, which is only used in very small power supplies for this and other reasons. A full wave rectifier achieves two peaks per cycle, which is the best to do with single-phase input. For three-phase inputs, a three-phase bridge will give six peaks per cycle, and even higher numbers of peaks can be achieved by using transformer networks placed before the rectifier to convert to a higher phase order.

To further reduce this ripple, a capacitor-input filter can be used. This complements the reservoir capacitor with an inductor and a second filter capacitor so that a steadier dc output can be obtained across the terminals of the filter capacitor. The inductor presents a high impedance to the ripple current. Figure 11.13 shows the filtration process.

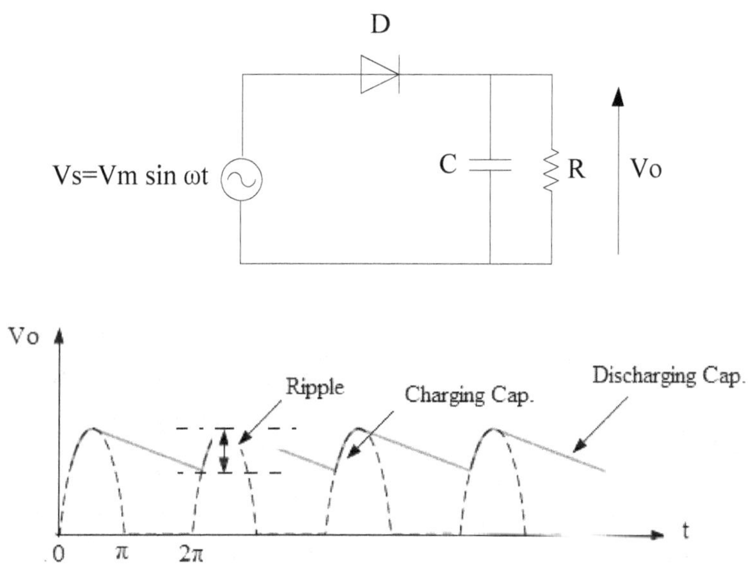

(a)

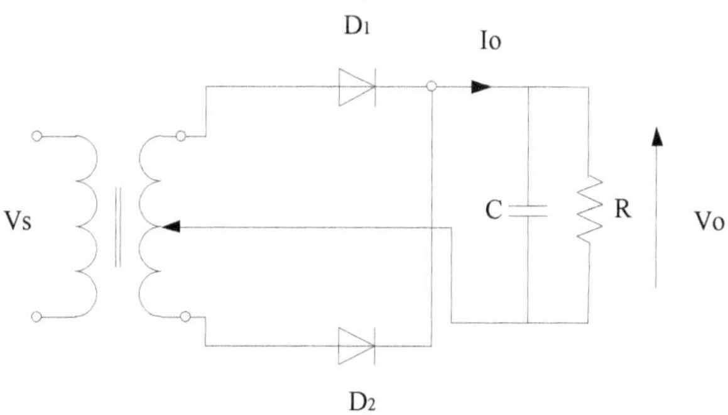

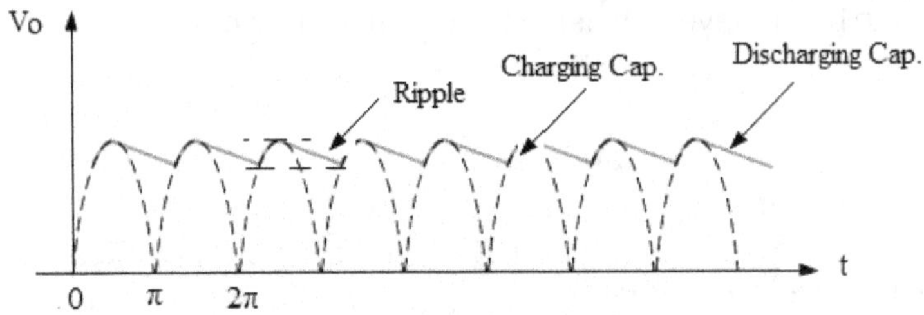

(b)

Figure 11.13 Filtration process (a)Half Wave Rectifier. (b)Full Wave Rectifier.

11.1.5 Controlled Rectifiers

These rectifiers use the thyristors and the delay angle to control the output voltage and current.

Thyristor Firing Circuits

The thyristor becomes a conductor if a positive voltage is found between the anode and the cathode and a pulse trigger on the gate. The trigger is a signal on the gate in the form of a pulse that takes time to operate the thyristor.

Shockley diodes are curious devices but rather limited in application. However, their usefulness may be expanded by equipping them with another latching means. In doing so, they become true amplifying devices (if only in an on/off mode), and we refer to them as silicon-controlled rectifiers or SCRs.

The progression from Shockley diode to SCR is achieved with one small addition, nothing more than a third wire connecting to the existing PNPN structure:

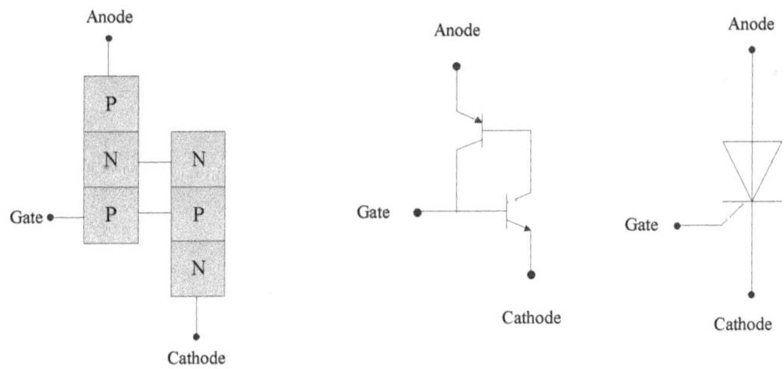

Figure 11.14 The silicon-controlled rectifier. A-Physical diagram, B-Equivalent schematic, C-Schematic. symbol

If SCR's gate is left floating (disconnected), it behaves exactly as a Shockley diode. It may be latched by brake overvoltage or by exceeding the critical rate of voltage rise between the anode and cathode, just as with the Shockley diode. Dropout is accomplished by reducing current until one or both internal transistors fall into the cutoff mode, like the Shockley diode.

However, because the gate terminal connects directly to the lower transistor's base, it may be an alternative means to latch the SCR. By applying a small voltage between the gate and cathode, the lower transistor will be forced on by the resulting base current, which will cause the upper transistor to conduct, which then supplies the lower transistor's base with the current so that it no longer needs to be activated by a gate voltage. The necessary gate current to initiate latch-up will be much lower than the current through the SCR from the cathode to the anode, so the SCR achieves amplification.

This method of securing SCR conduction is called triggering, and it is by far the most common way that SCRs are latched in actual practice. SCRs are usually chosen so that their brake overvoltage is far beyond the greatest voltage expected to be experienced from the power source so that it can be turned on only by an intentional voltage pulse applied to the gate.

The Use of a Thyristor in the Controlled Rectifier Circuits

The thyristors are used in the circuits of the controlled rectifier to convert the alternating current to a constant current and voltage that can be controlled by the angle of the trigger of the thyristor (α) and called (trigger angle). The control rectifiers are characterized by simplicity and high efficiency, as well as the low cost of manufacturing, which requires a voltage change to control the speed of the motors.

Single Phase Half Wave Controlled Rectifier with Resistive Load

The circuit in Figure 11.15 is one of the simplest applications on the controlled rectifier circuits, consisting of the alternating current source, thyristor, and resistance load.

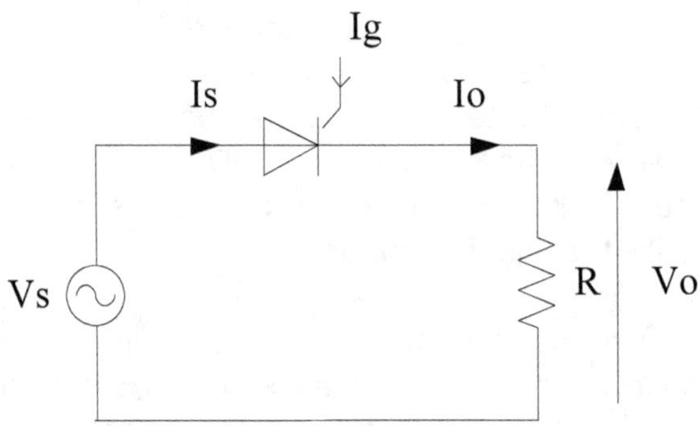

Figure 11.15 Half-wave controlled rectifier.

During the positive half of the wave, the anode voltage is higher than the cathode voltage, and the thyristor is in the forward bias state. When the gate's current triggers the thyristor at the firing angle, the input voltage will appear on the load. When the current follows the load to be zero, the thyristor becomes off, and when the negative half of the input wave is, the anode voltage is lower than the cathode voltage, and the thyristor is in reverse bias and repeated with each cycle (2 π). Figure 11.16 shows the input wave, gate current, load current, and voltage. The average value of the mean output voltage can be found in the following equation.

$$V_o = \frac{Vm}{2\pi}\,(1 + \cos \alpha)$$

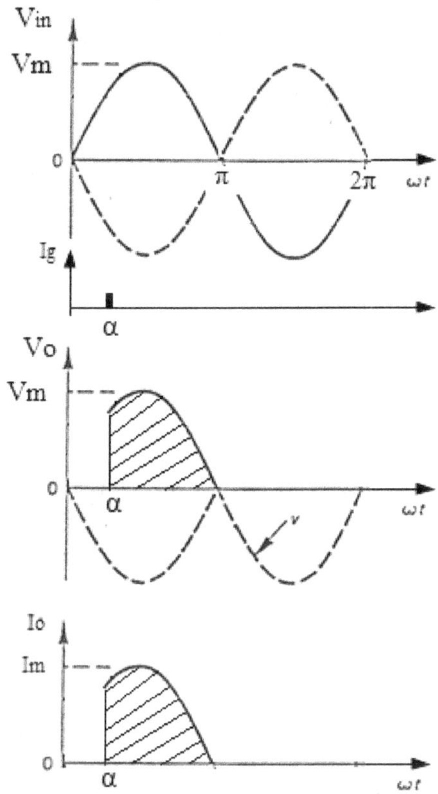

Figure 11.16 Input, output voltage, and output current waveforms.

Single Phase Half Wave Control Rectifier with R-L Load

The importance of this circuit lies in the fact that most industrial loads are inductive loads containing resistance and inductive.

Figure 11.17 shows a circuit with a resistive and inductive load, and Figure 11.18 shows the input and output currents of the voltages and current of this circuit.

The current will start to flow in the thyristor when it is triggered, and this current lags the negative voltage due to the transient state of the current caused by the inductor and sometimes freewheeling diode to get rid of the negative part of the load voltages.

The average value of the output voltage without the freewheeling diode can be found in the following formula:

$$V_o = \frac{Vm}{2\pi}(\cos\alpha - \cos\beta)$$

Where β: is the angle at which the thyristor turns OFF.

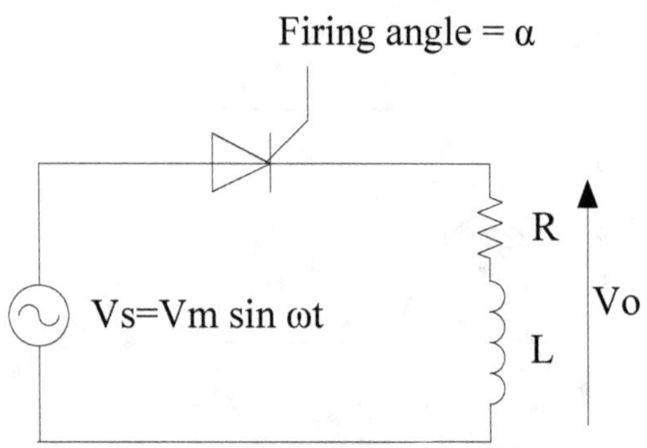

Figure 11.17 Single phase half wave control rectifier with R-L load Circuit.

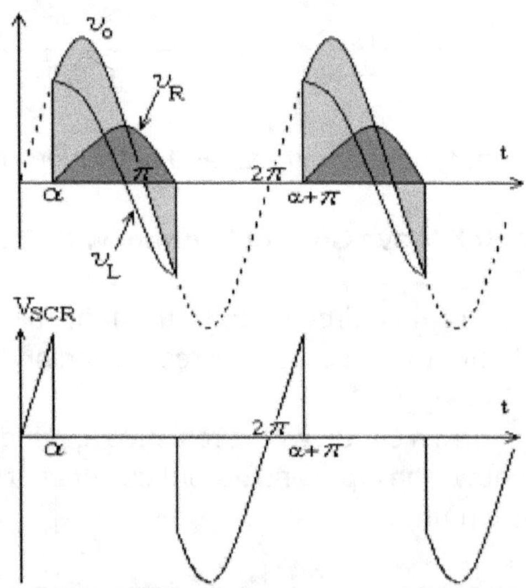

Figure 11.18 Output voltage and current and SCR voltage of single phase half wave control rectifier with the R-L load.

Example

A single-phase half-wave controlled rectifier is used to feed a pure resistance load of $R = 10\Omega$ and ac voltage of 220 V at frequency f= 60 Hz. Calculate:

(a) Output voltages at delay angle $\alpha = 90°$.
(b) The output and thyristor current at the angle of the $\alpha=90°$.

Solution

(a)

$$Vo = \frac{Vm}{2\pi}\ (1 + \cos \propto)$$

$$= \frac{220}{2\pi}\sqrt{2}(\ 1 + \cos 90^{o}\ )$$

$$V_o = 49.51\ \ V$$

(b)

$$I_o = V_o/R = 49.51/10 = 4.951\ A$$

$$I_{av} = I_o$$

$$= 4.951\ A$$

Example

A single-phase half-wave controlled rectifier with a resistive load and high inductance with a supply voltage of 120V and frequency of 60Hz

(a) Calculate the output voltage at $\alpha = 60^{o}$ and $\beta = 200^{o}$.
(b) Calculate the output current at the two angles above.

Solution

(a)

$$V_o = (\ Vm/2\pi)\ .\ (\ \cos \propto - \cos \beta\ )$$

$$= ((\sqrt{2} * 120)/2\,\pi)\,.\,(\cos 60^{o} - \cos 200^{o})$$

$$V_o = 38.88\text{ V}$$

(b)

$$I_o = V_o / R = 38.88/5$$

$$= 7.77\text{ A}$$

Example

A single-phase half-wave controlled rectifier with a resistance R = 2 Ω in series with a very high inductance. The supply voltage of 200 V and f = 60 Hz. Firing angle for each thyristor $\alpha = 60^{o}$. Calculate:
(a) The Output voltage V_o.
(b) The Output current I_o.

Solution

(a)

$$V_o = (2.Vm/\pi)\cos \propto$$

$$= ((2\sqrt{2} \times 200)/\pi)\,.\,\cos 60^{o}$$

$$= 90\text{ V}$$

(b)

$$I_o = V_o / R = 90/2 = 45\text{ A}$$

Example

The single phase half wave-controlled rectifier circuit, shown in Figure, has a purely resistive load of R and firing angle $\alpha = 50\%$ of total power. Determine

i. Derive an expression for load voltage in terms of α.
ii. Efficiency

iii. FF

iv. RF

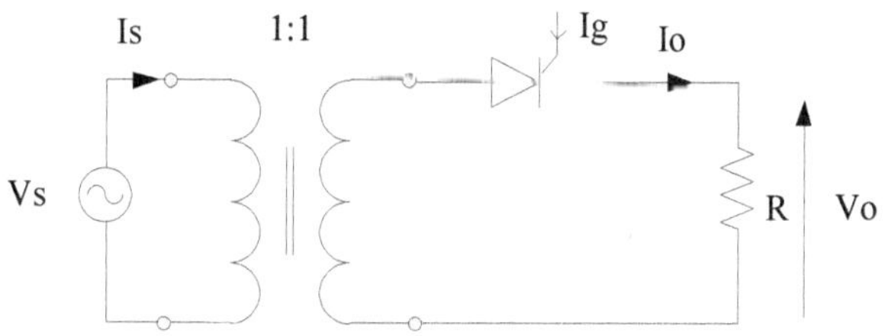

Solution

i.

$$V_{lmean} = \frac{1}{2\pi}\int_{\alpha}^{\pi} V_{s\,max} sin\theta d\theta$$

$$= \frac{V_{s\,max}}{2\pi}\int_{\alpha}^{\pi} sin\theta d\theta$$

$$= \frac{V_{s\,max}}{2\pi}[-cos\theta]_{\alpha}^{\pi}$$

$$= -\frac{V_{s\,max}}{2\pi}[cos\pi - cos\alpha]$$

$$\therefore \; V_{lmean} = \frac{V_{s\,max}}{2\pi}[1 + cos\alpha]$$

ii.

$$V_{lmean} = \frac{V_{s\,max}}{2\pi}[1 + \mathrm{COS}90] = \frac{V_{s\,max}}{2\pi} = 0.1592\, V_{s\,max}$$

Average power $(P_{av}) = \frac{V_{lmean}{}^{2}}{R} = \frac{(0.1592V_{s\,max})^2}{R}$

$$v_{0\,rms} = \sqrt{\frac{1}{2\pi}\int_{\alpha}^{\pi} V_{smax}{}^{2} sin^2\theta d\theta}$$

$$= V_{smax}\sqrt{\frac{1}{2\pi}\int_{\alpha}^{\pi}\frac{1}{2}(1-cos2\theta)d\theta}$$

$$= V_{smax}\sqrt{\frac{1}{4\pi}\int_{\alpha}^{\pi}(1-cos2\theta)d\theta}$$

$$=V_{smax}\sqrt{\frac{1}{4\pi}\int_{\alpha}^{\pi}1d\theta - \frac{1}{2}\int_{\alpha}^{\pi}cos2\theta\, d\theta}$$

$$= V_{smax}\sqrt{\frac{1}{4\pi}[\,\theta + \frac{sin2\theta}{2}\,]_{\alpha}^{\pi}}$$

$$]= V_{smax}\sqrt{\frac{1}{4\pi}[\,\pi - \frac{sin2\pi}{2} - \alpha + \frac{sin2\alpha}{2}}$$

$$= V_{smax}\sqrt{\frac{1}{4\pi}[\,\pi - \frac{sin2\pi}{2} - \frac{\pi}{2} + \frac{sin2\times\frac{\pi}{2}}{2}\,]}$$

$$= V_{smax}\sqrt{\frac{1}{4\pi}[\frac{\pi}{2}]} = V_{smax}\sqrt{\frac{1}{8}} = 0.3536\, V_{smax}$$

AC power (P_{ac}) $= \frac{{v_{0\,rms}}^2}{R} = \frac{(0.3536V_{s\,max})^2}{R}$

$$efficiency\,(\eta) = \frac{P_{av}}{P_{ac}} = \frac{\frac{(0.1592V_{s\,max})^2}{R}}{\frac{(0.3536V_{s\,max})^2}{R}} = 20.27\%$$

iii.

$$Form\ factor\ (FF) = \frac{v_{o\ rms}}{V_{lmean}} = \frac{0.3536\ V_{s\ max}}{0.1592\ V_{s\ max}} = 2.221$$

iv. $Ripple\ Factor\ (RF) = \sqrt{FF^2 - 1} = \sqrt{(2.221)^2 - 1} = 1.9831$

Single Phase Full Wave Control Rectifier with RL Load

The fully controlled rectifier contains four thyristors. We will assume that the load here contains resistance and high inductance to filter the output current.

The thyristors (T_1, T_2) are connected to the positive part of input voltage after being triggered at angle α, and the work continues to the next firing angle for (T_3, T_4) at α + π. The thyristors (T3, T4) occurrence in the negative part of the input wave is due to the transient condition of the presence of inductance in the load, as shown in Figure 11.19. Figure 11.20 shows input and output voltages and source currents. The following equation can find the average output voltage:

$$V_o = \frac{2Vm}{\pi} \cos \propto$$

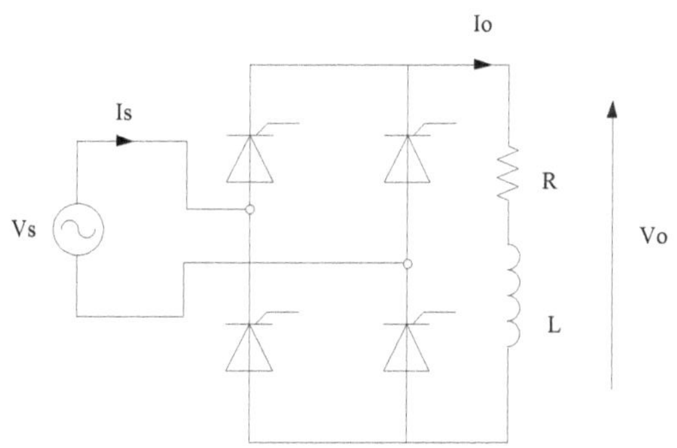

Figure 11.19 Single phase full wave control rectifier with R-L load circuit.

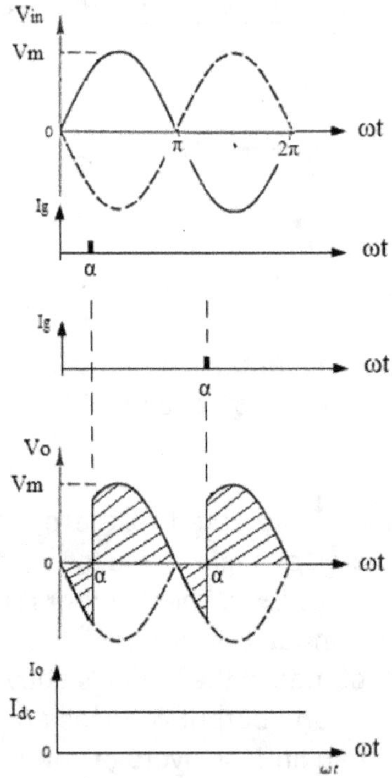

Figure 11.20 Input, output voltages, load, SCR, and source currents.

Formula Derivative:

a. Resistive load:

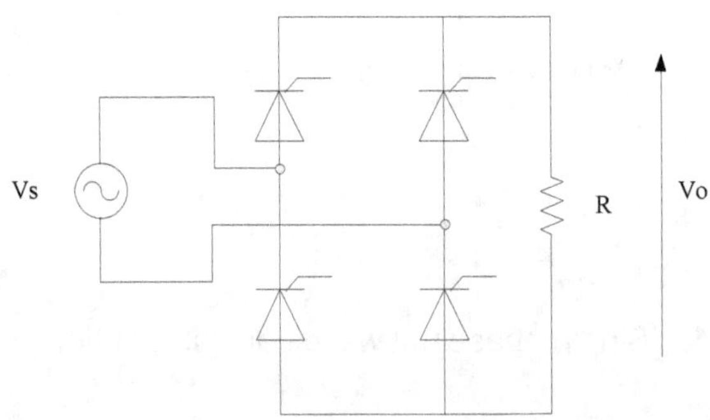

Figure 11.21 Single phase full wave control rectifier with R load circuit.

$$V_{lmean} = \frac{1}{2\pi}\int_{\alpha}^{\pi} V_{s\,max} sin\theta d\theta \times 2$$

$$= \frac{V_{s\,max}}{\pi}\int_{\alpha}^{\pi} sin\theta d\theta$$

$$= \frac{V_{s\,max}}{\pi}[-cos\theta]_{\alpha}^{\pi}$$

$$= -\frac{V_{s\,max}}{\pi}[cos\pi - cos\alpha]$$

$$\therefore \; V_{lmean} = \frac{V_{s\,max}}{\pi}[1 + \cos\alpha]$$

$$\therefore \; I_{lmean} = \frac{V_{lmean}}{R}$$

or $\quad I_{lmean} = \frac{1}{2\pi}\int_{\alpha}^{\pi} I_{s\,max} sin\theta d\theta \times 2$

$$v_{s\,rms} = \frac{V_{s\,max}}{\sqrt{2}}$$

$$v_{0\,rms} = \sqrt{\frac{1}{2\pi}\int_{\alpha}^{\pi} {V_{smax}}^{2} sin^{2}\theta d\theta \times 2}$$

$$= V_{smax}\sqrt{\frac{1}{\pi}\int_{\alpha}^{\pi}\frac{1}{2}(1 - cos2\theta)d\theta}$$

$$= V_{smax}\sqrt{\frac{1}{2\pi}\int_{\alpha}^{\pi}(1 - cos2\theta)d\theta} = V_{smax}\sqrt{\frac{1}{2\pi}\int_{\alpha}^{\pi} 1d\theta - \frac{1}{2}\int_{\alpha}^{\pi} cos2\theta\, d\theta}$$

$$v_{0rms} = V_{smax}\sqrt{\frac{1}{2\pi}[\,\pi - \frac{sin2\pi}{2} - \alpha + \frac{sin2\alpha}{2}\,]}$$

$$\therefore\ i_{orms} = \frac{v_{orms}}{R}$$

Example

For the fully-controlled full wave rectifier circuit shown in Figure 11.22, firing angle $\alpha = 90^0$, Calculate the efficiency.

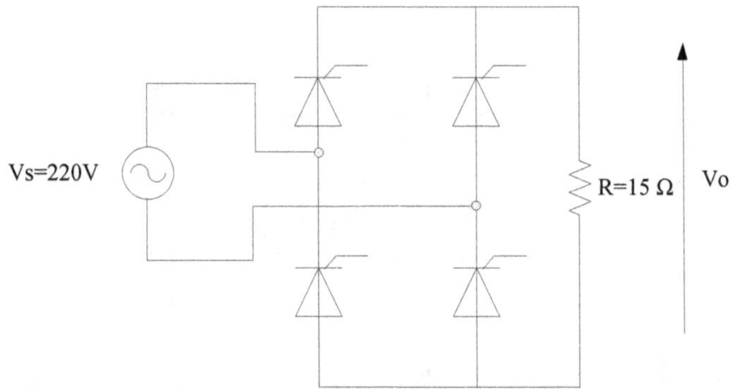

Figure 11.22 Circuit diagram.

Solution

$$V_{lmean} = \frac{V_{s\,max}}{\pi}[1+\cos\alpha] = \frac{220\times\sqrt{2}}{\pi}[1+\cos 90^\circ]$$
$$= 99.08\ V$$

$$I_{lmean} = \frac{V_{lmean}}{R} = \frac{99.08}{15} = 6.6\ A$$

$v_{0\,rms}$

$$= \sqrt{\frac{1}{2\pi}\int_{\alpha}^{\pi} V_{smax}{}^2 \sin^2\theta d\theta \times 2} = V_{smax}\sqrt{\frac{1}{2\pi}[\pi - \frac{sin2\pi}{2} - \alpha + \frac{sin2\alpha}{2}]}$$
$$= 155.5\ V$$

$$i_{orms} = \frac{v_{orms}}{R} = \frac{155.5}{15} = 10.36\ A$$

$$efficiency\ (\eta) = \frac{P_{av}}{P_{ac}} = \frac{V_{lmean}\times I_{lmean}}{v_{o\,rms}\times i_{o\,rms}} = \frac{99.08\times 6.6}{155.5\times 10.36}$$

$$= 40.59\%$$

Single Phase Full Wave Half Control Rectifier with R-L Load

A converter is called half-controlled when it contains both a diode and thyristors. These three circuits produce identical load waveforms, neglecting any differences in the number of types of semiconductor voltage drops. The power to the load is varied by controlling the angle α, at which the load voltage from going negative, extending the conduction period and reducing the a.c. Ripple.

$$Average\ output\ voltage = V_{dc} = \frac{1}{\pi}\int_{\alpha}^{\pi} V_m \sin\omega t \, d\omega t$$

$$= \frac{V_m}{\pi}(1 + cos\alpha)$$

$$\text{RMS output voltage} = V_{rms} = \sqrt{\frac{1}{\pi}\int_{\alpha}^{\pi}(V_m \sin\omega t)^2 \, d\omega t}$$

$$= V_m\sqrt{\frac{\pi - \alpha + \frac{1}{2}\sin 2\alpha}{2\pi}}$$

Inversion is not possible.

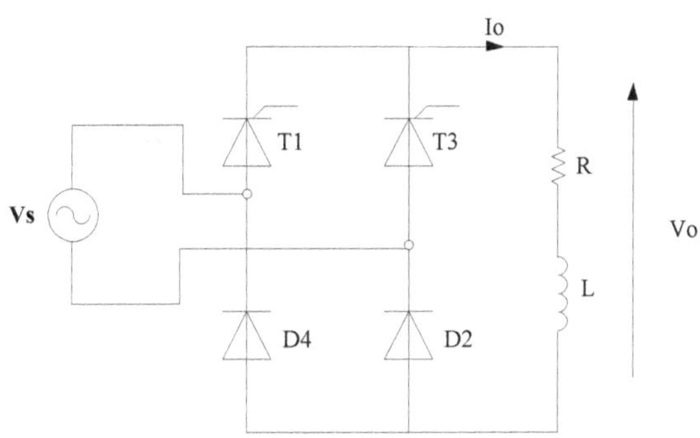

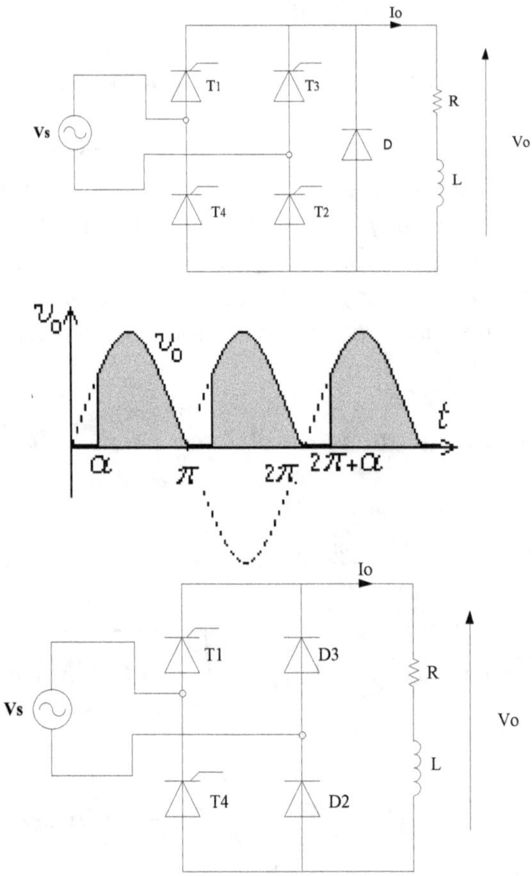

Figure 11.23 Half-controlled rectifiers and output voltage.

Example

For the fully-controlled full wave rectifier circuit shown in Figure 11.24, firing angle $\alpha = 45^0$. Calculate efficiency.

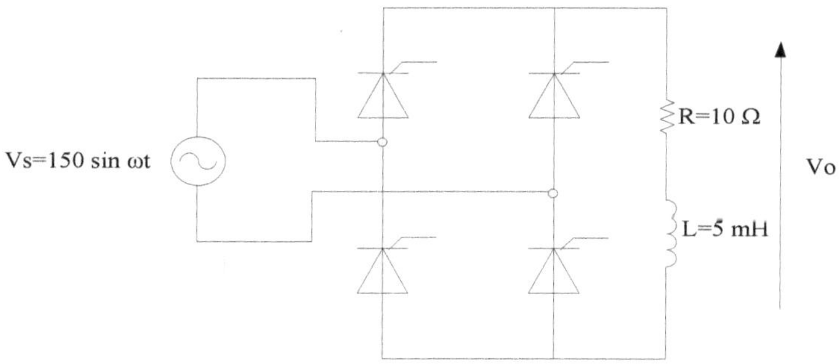

Figure 11.24 Fully-controlled full wave rectifier circuit diagram.

Solution

$$\varphi = \tan^{-1}\frac{\omega L}{R} = \tan^{-1}\frac{2\pi \times 50 \times 5 \times 10^{-3}}{10} = 11.9^{0}$$

$$V_{lmean} = \frac{V_{s\,max}}{\pi}[cos\varphi + cos\alpha]$$

$$= \frac{150}{3.14}[cos11.9^{\circ} + cos45^{\circ}] = 81\,V$$

$$I_{lmean} = \frac{V_{lmean}}{R} = \frac{81}{10} = 11.1\,A$$

$$v_{0\,rms} = \sqrt{\frac{1}{2\pi}\int_{\alpha}^{\pi+\varphi} V_{smax}{}^{2} sin^{2}\theta d\theta \times 2} \approx 101.1\,V$$

$$i_{orms} = \frac{v_{orms}}{R} = \frac{101.1}{10} = 10.11\,A$$

$$efficiency\,(\eta) = \frac{P_{av}}{P_{ac}} = \frac{V_{lmean} \times I_{lmean}}{v_{o\,rms} \times i_{o\,rms}}$$

$$= \frac{81 \times 11.1}{101.1 \times 10.11} = 64.25\%$$

Example

For the fully controlled full wave rectifier circuit shown in Figure 11.25, firing angle $(\alpha) = 45^{0}$. Calculate the efficiency.

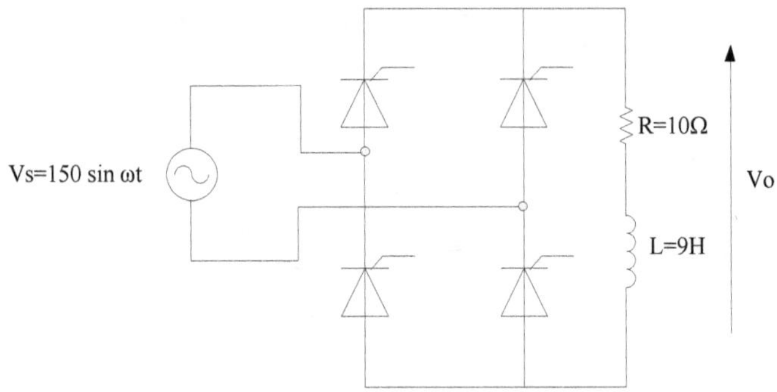

Figure 11.25 Fully-controlled full wave rectifier circuit diagram.

Solution

$$\varphi = \tan^{-1}\frac{\omega L}{R} = \tan^{-1}\frac{2\pi \times 50 \times 9}{10} = 89.79^0$$

$$V_{lmean} = \frac{2V_{s\,max}}{\pi}[cos\alpha] = \frac{2 \times 150}{3.14}[cos45°]$$
$$= 67.5\ V$$

$$I_{lmean} = \frac{V_{lmean}}{R} = \frac{67.5}{10} = 6.75\ A$$

$$v_{0\,rms} = \sqrt{\frac{1}{2\pi}\int_{\alpha}^{\pi+\alpha} V_{smax}{}^2 sin^2\theta d\theta \times 2} \approx 105.8\ V$$

$$i_{0\,rms} = \sqrt{\frac{1}{2\pi}\int_{\alpha}^{\pi+\alpha} I_{lmean}{}^2 \theta d\theta \times 2} = I_{lmean} = 6.75\ A$$

$$efficiency\ (\eta) = \frac{P_{av}}{P_{ac}} = \frac{V_{lmean} \times I_{lmean}}{v_{o\,rms} \times i_{o\,rms}}$$

$$= \frac{67.5 \times 6.75}{105.8 \times 6.75} = 63.8\%$$

11.2 DC-DC Converter Basics

A dc-dc converter is a device that accepts a dc input voltage and produces a dc output voltage. Typically the output produced is at a different voltage level than the input. In addition, dc-dc converters are used to provide noise isolation, power bus regulation, etc. This section considers a summary of some of the popular dc-dc converter topologies.

11.2.1 Step-Down (Buck) Converter

In the circuit in Figure 11.26, the transistor turning ON will put voltage V_{in} on one end of the inductor. This voltage will tend to cause the inductor current to rise. When the transistor is OFF, the current will continue flowing through the inductor but now flow through the diode. We initially assume that the current through the inductor does not reach zero. Thus the voltage at V_x will now be only the voltage across the conducting diode during the full OFF time. The average voltage at V_x will depend on the average ON time of the transistor, provided the inductor current is continuous, as in Figure 11.27.

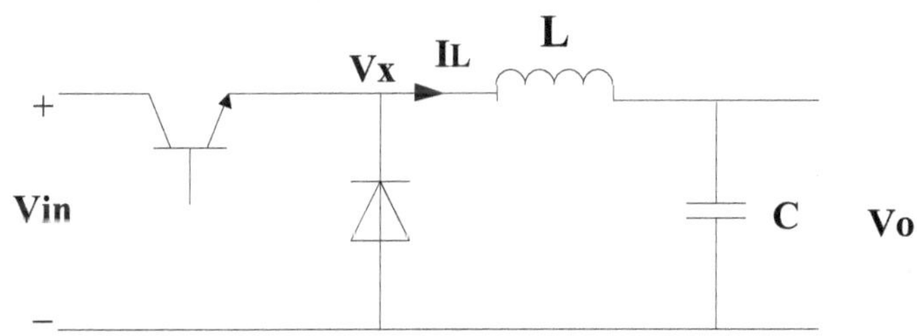

Figure 11.26 Buck converter.

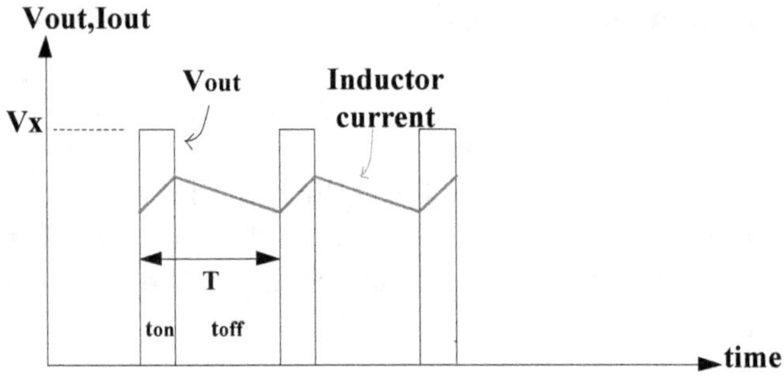

Figure 11.27 Voltage and current changes.

To analyze the voltages of this circuit, let us consider the changes in the inductor current over one cycle. From the relation

$$V_x - V_0 = L\,\frac{di}{dt}$$

The change of current satisfies.

$$di = \int_{ON} (V_x - V_0)\,dt + \int_{OFF} (V_x - V_0)\,dt$$

For steady state operation, the current at the start and end of a period T will not change. We assume no voltage drop across the transistor or diode during ON and a perfect switch change to get a simple relation between voltages. Thus during the ON time V_x=V_{in} and in the OFF, V_x=0.

So $\quad 0 = di = \int_0^{t_{on}} (V_{in} - V_0)\,dt + \int_{t_{on}}^{T} (-V_0)\,dt$

Which simplifies to

$$(V_{in} - V_0).\,t_{on} + (-V_0).\,t_{off} = 0$$

or

$$\frac{V_o}{V_{in}} = \frac{t_{on}}{T}$$

and defining "duty ratio" as

$$D = \frac{t_{on}}{T}$$

The voltage relationship becomes V_o=D V_{in} Since the circuit is lossless and the input and output powers must match on the average V_o* I_o = V_{in}* I_{in}. Thus the average input and output current must satisfy I_{in} =D I_o These relations are based on the assumption that the inductor current does not reach zero.

Transition between Continuous and Discontinuous

When the current in the inductor L always remains positive, either the transistor T_1 or the diode D_1 must be conducting. For continuous conduction, the voltage V_x is either V_{in} or 0. If the inductor current ever goes to zero, the output voltage will not be forced to either of these conditions. The current reaches zero at this transition point, as seen in Figure 11.28. During the ON time V_{in} -V_{out} is across the inductor, thus

$$I_L(peak) = (V_{in} - V_{out}).\frac{t_{on}}{L}$$

The average current, which must match the output current, satisfies

$$I_L(av) = \frac{I_L(peak)}{2} = (V_{in} - V_{out}).\frac{dT}{2\,L}$$

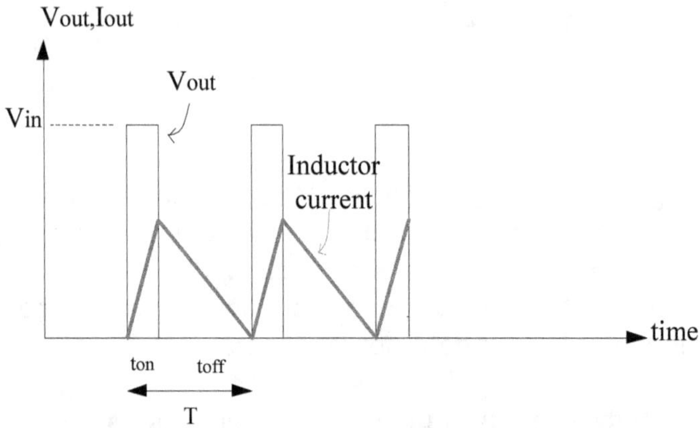

Figure 11.28 Buck converter at the boundary.

If the input voltage is constant, the output current at the transition point satisfies.

$$I_L(av) = I_{out} = V_{in}.\frac{(1-d)}{2L}.T$$

Voltage Ratio of Buck Converter (Discontinuous Mode)

As for the continuous conduction analysis, we use the fact that the integral of the voltage across the inductor is zero over a cycle of switching T. The transistor OFF time is now divided into segments of diode conduction $\delta_d T$ and zero conduction $\delta_o T$. as shown in Figure 11.29. The average inductor voltage thus gives

$$(V_{in} - V_o)\,DT + (-V_o)\,\delta.dT = 0$$

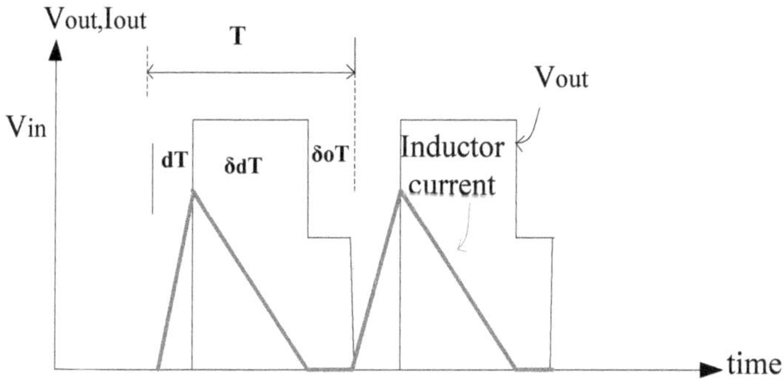

Figure 11.29 Buck converter - discontinuous conduction.

$$\frac{V_{out}}{V_{in}} = \frac{d}{d + d_\delta}$$

for the case d+δ_d <1. To resolve the value of δ_d, consider the output current, which is half the peak when averaged over the conduction times d+δ_d

$$I_{out} = \frac{I_{L(peak)}}{2} . d + \delta_d$$

Considering the change of current during the diode conduction time

$$I_{L(peak)} = \frac{V_o(\delta_d T)}{L}$$

Thus from Equations 11.33 and 11.34, we can get

$$I_{out} = \frac{V_o \delta_d T(d + d_\delta)}{2L}$$

using the relationship in Equation 11.32

$$I_{out} = \frac{V_{in} d T d_\delta}{2L}$$

And solving for the diode conduction

$$\delta_d = \frac{2LI_{out}}{V_{in}dT}$$

The output voltage is thus given as

$$\frac{V_{out}}{V_{in}} = \frac{d^2}{d^2 + (\frac{2LI_{out}}{V_{in}T})}$$

Defining k* = 2L/(Vin T), we can see the effect of discontinuous current on the voltage ratio of the converter. As seen in Figure 11.30, once the output current is high enough, the voltage ratio depends only on the duty ratio "d." At low currents, the discontinuous operation increases the converter's output voltage towards Vin.

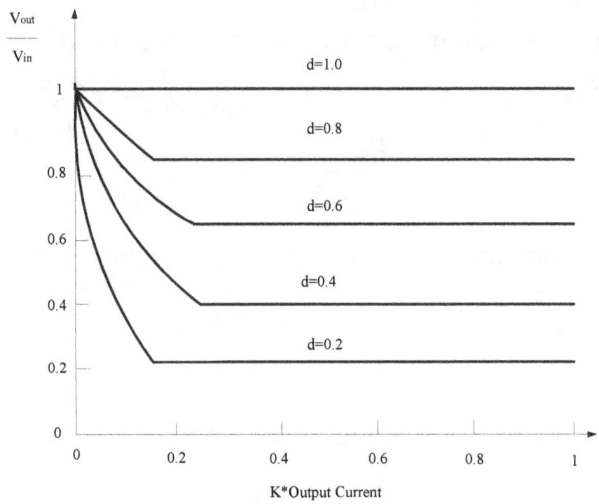

Figure 11.30 Output voltage vs. current.

11.2.2 Step-Up (Boost) Converter

The schematic in Figure 11.31 shows the basic boost converter. This circuit is used when a higher output voltage than input is required.

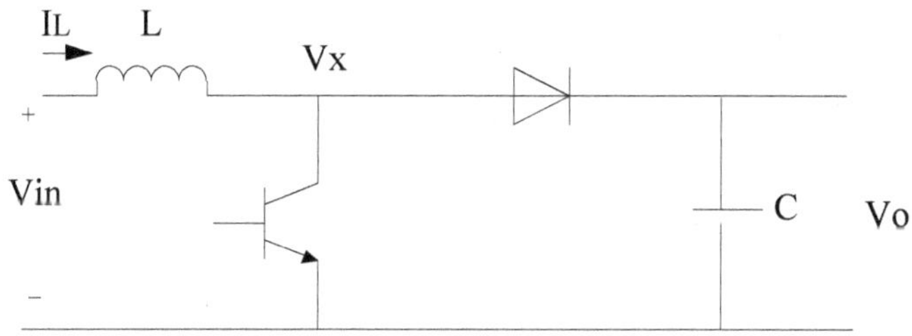

Figure 11.31 Boost converter circuit.

While the transistor is ON $V_x = V_{in}$ and the OFF state, the inductor current flows through the diode giving $V_x = V_o$. For this analysis, it is assumed that the inductor current always remains flowing (continuous conduction). The voltage across the inductor is shown in Figure 11.32, and the average must be zero for the average current to remain in a steady state

$$V_{in} . t_{on} + (V_{in} - V_o) t_{off} = 0$$

This can be rearranged as

$$\frac{V_{out}}{V_{in}} = \frac{T}{t_{off}} = \frac{1}{(1-D)}$$

And for a lossless circuit, the power balance ensures

$$\frac{I_o}{I_{in}} = (1-D)$$

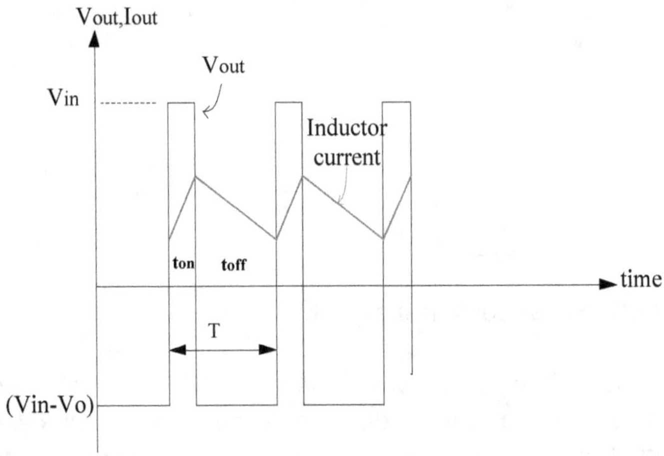

Figure 11.32 Voltage and current waveforms (boost converter).

Since the duty ratio "D" is between 0 and 1, the output voltage must always be higher than the input voltage in magnitude. The negative sign indicates a reversal of the sense of the output voltage.

11.2.3 Buck-Boost Converter

With continuous conduction for the Buck-Boost converter in Figure 11.33, $V_x = V_{in}$ when the transistor is ON and $V_x = V_o$ when the transistor is OFF. The average voltage across the inductor is zero for zero net current change over a period.

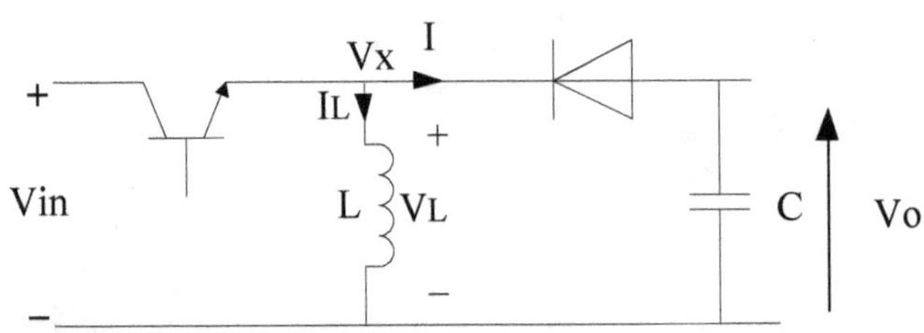

Figure 11.33 Schematic for the buck-boost converter.

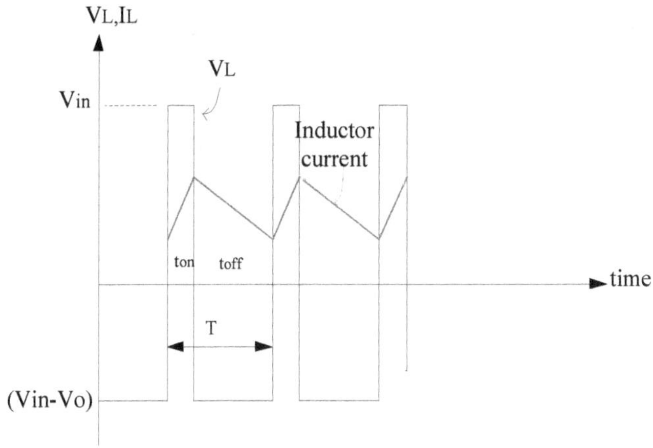

Figure 11.34 Waveforms for the buck-boost converter.

From Figure 11.34

$$V_{in}\,.t_{on} + V_o\,.t_{off} = 0$$

Which gives the voltage ratio

$$\frac{V_{out}}{V_{in}} = \frac{T}{t_{off}} = \frac{D}{(1-D)}$$

and the corresponding current

$$\frac{I_{out}}{I_{in}} = \frac{(1-D)}{D}$$

Since the duty ratio D is between 0 and 1, the output voltage can vary lower or higher than the input voltage in magnitude. The negative sign indicates a reversal of the sense of the output voltage.

11.2.4 Converter Comparison

The voltage ratios achievable by the dc-dc converters are summarized in Figure 11.35. Notice that only the buck converter shows a linear relationship between the control (duty ratio) and output voltage. The buck boost can reduce or increase the voltage ratio with rectifier gain for a duty ratio of 50%.

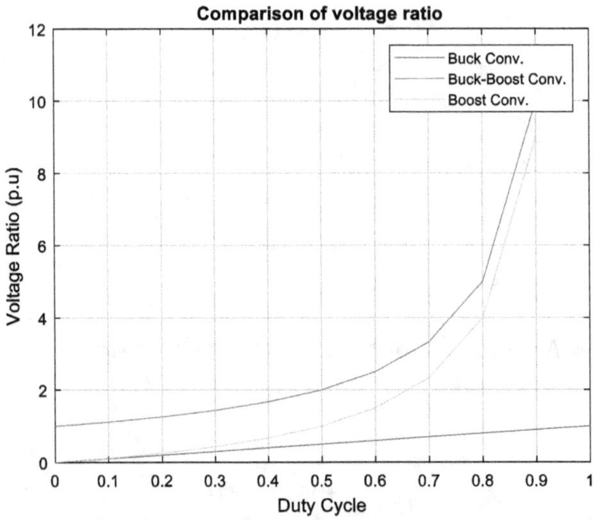

Figure 11.35 Comparison of the voltage ratio.

Example

A step-down (buck) converter is used to supply an inductive load of $R = 3\Omega$ and $L = 9$ mH, with a dc voltage of 60V, as shown in Figure 11.36. If $D = 0.4$, $F_s = 2\ KH_z$, calculate:

a) I_{Lmean} and V_{Lmean}.
b) Trace the output voltage and current.

Solution

a.

$$Ts = 1/2000 = 5 \times 10^{-4}$$

$$D = \frac{a}{Ts}$$

$$a = D \times Ts = 0.4 \times 5 \times 10^{-4} = 2 \times 10^{-4}$$

$$Ts = a + b$$

$$b = Ts - a = 5 \times 10^{-4} - 2 \times 10^{-4} = 3 \times 10^{-4}$$

$$I_{max} = \frac{Ed}{R}\left[\frac{e^{\frac{RTs}{L}} - e^{\frac{Rb}{L}}}{e^{\frac{RTs}{L}} - 1}\right] = 11.4A$$

$$I_{min} = I_{max}e^{\frac{-Rb}{L}} = 7.6A$$

$$I_{Lmean} = \frac{I_{max} + I_{min}}{2} = 7.8A$$

$$V_{max} = I_{max} \times R = 25.2V$$

$$V_{min} = I_{min} \times R = 22.8V$$

$$V_{Lmean} = \frac{V_{max} + V_{min}}{2} = 24V$$

b.

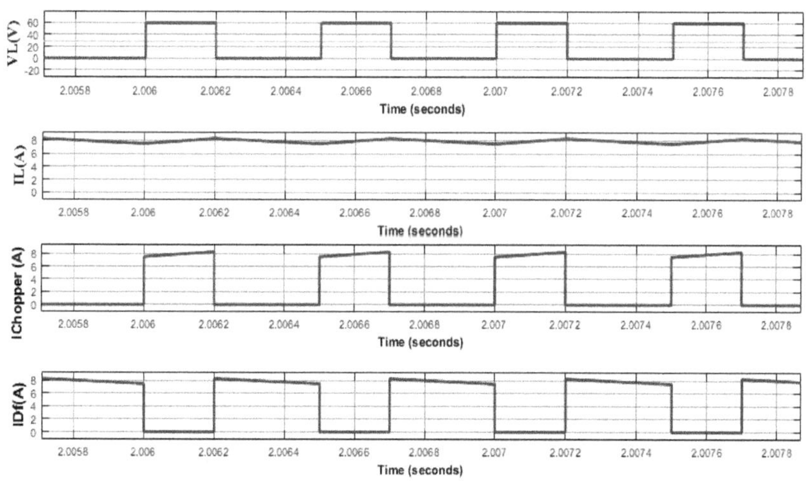

Figure 11.36 Waveforms.

11.3 Harmonic in Single–Phase AC Circuit

If an alternating voltage containing various harmonics is applied to a single–phase circuit containing linear circuit elements, then the current produced also contains harmonics. Each harmonic voltage will produce its current independent of others. By the principle of superposition, the combined current can be found. We will now consider well-known elements like pure resistance, pure inductance, and pure capacitance.

In each case, we will assume that the applied complex voltage is represented by the following:

$$e = E_{1m} \sin \omega t + E_{2m} \sin 2\omega t + \dots\dots + E_{nm} \sin n\omega t$$

11.3.1 Pure Resistance

Let the circuit have a resistance of R independent of frequency. The instantaneous current i_1 due to fundamental voltage is:

$$i_1 = \frac{E_{1m} \sin \omega t}{R}$$

similarly,

$i_2 = \frac{E_{1m} \sin 2\omega t}{R}$ - for 2nd harmonic

and

$i_n = \frac{E_{nm} \sin n\omega t}{R}$ - for the nth harmonic

total current $i = i_1 + i_2 + i_3 + \dots\dots + i_n$

$$i = \frac{E_{1m} \sin \omega t}{R} + \frac{E_{2m} \sin 2\omega t}{R} + \dots + \frac{E_{nm} \sin n\omega t}{R}$$

$$= I_{1m} \sin \omega t + I_{2m} \sin 2\omega t + \dots + I_{nm} \sin n\omega t$$

It shows that:

(i) The waveform of the resulting current is similar to that of the applied voltage, i.e., the two waves are identical.

(ii) The percentage of harmonic content in the current wave is the same as in the applied voltage.

11.3.2 Pure Inductance

Let the circuit's inductance be L Henry, whose reactance varies directly with the frequency of the applied voltage.
Its reactance for the fundamental would be X1 =ωL; for the second harmonic, $X_2 = 2\omega L$. However, for every harmonic term, the current will lag behind the voltage by 90°.

Current due to fundamental:

$$i_1 = \frac{E_{1m}}{\omega L} sin(\omega t - {}^{\pi}/_{2})$$

Current due to 2nd harmonic :

$$i_2 = \frac{E_{2m}}{2\omega L} sin(2\omega t - {}^{\pi}/_{2})$$

Current due to 3rd harmonic :

$$i_3 = \frac{E_{3m}}{3\omega L} sin(3\omega t - {}^{\pi}/_{2})$$

Current due to nth harmonic :

$$i_n = \frac{E_{nm}}{n\omega L} sin(n\omega t - {}^{\pi}/_{2})$$

∴ Total current:

$$i(t) = i_1 + i_2 + i_3 + \ldots\ldots + i_n = \frac{E_{1m}}{\omega L} sin(\omega t - {}^{\pi}/_{2}) + \frac{E_{2m}}{2\omega L} sin(2\omega t - {}^{\pi}/_{2}) + \frac{E_{3m}}{3\omega L} sin(3\omega t - {}^{\pi}/_{2}) + \cdots + \frac{E_{nm}}{n\omega L} sin(n\omega t - {}^{\pi}/_{2})$$

It can be seen from the above equation that:

(i) The current waveform differs from the applied voltage's waveform.

(ii) for the nth harmonic, the percentage harmonic content in the current wave is 1/n of the corresponding harmonic content in the voltage wave. In an inductive circuit, the current waveform shows less distortion than the voltage waveform .in this case, current more nearly approaches a sine wave than it does in a circuit containing resistance.

11.3.3 Pure Capacitance

In this case:

$$X_1 = \frac{1}{\omega C} \text{ -for fundamental}$$

$$X_2 = \frac{1}{2\omega C} \text{ -for 2}^{\text{nd}} \text{ harmonic}$$

$$X_3 = \frac{1}{3\omega C} \text{ -for 3}^{\text{rd}} \text{ harmonic}$$

$$X_n = \frac{1}{n\omega C} \text{ -for the n}^{\text{th}} \text{ harmonic}$$

$$i_1 = \frac{E_{1m}}{1/_{\omega C}} sin(\omega t + \pi/_2) = \omega C E_{1m} sin(\omega t + \pi/_2) \quad \text{..........(2.16)}$$

$$i_2 = \frac{E_{2m}}{1/_{2\omega C}} sin(2\omega t + \pi/_2) = 2\omega C E_{2m} sin(2\omega t + \pi/_2) \text{ ....(2.17)}$$

$$i_n = \frac{E_{nm}}{1/_{n\omega C}} sin(n\omega t + \pi/_2) = n\omega C E_{1m} sin(n\omega t + \pi/_2) \text{ ....(2.18)}$$

For every harmonic term, the current will lead the voltage by 90°.
Now:

$$i_{\square} = i_1 + i_2 + i_3 + \ldots\ldots + i_n$$

$$= \omega C E_{1m} sin(\omega t + \pi/2) + 2\omega C E_{2m} sin(2\omega t + \pi/2) + \ldots. + n\omega C E_{nm} sin(n\omega t + \pi/2)$$

This equation shows that :

(i) The current and voltage waveforms are dissimilar.

(ii) The percentage harmonic content of the current is larger than that of the applied voltage wave. For example, it would be n times larger for the nth harmonic.
(iii) As a result, the current wave is more distorted than the voltage wave.
(iv) The effect of the capacitor on distortion is just the reverse of inductance.

11.4 Stepped Inverter Voltage Waveforms

11.4.1 The stepped waveform

For voltage waveform fabrication, switching the thyristors of FIGURE 11.37 sequentially at intervals of 60o electrical or 1/6th of the period is convenient. The use of a d.c. Supply having equal positive and negative voltage values ± V_{dc} is common.

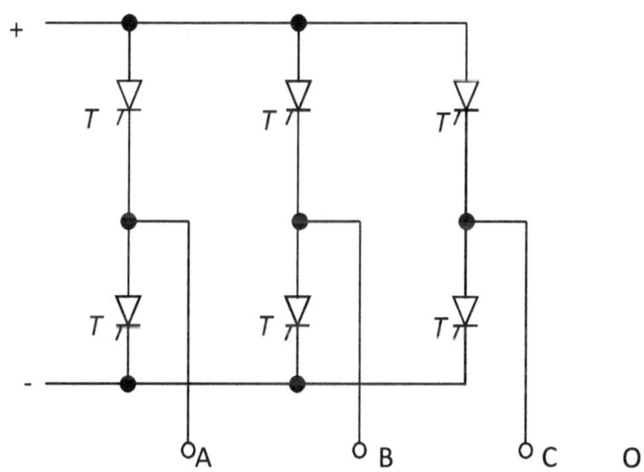

Figure 11.37 Skeleton switching circuit of voltage source inverter.

11.4.2 Two simultaneously conducting thyristors

If two thyristors conduct at any instant, a suitable switching pattern for no-load operation is defined in Figure 11.37.

The thyristors are switched numerically, each remaining in conduction for 120° electrical. Phase voltage v_{AN} , v_{BN} , v_{CN} consist of rectangular pulses of height ± V_{dc}, which sum to zero at every instant. The corresponding line voltages are six-stepped waves of height ± 2Vdc. If equal resistors R are now connected in star to the load terminals A,B, and C of Figure 11.37, then interval 0<ωt<π/3 :

$$v_{AN} = -I_L R = \frac{-2V_{dc}}{2R} R = -V_{dc}$$

$$v_{BN} = 0$$

$$v_{CN} = I_L R = \frac{2V_{dc}}{2R} . R = +V_{dc}$$

$$v_{AB} = v_{AN} + v_{NB} = v_{AN} - v_{BN} = -V_{dc}$$

In the interval π/3 <ωt <2 π/3 :

$$V_{AN}=0$$

$$V_{BN}=-I_L R=-V_{dc}$$

$$V_{CN}=+I_L R=+V_{dc}$$

$$V_{AB}=+V_{dc}$$

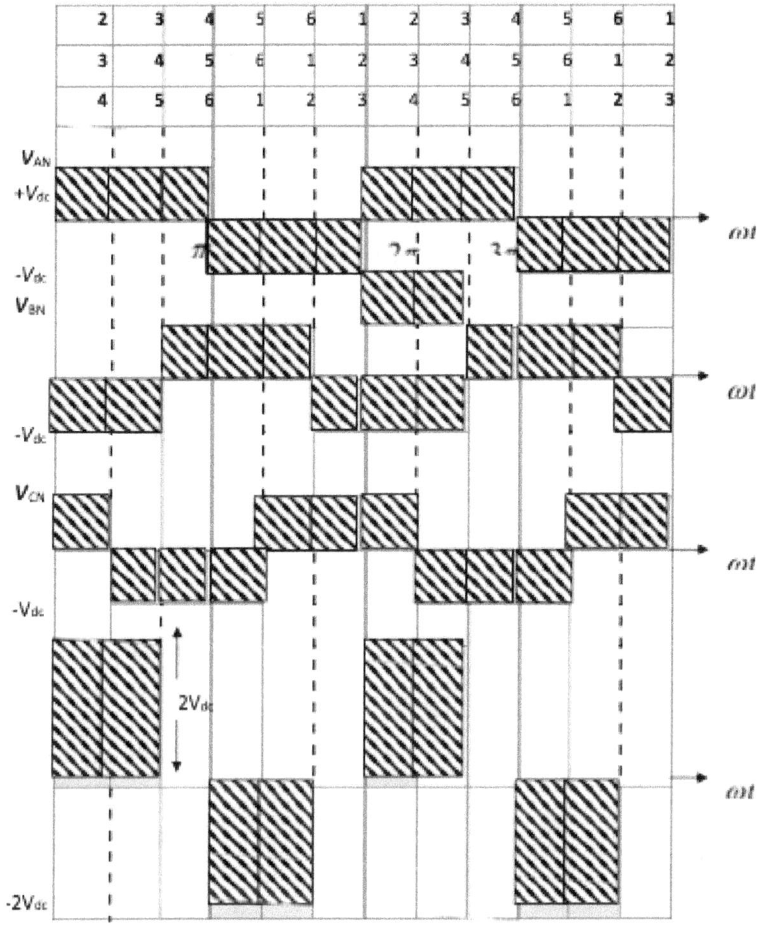

Figure 11.38 Voltage wave forms with two simultaneously conducting thyristors. No- load and resistive load.

In the interval 2π/3 <ωt < π :

$$v_{AN} = +I_L R = +V_{dc}$$

$$v_{BN} = -I_L R = -V_{dc}$$

$$v_{CN} = 0$$

$$v_{AB} = +2Vdc$$

For each interval, it is seen that the load current during conduction is :

$$I_L = \frac{\pm 2V_{dc}}{2R} = \pm\frac{V_{dc}}{R}$$

for this particular mode of switching, the load voltage and current waveforms with star-connected resistive load are, therefore, identical to the pattern of the open-circuit voltages. Phase voltage waveform v_{AN} in Figure 11.38 has the r.m.s value given by:

$$V_{AN} = \sqrt{\frac{1}{2\pi}\int_0^{2\pi} v_{AN}^2\,(\omega t)d\omega t} = \sqrt{\frac{2}{3}}V_{dc} = 0.816V_{dc}$$

The fundamental Fourier coefficients of waveform $v_{AN}\,(\omega t)$ are found to be:

$$a_1 = \frac{1}{\pi}\int_0^{2\pi} v_{AN}\,(\omega t)\cos\omega\, t d\omega t = -\frac{2\sqrt{3}}{\pi}V_{dc}$$

$$b_1 = \frac{1}{\pi}\int_0^{2\pi} v_{AN}\,(\omega t)\sin\omega\, t d\omega t = 0$$

Line voltage $v_{AB}(\omega t)$ in Figure (3.2) has fundamental Fourier coefficients of value :

$$a_1 = -\frac{3\sqrt{3}}{\pi}V_{dc}$$

$$b_1 = +\frac{3}{\pi}V_{dc}$$

$$\therefore c_1 = \frac{6}{\pi}V_{dc}$$

So the fundamental phase voltage has a peak value $(\frac{2\sqrt{3}}{\pi})$ V_{dc} or 1.1 V_{dc}, and this is sketched in Figure 11.38. The distortion factor of the phase voltage is given by :

$$distortionfactor = \frac{V_{AN1}}{V_{AN}} = \frac{\frac{C_1}{\sqrt{2}}}{V_{AN}} = 1.65$$

Summary of 120° Conduction mode

1. Each thyristor conduction for 120°.
2. Two thyristors remain ON at any instant of time.
3. There are three modes of operation in one half-cycle.
4. The duration of each mode is 120°.
5. Gating signals are shifted from each other by 120°.
6. The load is star connected.

11.4.3 Three simultaneously conducting thyristors

A further load voltage waveform is generated if a switching mode is used where three thyristors conduct at any instant. Once again, the thyristors conduct in numerical sequence but each with a conduction angle of 180° electrical. Three thyristors with consecutive numbering are conducted simultaneously at any instant of the cycle.

The pattern of waveforms obtained on no-load is shown in Figure 11.39.

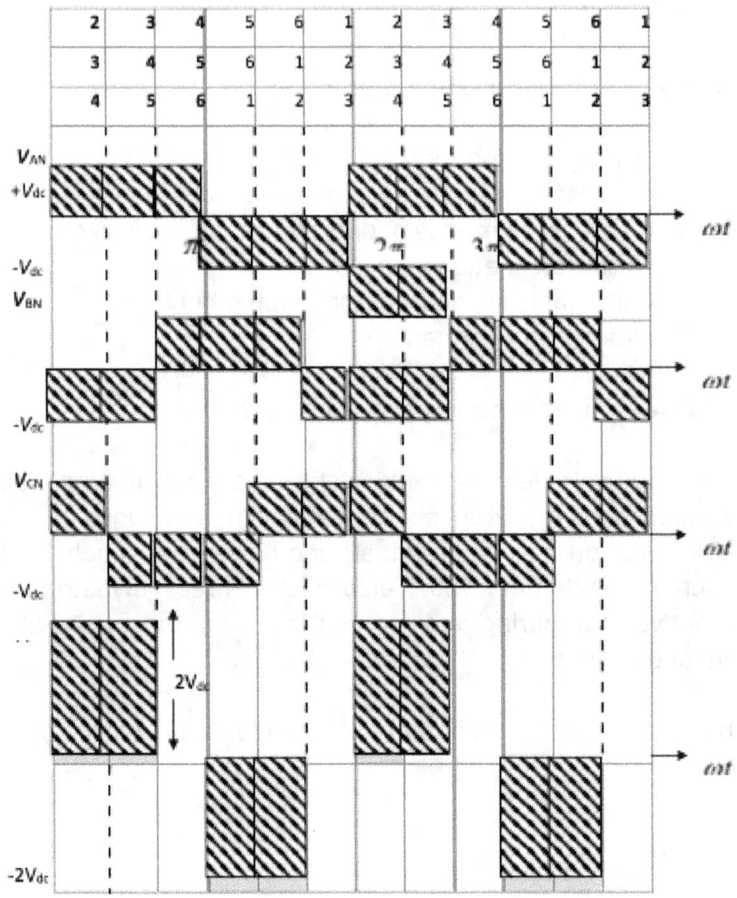

Figure 11.39 No-load voltage wave forms with free simultaneously conducting thyristor

For each interval:

$$I_L = \frac{2V_{dc}}{R + \frac{R}{2}} = \frac{2V_{dc}}{R(1 + 0.5)} = \frac{2V_{dc}}{1.5R} = \frac{4V_{dc}}{3R}$$

In the interval 0<ωt<π/3 :

$$V_{AN} = V_{CN} = \frac{I_L R}{2} = \frac{2}{3} V_{dc}$$

$$V_{BN} = -I_L R = -\frac{4}{3}V_{dc}$$

In the interval 2π/3<ωt<π :

$$V_{AN} = V_{BN} = \frac{I_L R}{2} = \frac{2}{3}V_{dc}$$

$$V_{CN} = -I_L R = -\frac{4}{3}V_{dc}$$

$$V_{AB} = 0$$

The fundamental component of V_{AN} (ωt) can be seen by inspection. Is in time –phase with it, so that

$$a_1=0$$

fundamental Fourier coefficient b_1 for the load peak phase voltage is found to be :

$$b_1 = c_1 = \frac{4}{\pi}V_{dc}$$

T

he corresponding fourier coefficients for line voltage v_{AB}(ωt) are given by:

$$a_1 = \frac{2\sqrt{3}}{\pi}V_{dc}$$

$$b_1 = -\frac{6}{\pi}V_{dc}$$

$$c_1 = \frac{4}{\pi}\sqrt{3}V_{dc}$$

$$\therefore \psi_1 = -\tan^{-1}\frac{1}{\sqrt{3}} = -30^o$$

the fundamental components of the load voltages for athree–phase sinusoidal system, the line voltage leads its corresponding phase voltage by 30°.

The r.m.s.value of phase voltage $v_{AN}(\omega t)$ is found to be:

$$V_{AN} = \sqrt{\frac{1}{\pi}\int_0^{\pi} v_{AN}^2\,(\omega t)d\omega t} = \frac{2\sqrt{2}}{3}V_{dc} = 0.943V_{dc}$$

the distortion factor of the phase voltage:

$$distortionfactorDF = \frac{V_{AN1}}{V_{AN}} = \frac{{c_1}/{\sqrt{2}}}{V_{AN}} = 1.65$$

Summary of 180° Conduction mode

1. Each thyristor conduts for 180° .
2. Three thyristors remain ON at any instant time.
3. There are six- modes of operation in a cycle.
4. The duration of each mode is 60°.
5. Gating signals are shifted from each other by 60° to obtain three–phase balance (fundamental) voltages.
6. The load may be connected in a star or delta.

11.5 Summary of Power Electronics Formulas

Single phase uncontrolled Rectifier

Circuit	Load	V_{dc}	I_{dc}	$V_{o\ rms}$	$I_{o\ rms}$	$I_{D\ dc}$	$I_{s\ rms}$
Half wave	R	$\frac{V_m}{\pi}$	$\frac{V_{dc}}{R}$	$\frac{V_m}{2}$	$\frac{V_{o rms}}{R}$	I_{dc}	$I_{o\ rms}$
Full wave (bridge)	R	$\frac{2V_m}{\pi}$	$\frac{V_{dc}}{R}$	$\frac{V_m}{\sqrt{2}}$	$\frac{V_{o rms}}{R}$	$\frac{I_{dc}}{2}$	$I_{o\ rms}$
	R+L	$\frac{2V_m}{\pi}$	$\frac{V_{dc}}{R}$	$\frac{V_m}{\sqrt{2}}$	I_{dc}	$\frac{I_{dc}}{2}$	I_{dc}

Single phase-controlled Rectifier

Circuit	Load	V_{dc}	I_{dc}	$V_{o\,rms}$	$I_{o\,rms}$	$I_{D\,dc}$	$I_{s\,rms}$
Half wave	R	$\frac{V_m}{2\pi}(1+cos\alpha)$	$\frac{V_{dc}}{R}$	$\frac{V_m}{2} * \sqrt{\frac{1}{\pi}\left[\pi-\alpha+\frac{1}{2}\sin(2\alpha)\right]}$	$\frac{V_{o\,rms}}{R}$	I_{dc}	$I_{o\,rms}$
Full wave (bridge)	R	$\frac{V_m}{\pi}(1+cos\alpha)$	$\frac{V_{dc}}{R}$	$\frac{V_m}{\sqrt{2}} * \sqrt{\frac{1}{\pi}\left[\pi-\alpha+\frac{1}{2}\sin(2\alpha)\right]}$	$\frac{V_{o\,rms}}{R}$	$\frac{I_{dc}}{2}$	$I_{o\,rms}$
	R+L	$\frac{2\,V_m}{\pi}cos\alpha$	$\frac{V_{dc}}{R}$	$\frac{V_m}{\sqrt{2}}$	I_{dc}	$\frac{I_{dc}}{2}$	I_{dc}

Three phase Uncontrolled Rectifier

Circuit	Load	V_{dc}	I_{dc}	$V_{o\,rms}$	$I_{o\,rms}$	$I_{D\,dc}$	$I_{D\,rms}$	$I_{s\,rms}$	PIV
Half wave	R	$\frac{3\,V_{line\,m}}{2\pi}$	$\frac{V_{dc}}{R}$	$0.84\,V_{ph\,m}$	$\frac{V_{orms}}{R}$	$\frac{I_{dc}}{3}$	$\frac{I_{o\,rms}}{\sqrt{3}}$	$\frac{I_{o\,rms}}{\sqrt{3}}$	$\sqrt{3}\,V_{ph.m}$
	R+L	$\frac{3\,V_{line\,m}}{2\pi}$	$\frac{V_{dc}}{R}$	$0.84\,V_{ph\,m}$	I_{dc}	$\frac{I_{dc}}{3}$	$\frac{I_{dc}}{\sqrt{3}}$	$\frac{I_{dc}}{\sqrt{3}}$	$\sqrt{3}\,V_{ph.m}$
Full wave (bridge)	R	$\frac{3\,V_{line\,m}}{\pi}$	$\frac{V_{dc}}{R}$	$1.655\,V_{ph.m}$	$\frac{V_{orms}}{R}$	$\frac{I_{dc}}{3}$	$\frac{I_{o\,rms}}{\sqrt{3}}$	$\frac{I_{o\,rms}}{\sqrt{3}}$	$\sqrt{3}\,V_{ph.m}$
	R+L	$\frac{3\,V_{line\,m}}{\pi}$	$\frac{V_{dc}}{R}$	$1.655\,V_{ph.m}$	I_{dc}	$\frac{I_{dc}}{3}$	$\frac{I_{dc}}{\sqrt{3}}$	$\frac{I_{dc}}{\sqrt{3}}$	$\sqrt{3}\,V_{ph.m}$

Three phase-controlled Rectifier

Circuit	Load	α	V_{dc}	I_{dc}	$V_{o,rms}$	$I_{o,rms}$	$I_{Th,dc}$	$I_{Th,rms}$	$I_{s,rms}$	PIV
Half wave	R	$0° \leq \alpha \leq 30°$	$\frac{3\,V_{line\,m}}{2\pi}\cos\alpha$	$\frac{V_{dc}}{R}$	$\sqrt{3}V_{m\,ph}\sqrt{\frac{1}{6}+\frac{\sqrt{3}}{8\pi}\cos(2\alpha)}$	$\frac{0.84\,V_{ph,m}}{R}$	$\frac{I_{dc}}{3}$	$\frac{0.4854\,V_{ph,m}}{R}$	$\frac{0.4854\,V_{ph,m}}{R}$	$\sqrt{3}\,V_{ph,m}$
		$30° < \alpha < 150°$	$\frac{3\,V_{line\,m}}{\pi}\left[1+\cos\left(\frac{\pi}{6}+\alpha\right)\right]$	$\frac{V_{dc}}{R}$	$\frac{\sqrt{3}V_{m\,ph}}{2\sqrt{\pi}}\sqrt{\frac{5\pi}{6}-\alpha+\frac{1}{2}\sin(\frac{\pi}{6}+2\alpha)}$	$\frac{0.84\,V_{ph,m}}{R}$	$\frac{I_{dc}}{3}$	$\frac{0.4854\,V_{ph,m}}{R}$	$\frac{0.4854\,V_{ph,m}}{R}$	$\sqrt{3}\,V_{ph,m}$
		$150° < \alpha$	0	0	0	0	0	0	0	$\sqrt{3}\,V_{ph,m}$
	R+L	Any value of α	$\frac{3\,V_{line\,m}}{\pi}\cos\alpha$	$\frac{V_{dc}}{R}$	$\sqrt{3}V_{m\,ph}\sqrt{\frac{1}{6}+\frac{\sqrt{3}}{8\pi}\cos(2\alpha)}$	I_{dc}	$\frac{I_{dc}}{3}$	$\frac{I_{dc}}{\sqrt{3}}$	$\frac{I_{dc}}{\sqrt{3}}$	$\sqrt{3}\,V_{ph,m}$
Full wave (bridge)	R	$0° \leq \alpha \leq 60°$	$\frac{3\,V_{line\,m}}{\pi}\cos\alpha$	$\frac{V_{dc}}{R}$	$\sqrt{3}V_{m\,ph}\sqrt{\frac{1}{2}+\frac{3\sqrt{3}}{4\pi}\cos(2\alpha)}$	$\frac{V_{o\,rms}}{R}$	$\frac{I_{dc}}{3}$	$\frac{0.582\,V_{line\,m}}{R}$	$\frac{0.78\,V_{line\,m}}{R}$	$\sqrt{3}\,V_{ph,m}$
		$60° < \alpha < 120°$	$\frac{3\,V_{line\,m}}{\pi}\left[1+\cos\left(\frac{\pi}{3}+\alpha\right)\right]$	$\frac{V_{dc}}{R}$	$\frac{\sqrt{3}V_{m\,ph}}{2\sqrt{\pi}}\sqrt{4\pi-6\alpha+3\cos(\frac{\pi}{6}+2\alpha)}$	$\frac{V_{o\,rms}}{R}$	$\frac{I_{dc}}{3}$	$\frac{0.582\,V_{line\,m}}{R}$	$\frac{0.78\,V_{line\,m}}{R}$	$\sqrt{3}\,V_{ph,m}$
		$120° < \alpha$	0	0	0	0	0	0	0	$\sqrt{3}\,V_{ph,m}$
	R+L	Any value of α	$\frac{3\,V_{line\,m}}{\pi}\cos\alpha$	$\frac{V_{dc}}{R}$	$\sqrt{3}\,V_{m\,ph}\sqrt{\frac{1}{2}+\frac{3\sqrt{3}}{4\pi}\cos(2\alpha)}$	I_{dc}	$\frac{I_{dc}}{3}$	$\frac{I_{dc}}{\sqrt{3}}$	$\sqrt{\frac{2}{3}}\,I_{dc}$	$\sqrt{3}\,V_{ph,m}$

Single Phase Full Wave Voltage Controller

$$i_{oN} = i_{A1N} = [\sin(\omega t - \phi] - \sin(\alpha - \phi)e^{(R/L)\left(\frac{\alpha}{\omega} - t\right)}$$

$$Z = \left[R^2 + (\omega L)^2\right]^{1/2} \quad \beta = \gamma + \alpha \quad \phi = \tan^{-1}\left(\frac{\omega L}{R}\right)$$

Thyristor Norm Currents:

$$I_N = \frac{1}{2\pi}\int_{\alpha}^{\alpha+\gamma} i_{A1N}\, d(\omega t) \quad I_{RN} = \left[\frac{1}{2\pi}\int_{\alpha}^{\alpha+\gamma} i_{A1N}^2\, d(\omega t)\right]^{1/2}$$

RMS Norm Output Current:

$$I_{RoN} = \left[I_{RN}^2 + I_{RN}^2\right]^{1/2} = \sqrt{2} I_{RN}$$

RMS Norm Voltage:

$$V_{RoN} = \left[\frac{1}{\pi}\int_{\alpha}^{\alpha+\gamma} \sin^2 \omega t \cdot d(\omega t)\right]^{1/2}$$

$$= \left\{\frac{1}{2\pi}\left[\gamma + \frac{1}{2}\sin 2\alpha - \frac{1}{2}\sin 2(\alpha + \gamma)\right]\right\}^{1/2}$$

RMS Ouputs:

$$I_{Ro} = \frac{\sqrt{2}V}{Z} I_{RoN} \quad [A]; \quad V_{Ro} = \sqrt{2}V \cdot V_{RoN} \quad [V$$

Harmonic Analysis

$$i_o = \sum_{n=1}^{\infty} a_n \cos n\omega t + b_n \sin n\omega t$$

$$a_n = \frac{1}{\pi}\int_0^{2\pi} i_o \cos n\omega t\, d(\omega t) = \frac{2}{\pi}\int_0^{\pi} i_o \cos n\omega t\, d(\omega t)$$

$$b_n = \frac{1}{\pi}\int_0^{2\pi} i_o \sin n\omega t\, d(\omega t) = \frac{2}{\pi}\int_0^{\pi} i_o \sin n\omega t\, d(\omega t)$$

$$i_o = i_{A1}$$

$$I_{nR} = \frac{1}{\sqrt{2}}\left[a_n^2 + b_n^2\right]^{1/2}$$

$$I_{nR} = H_{n\alpha} I_R\big|_{\alpha=0}$$

$$I_{nRpeak} = \sqrt{2} I_{nR}$$

$$H_{n\alpha} = \frac{RMS\ value\ of\ n^{th}\ harmonic\ at\ angle\ \alpha}{RMS\ value\ of\ line\ current\ at\ angle\ \alpha = 0} = \frac{I_{nR}}{I_R\big|_{\alpha=0}}$$

$$V_{nR} = I_{nR}\left[R^2 + (n\omega L)^2\right]^{1/2}$$

$$P_o = R_o I_{Ro}^2 \qquad p.f. = \frac{I_{Ro}^2 R}{V I_{Ro}}$$

Resistive: $\beta = \pi = \gamma$

Inductive: $\gamma = 2(\pi - \alpha)$

RL $\alpha_{\min} = \phi = \tan^{-1}\left(\frac{\omega L}{R}\right)$

when $\alpha = \phi$ treat as complete sinusoidal wave

$$\therefore I_{Ro} = \frac{V}{Z}$$

$$I_Q = I_N \cdot I_B$$

$$I_{QR} = I_{RN} \cdot I_B \quad I_{RoN} = \sqrt{2} \cdot I_{RN}$$

Single Phase Full-Wave Controlled Rectifier

- Determine what modes its operating in
- If discontinuous use graphs and theory (the formulas are for only one thyristor so for, where $I_{RN} \cdot I_B$ off graph $= I_R = I_{QR}$ and $I_N \cdot I_B$ off graph $= I_Q$ $\therefore I_{Ro} = \sqrt{2} I_{QR}$ and $I_o = 2I_Q$) $\quad p.f. = \frac{V_o I_o}{V_{sec} I_{SR}}$
- gamma from graph then find beta else continue with following formulae:

$$m = \frac{V_c}{\sqrt{2}V_{sec}} \quad ; \quad \phi = \tan^{-1}\left(\frac{\omega L}{R}\right) \quad ; \quad i_N = \sin(\omega t - \phi)$$

Fourier Analysis

$$v_o = V_o + \sum_{n=1}^{\infty} c_n \cos(n\omega t - \theta_n)$$

$$c_n \left[a_n^2 + b_n^2\right]^{1/2} \qquad \theta_n = \tan^{-1}\left(\frac{b_n}{a_n}\right)$$

$$V_o = \frac{1}{\pi}\int_{\alpha}^{\pi+\alpha} \sqrt{2}V \sin\omega t \; d(\omega t) = \frac{2\sqrt{2}}{\pi} V \cos\alpha \quad [V]$$

$$a_n = \frac{2}{\pi}\int_{\alpha}^{\alpha+\pi} v_o \cos n\omega t \; d(\omega t) = \left[\frac{\cos(n+1)\alpha}{(n+1)} - \frac{\cos(n-1)\alpha}{(n-1)}\right]$$

$$b_n = \frac{2}{\pi}\int_{\alpha}^{\alpha+\pi} v_o \sin n\omega t \; d(\omega t) = \left[\frac{\sin(n+1)\alpha}{(n+1)} - \frac{\sin(n-1)\alpha}{(n-1)}\right]$$

RMS Voltage

$$V_R = \left[\frac{1}{\pi}\int_{\alpha}^{\alpha+\pi} \left[\sqrt{2}V \sin\omega t\right]^2 d(\omega t)\right]^{1/2} = V \quad [V]$$

Ripple Voltage:

$$V_{RI} = \left[V_R^2 - V_o^2\right]^{1/2} = V\left[1 - \frac{8\cos^2\alpha}{\pi^2}\right]$$

Voltage Ripple Factor:

$$K_v = \frac{V_{RI}}{V_o}$$

Load circuit Current

$$i_o = I_o + \sum d_n \cos(n\omega t - \theta_n - \phi_n) \quad [A]$$

avg output current:

$$I_o = \frac{V_o - V_c}{R} \quad [A]$$

get this: $\frac{c_n}{\sqrt{2}}$ off graph and then solve for c_n

$$d_n = \frac{c_n}{\left[R^2 + (n\omega L)^2\right]^{1/2}} \quad [A] \;;$$

$$\phi_n = \tan^{-1}(n\omega L/R) \quad [rad]$$

nth harmonic component of current:

$$I_{nR} = \frac{d_n}{\sqrt{2}} \quad [A]$$

ripple current:

$$I_{RI} = \left[\sum I_{nR}^2\right]^{1/2} \quad [A]$$

rms output current:

$$I_{Ro} = \left[I_o^2 + I_{RI}^2\right]^{1/2} \quad [A]$$

just calculate the second harmonic for I_{RI} to get the *rms* output calculation

current ripple factor:

$$K_i = \frac{I_{RI}}{I_o}$$

Thyristor Ratings

avg thyristor current:

$$I_Q = \frac{I_o}{2} \quad [A]$$

rms thyristor current:

$$I_{QR} = \frac{I_R}{\sqrt{2}} \quad [A]$$

max. forward or rev. voltage across thyristor:

$$v_{AK\,\max} = \pm 2\sqrt{2}V \quad [V]$$

turn off time available:

$$t_q = \frac{\pi - \alpha}{\omega} \quad [s]$$

$$p.f. = \frac{I_{Ro}^2 R + V_c I_o}{V_{sec} I_{Ro}}$$

($-V_o$)then DC delivers power

(V_o) then AC delivers power

if current totally constant(no ripple):

$$p.f. = \frac{V_o I_o}{V_{sec} I_o} \quad because \quad I_{srms} = I_o$$

<u>with Freewheeling Diode circuit</u>

$$V_o = v_o = \frac{1}{\pi}\int_{\alpha}^{\pi} \sqrt{2}V \sin \omega t \, d(\omega t)$$

because the source current is not constant must calculate with:

$$I_{SR} = \left[\frac{1}{\pi}\int_{\alpha}^{\pi}(I_o)^2 \, d(\omega t)\right]^{\frac{1}{2}}$$ for p.f @ source

BASE VALUES

$$V_B = \sqrt{2}V, \quad Z_B = Z, \quad I_B = \frac{\sqrt{2}V}{Z}$$

$$X_o = X_N \cdot X_B$$

<u>Circuits with Switches and Diodes</u>

<u>First Principles:</u>

$$v_R = Ri$$

$$v_c = \frac{1}{C}\int_0^t i \, dt + v_c(0)$$

$$v_L = L\frac{di}{dt}$$

<u>Half-Wave Rectifier:</u>

$$v = \sqrt{2}V \sin \omega t$$

$$v_o = V_o + \sum(a_n \cos n\omega t + b_n \sin n\omega t)$$

$\alpha \rightarrow$ firing angle

$\beta \rightarrow$ extinction angle

$\gamma = \beta - \alpha \rightarrow$ conduction angel

$$\eta = \sin^{-1} m = \sin^{-1}\left(\frac{V_c}{\sqrt{2}V}\right) \quad \therefore m = \frac{V_c}{\sqrt{2}V}$$

$$V_o = \frac{1}{2\pi}\int_0^{2\pi} v_o \, d(\omega t) = \frac{1}{2\pi}\int_{\alpha}^{\beta} v_o \, d(\omega t)$$

$$a_n = \frac{1}{\pi}\int_{\alpha}^{\beta} v_o \cos n\omega t \, d(\omega t)$$

$$b_n = \frac{1}{\pi}\int_{\alpha}^{\beta} v_o \sin n\omega t \, d(\omega t)$$

n[th] *rms* harmonic voltage:

$$V_{nR} = \frac{1}{\sqrt{2}}\left[a_n^2 + b_n^2\right]^{1/2}$$

$$V_R = \left[\frac{1}{2\pi}\int_{\alpha}^{\beta} v_o^2 \, d(\omega t)\right]^{1/2} = \left[V_o^2 + \sum V_{nR}^2\right]^{1/2}$$

rms ripple voltage:

$$V_{RI} - \left[V_R^2 - V_o^2\right]^{1/2}$$

voltage ripple factor:

$$K_v = \frac{V_{RI}}{V_o}$$

$$i = I_o + \sum\left[c_n \cos(n\omega t - \phi_n) + d_n \sin(n\omega t - \phi_n)\right]$$

$$I_0 = \frac{V_0}{R}$$

$$c_n = \frac{a_n}{|Z_n|} \qquad d_n = \frac{b_n}{|Z_n|} \qquad \phi_n = \tan^{-1}\left[\frac{n\omega L}{R}\right]$$

n[th] *rms* harmonic current:

$$I_{nR} = \frac{1}{\sqrt{2}}\left[c_n^2 + d_n^2\right]^{1/2}$$

rms ripple current:

$$I_{RI} = \left[I_R^2 - I_o^2\right]^{1/2}$$

current ripple factor: $I_v = \dfrac{I_{RI}}{I_o}$

<u>Single Phase Half Wave Controlled Rectifier</u>

mainly done through graphs

RL Load Circuit

$$L\frac{di}{dt}+R_i=\sqrt{2}V\sin\omega t$$

$\Downarrow$

$$i=\frac{\sqrt{2}}{Z}\left[\sin(\omega t-\phi)-\sin(\alpha-\phi)e^{(R/L)\left(\alpha/\omega-t\right)}\right]$$

$$Z=\left[R^2+(\omega L)^2\right]^{1/2}\qquad \phi=\tan^{-1}\left(\frac{\omega L}{R}\right)$$

$$I_N=\frac{1}{2\pi}\int_{\alpha}^{\alpha+\gamma}i_n\,d(\omega t)\rightarrow \text{avg. norm}$$

$$I_{RN}=\left[\frac{1}{2\pi}\int_{\alpha}^{\alpha+\gamma}i_N^2\,d(\omega t)\right]^{1/2}\rightarrow \text{RMS}$$

norm

R Load Circuit

$$Z=R\quad \phi=0\quad \beta=\pi$$

$$i=\frac{\sqrt{2}V}{Z}\sin\omega t$$

$$I_N=\frac{1}{2\pi}\int_{\alpha}^{\pi}\sin\omega t\,d(\omega t)=\frac{1+\cos\alpha}{2\pi}$$

$$I_{RN}=\left[\frac{1}{2\pi}\int_{\alpha}^{\pi}\sin^2\omega t\,d(\omega t)\right]^{1/2}=\left[\frac{1}{4\pi}\left(\pi-\alpha+\frac{\sin 2\alpha}{2}\right)\right]$$

L Load Circuit

$$Z=\omega L\quad \varphi=\frac{\pi}{2}\quad \beta=2\pi-\alpha$$

$$i=\frac{\sqrt{2}V}{\omega L}(\cos\alpha-\cos\omega t)$$

$$I_N=\frac{1}{2\pi}\int_{\alpha}^{2\pi-\alpha}(\cos\alpha-\cos\omega t)d(\omega t)=\frac{1}{\pi}[(\pi-\alpha)\cos\alpha+\sin\alpha]$$

$$I_{RN}=\left[\frac{1}{2\pi}\int_{\alpha}^{2\pi-\alpha}(\cos\alpha-\cos\omega t)^2d(\omega t)\right]^{1/2}$$

$$V_{RN}=\left[\frac{1}{2\pi}\int_{\alpha}^{2\pi-\alpha}\sin^2\omega t\ d(\omega t)\right]^{1/2}=\left[\frac{1}{2}-\frac{\alpha}{2\pi}+\frac{1}{4\pi}\sin 2\alpha\right]$$

RL Load with EMF Circuit

$$\eta=\sin^{-1}\left(\frac{V_c}{\sqrt{2}V}\right)=\sin^{-1}m$$

$$\beta_{max}=2\pi-\sin^{-1}\frac{|V_c|}{\sqrt{2}V}$$

$$i_N=\sin(\omega t-\varphi)-\left[\frac{m}{\cos\varphi}-Be^{(\alpha-\omega t)/\tan\varphi}\right]$$

$$B=\frac{m}{\cos\varphi}-\sin(\alpha-\varphi)$$

$$I_N=\frac{1}{2\pi}\int_{\alpha}^{\alpha+\gamma}i_N d(\omega t)\rightarrow \text{avg. norm}$$

$$I_{RN}=\left[\frac{1}{2\pi}\int_{\alpha}^{\alpha+\gamma}i_N^2 d(\omega t)\right]^{1/2}\rightarrow \text{rms norm}$$

R Load with EMF Circuit

$$\eta=\sin^{-1}\left(\frac{V_c}{\sqrt{2}V}\right)=\sin^{-1}m$$

$$\beta_{max}=2\pi-\sin^{-1}\frac{|V_c|}{\sqrt{2}V}$$

$$i_N=\sin\omega t-m$$

$$\gamma=(\pi-\eta)-\alpha$$

$$I_N = \frac{1}{2\pi}\int_{\alpha}^{\pi+\eta} i_N \, d(\omega t) \rightarrow avg\ norm$$

$$= \frac{1}{2\pi}\left[\cos\alpha - \sqrt{1-m^2} - m(\pi - \eta - \alpha)\right]$$

$$I_{RN} = \left[\frac{1}{2\pi}\int_{\alpha}^{\pi+\eta} i_N^{\,2} \, d(\omega t)\right]^{1/2} \rightarrow rms\ norm$$

L Load with EMF Circuit

$$\eta = \sin^{-1}\left(\frac{V_c}{\sqrt{2}V}\right) = \sin^{-1} m$$

$$\beta_{\max} = 2\pi - \sin^{-1}\frac{|V_c|}{\sqrt{2}V}$$

$$i_N = \cos\alpha - \cos\omega t - m(\omega t - \alpha)$$

$$m\gamma = \cos\alpha - \cos(\alpha - \gamma)$$

$$I_N = \frac{1}{2\pi}\left[\gamma\cos\alpha + \sin\alpha - \sin(\alpha + \gamma) - \frac{m\gamma^2}{2}\right]$$

$$I_{RN} = \left[\frac{1}{2\pi}\int_{\alpha}^{a+\gamma}[\cos\alpha - \cos\omega t - m(\omega t - \alpha)]^2 \, d(\omega t)\right]^{1/2}$$

One Quad Chopper DC to DC Converter

- Type A: V_o and I_o only positive find:

$$\omega = \frac{2\pi}{T}$$

$$m = \frac{V_c}{V} = \frac{e^{\rho\sigma}-1}{e^{\sigma}-1}, \quad \sigma = \frac{T}{\tau}, \quad \tau = \frac{L}{R}, \quad \rho = \frac{t_{ON}^{x}}{T}$$

(solve for ρ from m equation)

if $\rho < \frac{t_{ON}}{T}$ then CCM else DCM, if you don't have t_{ON} look on graph with all the variables, continue with respective formulae:

Continuous Current Mode

$$V_o = \frac{t_{ON}}{T}V \quad I_o = \frac{V_o - V_c}{R}$$

current at moment of commutation: I_{max}

$$I_{\max} = \frac{V}{R}\frac{\left(1-e^{-t_{ON}/\tau}\right)}{\left(1-e^{-T/\tau}\right)} - \frac{V_c}{R}$$

$$I_{\min} = \frac{V}{R}\frac{\left(e^{t_{ON}/\tau}-1\right)}{\left(e^{T/\tau}-1\right)} - \frac{V_c}{R}$$

$$v_o = V_o + \sum_{n=1}^{\infty} a_n \cos n\omega t + b_n \sin n\omega t$$

$$= V_o + \sum_{n=1}^{\infty} c_n \sin(n\omega t + \theta_n)$$

$$a_n = \frac{V}{n\pi} sin\, n\, \omega t_{ON}$$

$$b_n = \frac{V}{n\pi}[1 - cos\, n\, \omega t_{ON}]$$

$$c_n = \frac{\sqrt{2}V}{n\pi}[1 - cos\, n\, \omega t_{ON}]^{1/2}$$

$$\theta_n = tan^{-1}\left[\frac{sin\, n\omega t_{ON}}{1 - cos\, n\omega t_{ON}}\right]$$

the preceding 4 formulas only for CCM.

rms n[th] harmonic voltage:

$$V_{nR} = \frac{1}{\sqrt{2}}\left[a_n^2 + b_n^2\right]^{1/2}$$

rms output voltage:

$$V_R = \left[V_o^2 + \sum_{n=1}^{\infty} V_{nR}^2\right]^{1/2}$$

rms "ripple voltage":

$$V_{RI} = \left[\sum_{n=1}^{\infty} V_{nR}^2\right]^{1/2} = \left[V_R^2 - V_o^2\right]^{1/2}$$

voltage ripple factor:

$$K_v = \frac{V_{RI}}{V_o}$$

$$i_o = I_o + \sum_{n=1}^{\infty}\left[A_n \cos(n\omega t - \phi_n) + B_n \sin(n\omega t - \phi_n)\right]$$

$$A_n = \frac{a_n}{\sqrt{R^2 + (n\omega L)^2}}$$

$$B_n = \frac{b_n}{\sqrt{R^2 + (n\omega L)^2}}$$

$$\phi_n = \tan^{-1}(n\omega L/R)$$

rms nth harmonic current:

$$I_{nR}=\frac{1}{\sqrt{2}}\left[A_n^2+B_n^2\right]^{1/2}$$ or

$$Z_n=\left[R^2+(n\omega L)^2\right]^{1/2}$$ which leads to:

$$I_{nR}=\frac{V_{nR}}{Z_n}$$

rms ripple current:

$$I_{RI}=\left[\sum_{n=1}^{\infty}I_{nR}^2\right]^{1/2}=\left[I_R^2-I_o^2\right]^{1/2}$$

current ripple factor: $K_i=\frac{I_{RI}}{I_o}$

<u>Discontinuous Current Mode</u>

$$t_x=\tau\ln\left\{e^{t_{ON}/\tau}\left[1+\frac{V-V_c}{V_c}\left(1-e^{-t_{ON}/\tau}\right)\right]\right\}$$

$$V_o=\frac{t_{ON}}{T}V+\frac{(T-t_x)}{T}V_c \quad I_o=\frac{V_o-V_c}{R}$$

current at moment of commutation: I_{max}

$$I_{max}=\frac{V-V_c}{R}\left(1-e^{-t_{ON}/\tau}\right)$$

$$I_{min}=0$$

$$v_o=V_o+\sum_{n=1}^{\infty}a_n\cos n\omega t+b_n\sin n\omega t$$

$$=V_o+\sum_{n=1}^{\infty}c_n\sin(n\omega t+\theta_n)$$

$$a_n=\frac{2}{T}\int_0^T v_0\cos n\omega t\,dt$$

$$=\frac{2}{T}\left[\int_0^{t_{ON}}V\cos\frac{2n\omega t}{T}dt+\int_{t_x}^T V_c\cos\frac{2n\omega t}{T}dt\right]$$

$$=\frac{V}{n\pi}\sin n\omega t_{ON}-\frac{V_c}{n\pi}\sin n\omega t_x$$

$$b_n=\frac{2}{T}\int_0^T v_0\sin n\omega t\,dt$$

$$=\frac{V}{n\pi}\left[1-\cos n\omega t_{ON}\right]-\frac{V_c}{n\pi}\left[1-\cos n\omega t_x\right]$$

$$c_n=\left[a_n^2+b_n^2\right]^{1/2}$$

$$\theta_n=\tan^{-1}\left[a_n/b_n\right]$$

$$A_n=\frac{a_n}{\sqrt{R^2+(n\omega L)^2}}$$

$$B_n=\frac{b_n}{\sqrt{R^2+(n\omega L)^2}}$$

$$\phi_n=\tan^{-1}(n\omega L/R)$$

the preceding 4 formulas good for C.C.M. and D.C.M. but C.C.M. has simplified ones in its section

rms nth harmonic voltage:

$$V_{nR}=\frac{1}{\sqrt{2}}\left[a_n^2+b_n^2\right]^{1/2}$$

rms output voltage:

$$V_R=\left[V_o^2+\sum_{n=1}^{\infty}V_{nR}^2\right]^{1/2}$$

rms "ripple voltage":

$$V_{RI}=\left[\sum_{n=1}^{\infty}V_{nR}^2\right]^{1/2}=\left[V_R^2-V_o^2\right]^{1/2}$$

voltage ripple factor: $K_v=\frac{V_{RI}}{V_o}$

$$i_o=I_o+\sum_{n=1}^{\infty}\left[A_n\cos(n\omega t-\phi_n)+B_n\sin(n\omega t-\phi_n)\right]$$

rms nth harmonic current:

$$I_{nR}=\frac{1}{\sqrt{2}}\left[A_n^2+B_n^2\right]^{1/2}$$ or

$$Z_n=\left[R^2+(n\omega L)^2\right]^{1/2}$$ which leads

$$I_{nR}=\frac{V_{nR}}{Z_n}$$

to:

rms ripple current:

$$I_{RI}=\left[\sum_{n=1}^{\infty}I_{nR}^2\right]^{1/2}=\left[I_R^2-I_o^2\right]^{1/2}$$

current ripple factor:

$$K_i=\frac{I_{RI}}{I_o}$$

Current Commutation of Type 'A' Chopper

$I_{01} = I_{\max} = current\ at\ commutation$

$\omega_r = \frac{1}{\sqrt{L_1 C}}$

$t_1 = \frac{1}{\omega_r}\sin^{-1}\left(-\frac{\omega_r L_1 I_{01}}{V}\right) = \frac{\pi}{\omega_r} + \frac{1}{\omega_r}\sin^{-1}\frac{\omega_r L_1 I_{01}}{V}$

$t_2 = \frac{2\pi}{\omega_r} - \frac{1}{\omega_r}\sin^{-1}\left(\frac{\omega_r L_1 I_{01}}{V}\right) = \frac{2\pi}{\omega_r} - t_1 + \frac{\pi}{\omega_r} = \frac{3\pi}{\omega_r} - t_1$

$t_q = t_2 - t_1$

$t_1' = \frac{CV}{I_{01}}(1 - \cos\omega_r t_2) = \frac{x}{\omega_r}(1 - \cos\omega_r t_2)$

$t_1'' = \frac{\pi}{2\omega_r}$

$t_c = t_2 + t_1' + t_1'' \qquad x = \frac{V}{I_{01}}\left(\frac{C}{V}\right)^{1/2}$

$t_{q1} = \frac{1}{\omega_r}\left(\pi - 2\sin^{-1}\frac{1}{x}\right) \qquad t_{q2} = t_2 - \frac{\pi}{\omega_r}$

Single Phase Voltage-Source Inverter

for Half-Bridge: $E = V/2$

for Full-Bridge: $E = V$

using the following 2 equations, the nth harmonic voltage and current equations can be found:

$$v_o = \sum_{n=1,3}^{\infty}\frac{4E}{n\pi}\sin n\omega t = \quad [V]$$

$$i_o = \sum_{n=1,3}^{\infty}\frac{4E}{n\pi Z_n}\sin(n\omega t - \phi_n) \quad [A]$$

$$Z_n = \left[R^2 + \left(n\omega L - \frac{1}{n\omega C}\right)^2\right]^{1/2} \quad [\Omega]$$

$$\phi_n = \tan^{-1}\left[\frac{n\omega L - 1/(n\omega C)}{R}\right] \quad [rad]$$

$t_q = \frac{T}{2} - t_x > t_{off}$: has to be true for guaranteed commutation load current at instant of commutation sub this into the i_o *formula:*

$\omega t = \pi, \quad \therefore i_o = I_{01} \quad [A]$

$$I_{o_n} = \frac{1}{2\pi}\int_\alpha^\beta \frac{4E}{n\pi Z_n}\sin(n\omega t - \phi_n)d(\omega t)$$

$$I_{nRo} = \left[\frac{1}{2\pi}\int_\alpha^\beta \left(\frac{4E}{n\pi Z_n}\right)^2 \sin^2(n\omega t - \phi_n)d(\omega t)\right]^{1/2}$$

$$I_{nD} = \frac{1}{2\pi}\int_\beta^\pi \frac{4E}{n\pi Z_n}\sin(n\omega t - \phi_n)d(\omega t)$$

$$I_o = \sum_{n=1,3,..}^{\infty} I_{o_n} \qquad I_{Ro} = \sqrt{\sum_{n=1,3,..}^{\infty} I_{o_n}^2}$$

$P_o = RI_{1R}^2 \quad P_s = \frac{V}{2}\times\frac{\sqrt{2}I_{1R}}{\pi}$

if resistive:
the only difference use this formula for *rms* output current:

$$I_{Ro} = \frac{I_{o\max}}{\sqrt{2}} = \frac{i_{o1(peak)}}{\sqrt{2}}, \quad I_{QR} = \frac{I_{Ro}}{\sqrt{2}}$$

single pulse width:

$$v_{AB} = \sum_{n=1,3}^{\infty}\frac{4V_s}{n\pi}\sin\left(\frac{n\pi}{2}\right)\sin\left(\frac{n\delta}{2}\right)\sin(n\omega t)$$

where δ is pulse width

$$V_{ABn} = \frac{4V_s}{n\pi}\sin\left(\frac{n\pi}{2}\right)\sin\left(\frac{n\delta}{2}\right) = V_{nPeak}$$

$V_{ABnR} = V_{ABn}/\sqrt{2}$

$$i_{o_n} = \frac{|V_{nPeak}|}{|Z_n|}\sin(\omega t - \phi_n)$$

$$I_{o_n} = \frac{|V_{nPeak}|}{|Z_n|}\sin(\omega t - \phi_n)\Big|_{\omega t = \text{angle desired, i.e. commutation}}$$

Power Transistors BJT

t_d :

$P_i(t) = i_c(t)\cdot v_{ce}(t) = I_{ceo}V_{cc}$

$$P_d = \frac{1}{T}\int_0^{t_d} P_i(t)\,dt$$

t_r :

$$P_i(t) = i_c(t) \cdot v_{ce}(t) = I_{cs} \frac{t}{t_r}\left[V_{cc} + \left(V_{ce(sat)} - V_{cc}\right)\frac{t}{t_r}\right]$$

$$P_r = \frac{1}{T}\int_0^{t_r} P_i(t)dt$$

$t_n = kT - t_d - t_r$:

$$P_i(t) = i_c(t) \cdot v_{ce}(t) = V_{ce(sat)} I_{cs} \quad inst.\ power$$

$$P_n = \frac{1}{T}\int_0^{t_n} P_i(t)dt \quad avg\ power$$

t_s :

$$P_i(t) = i_c(t) \cdot v_{ce}(t) = V_{ce(sat)} I_{cs} \quad inst.\ power$$

$$P_s = \frac{1}{T}\int_0^{t_s} P_i(t)dt \quad avg\ power$$

t_f :

$$P_i(t) = i_{cs}(t) \cdot v_{ce}(t) = I_{cs}\left(1 - \frac{t}{t_f}\right)V_{cc}\left(\frac{t}{t_f}\right) \quad inst.\ power$$

$$P_f = \frac{1}{T}\int_0^{t_f} P_i(t)dt \quad avg\ power$$

$\bar{t}_o = (1-k)T - t_s - t_f$:

$$P_i(t) = i_{cs}(t) \cdot v_{ce}(t) = I_{ceo} V_{cc} \quad inst.\ power$$

$$P_o = \frac{1}{T}\int_0^{t_o} P_i(t)dt \quad avg\ power$$

General Equations:

$$t_{ON} = t_d + t_r \qquad t_{OFF} = t_s + t_r$$

$$P_{total} = P_T = P_d + P_r + P_n + P_s + P_f + P_o$$

$$\frac{dP_i(t)}{dt} = 0 \ for \max\ power$$

Thermal Evaluation

$$T_{JUNCT} - T_{AMB} = P_T(\theta_{JC} + \theta_{CS} + \theta_{SA})$$

Secondary Breakdown Derating Factor:

$$T_C = T_{JUNCT} - P_T\theta_{JC}$$

MOSFET

$$Q_x = V_x C_x$$

$$I_1 = \frac{Q_{gd}}{\Delta t}, \quad I_2 = \frac{Q_{gs}}{\Delta t}, \quad I_T = I_1 + I_2$$

$$V_{gd} = V_{gs} - (-V_{ds})$$

$$1 + \frac{V_{ds}}{V_{gs}} = miller\ effect$$

$$P_L = I_{Drms}^2 R_{DS(ON)}$$

DC to DC Converters

Buck

Rise:

$$V_o = \frac{V_s}{1-D} = \frac{V_o DT}{RC} = \frac{V_o D}{RCf}$$

$$V_s - V_o = L\frac{I_2 - I_1}{t_1} = L\frac{\Delta I}{t_1}$$

Fall:

$$V_o = L\frac{\Delta I}{t_2}$$

$$\Delta I = \frac{(V_s - V_o)t_1}{L} = \frac{V_o t_2}{L}$$

General for Buck:

$$T = \frac{1}{f} \quad T = t_1 + t_2 = \frac{\Delta I}{V_s - V_o}L + \frac{\Delta I}{V_o}L$$

$$V_o = kV_s \quad \frac{t_1}{T} = k = \frac{V_o}{V_s} = duty\ cycle$$

$$k = D \quad D = \frac{t_1}{T} \quad D' = \frac{t_2}{T}$$

$$\Delta I = \frac{V_o(V_s - V_o)}{fLV_s} = \frac{kV_s(1-k)}{fL}$$

$$|\Delta V_o| = \frac{\Delta q}{C} = \frac{1}{8}\frac{\Delta I}{fC} \qquad \frac{|\Delta V_o|}{V_o} = \frac{1}{8}\frac{D'}{fLC}$$

Boost:

$$T = T_s \quad D = k$$

$$T = \frac{1}{f} \quad T = t_1 + t_2 = \frac{\Delta I}{V_s - V_o}L + \frac{\Delta I}{V_o}L$$

$$t_1 = \frac{\Delta I \cdot L}{V_s} \qquad t_2 = \frac{\Delta I \cdot L}{V_0 - V_s}$$

$$t_1 = DT \qquad t_2 = (1-D)T$$

$$\Delta I = \frac{V_o(V_s - V_o)}{fLV_o} = \frac{V_s D}{fL}$$

$$t = DT_s$$

$$I_c = I_o$$

$$\Delta V_c = \frac{1}{C}\int_0^{DT_s} I_o\, dt = I_o\frac{DT_s}{C} = \frac{I_o D}{fC}$$

$$P_o = \frac{V_o^2}{R}$$

$$V_s I_s = \frac{V_o^2}{R} = \frac{\left(\frac{V_s}{1-D}\right)}{R} = \frac{V_s^2}{(1-D)^2 R}$$

$$I_L = \frac{V_s}{(1-D)^2 R} \qquad \frac{\Delta i_L}{2} = \frac{V_s DT}{2L}$$

$$I_{max} = I_L + \frac{\Delta i_L}{2} = \frac{V_s}{(1-D)^2 R} + \frac{V_s DT}{2L}$$

$$I_{min} = I_L - \frac{\Delta i_L}{2} = \frac{V_s}{(1-D)^2 R} - \frac{V_s DT}{2L}$$

$$L_{min} = \frac{D(1-D)^2 R}{2f}$$

$$|\Delta Q| = \left(\frac{V_o}{R}\right) DT = C\Delta V_o$$

$$\frac{\Delta V_o}{V_o} = \frac{D}{RCf} < given\, ripple\ percent$$

Select a suitable answer:

1- In a single-phase diode rectifier fed from Vm sin (wt), the lowest ripple frequency and PIV are, respectively

1. f /2, V m for half-wave circuit
2. f, Vm for the full-wave circuit.
3. f/ 2 Vm for M-2 circuit.
4. f, Vm for half-wave circuit

From these, the correct statements are :

(a) 1,2,3 (b) 2,4 (c) 3,4 (d) 2,3,4

Ans: c

2- A 3-phasehalf-wave diode rectifier feeds a load of R = 100 n. The power delivered to load is for an input supply of 400 V,50 Hz.

(a)753.73 W (b) 974.23 W (c) 376.98 W (d) 487.26 W .

Ans: a

3- Read the following statements:

1. If load resistance R is low, the ripple factor for L- filter is high
2. For C-filter, the ripple factor gets reduced if R is increased
3. L-filter is suited to low-current loads.
4. A better waveform is obtained if R is increased for C-filter.

From these, the correct statement is

(a) 1,4 (b) 2, 4 (c) 1,3 (d) 2,3

Ans: b

4. The number of p-n junctions in a thyristor is

(a) 1 (b) 2 (c) 3 (d) 4

Ans: c

5. When a thyristor is forward-biased, the number of blocked junctions is

(a) 1 (b) 2 (c) 3 (d) 4

Ans: a

5. When a thyristor is reverse biased, the number of blockQ<i p-n junctions is

(a) 1 (b) 2 (c) 3 (d) 4

Ans: b

6. In a thyristor, anode current is made up of :

a) Electronic only
b) electrons or holes
c) Electrons and holes
d) holes only.

Ans: c

7. A thyristor, when triggered, will charge from a forward-blocking state to a conduction state if its anode-to-cathode voltage is equal to

a) peak repetitive off-state forward voltage.
b) peak working off-state forward voltage.
c) peak working off-state reverse voltage.
d) peak non-repetitive off-state forward voltage.

Ans: b

8. Commutation or turn-off of a thyristor requires that:

1. anode current is reduced below holding current
2. anode voltage is reduced to zero
3. anode current is allowed to reverse
4. anode voltage gets reversed
5. reverse voltage is applied to it.

From these, the correct statements are

(a) all (b) 1, 3, 4 (c) 1, 3, 4, 5 (d) 1, 2, 4

Ans: c

9. In a thyristor, the ratio of holding current to latching current is

(a) 0.4 (b) 1.0 (c) 2.5 (d) 4.00
Ans: a

10 . When a thyristor gets turned on, the gate drive :

(a) should not be removed as it will turn off the SCR
(b) the mayor may not be removed
(c) should be removed
(d) should be removed to avoid increased losses and higher junction temperature.

Ans: d

10. In step down dc choppers, if Ton is the on-period and f is the chopping frequency, then output voltage in terms of input voltage Vs is given by:

(a) Vs . Ton/f (b) Vs . f/Ton . (c) V/f· Ton (d) Vs . f· Ton
Ans: d

11. In a thyristor

(a) latching current h is associated with the turn-off process and holding current IH. with turn-on process·
(b) both IL and IH are associated with the turn-off process
(c) IH is associated with the turn-off process and h with the turn-on process ·
(d) both hand IH are associated with the turn-on process.

Ans: c

12. For a triac and SCR,

a) both are unidirectional devices
b) Triac requires more current for turn-on than SCR at a particular voltage
c) a triac has less time forturn-offthan SCR
d) both are available with comparable voltage and current ratings.

Ans: b

13. The correct sequence of the given devices in the decreasing order of their speed of operation is:

    a) power BJT, PMOSFET, IGBT, SCR
    b) IGBT, PMOSFET, power BJ'T, SCR
    c) SCR, PBJT, IGBT, PMOSFET
    d) PMOSFET, IGBT, PBJT, SCR

Ans:d

14. In controlled rectifiers, the nature of load current, i.e., whether the load current is continuous or discontinuous

    a) does not depend on the type of load and firing angle delay
    b) depends both on the type of load and firing angle delay
    c) depends only on the type of load
    d) depends .only on the firing angle delay.

Ans: b

15. A freewheeling diode is placed across the de load

    1. to prevent reversal of load voltage
    2. to transfer the load current away from the source
    3. to transfer the load current away from conducting thyristor.

The correct statements are

(a) 1,3 (b) 2,3 (c) 1,2 (d) 1,2,3
Ans: d

16. A six-pulse thyristor rectifier bridge is connected to a balanced 50 Hz three-phase ac source. Assuming that the dc output current of the rectifier is constant, the lowest frequency harmonic component of the ac source line current is

(a) 100 Hz (b) 150 Hz (c) 150 Hz. (d) 300 Hz
Ans: d

17 . In dc choppers, the waveforms for input and output voltages are, respectively

a) discontinuous, continuous
b) both continuous
c) both discontinuous
d) continuous, discontinuous.

Ans: d

References

Fink, D. G., & Beaty, H. W. (Eds.). (2006). Standard Handbook for Electrical Engineers (16th ed.). McGraw-Hill Education.

Horowitz, P., & Hill, W. (2015). The Art of Electronics (3rd ed.). Cambridge University Press.

Purcell, E. M. (2013). Electricity and Magnetism (3rd ed.). Cambridge University Press.

Griffiths, D. J. (2017). Introduction to Electrodynamics (4th ed.). Cambridge University Press.

Beaty, H. W., & Santoso, S. (2006). Handbook of Electric Power Calculations (3rd ed.). McGraw-Hill Education.

Papoulis, A., & Pillai, S. U. (2002). Probability, Random Variables, and Stochastic Processes (4th ed.). McGraw-Hill Education.

Kreyszig, E. (2010). Advanced Engineering Mathematics (10th ed.). John Wiley & Sons.

Strang, G. (2005). Introduction to Linear Algebra (4th ed.). Wellesley-Cambridge Press.

Sauer, T. (2006). Numerical Analysis (2nd ed.). Pearson Education.

Bender, C. M., & Orszag, S. A. (1999). Advanced Mathematical Methods for Scientists and Engineers. Springer.

Ross, S. M. (2018). A First Course in Probability (10th ed.). Pearson.

DeGroot, M. H., & Schervish, M. J. (2011). Probability and Statistics (4th ed.). Pearson.

Devore, J. L., & Berk, K. N. (2018). Modern Mathematical Statistics with Applications (2nd ed.). Springer.

Rice, J. A. (2006). Mathematical Statistics and Data Analysis (3rd ed.). Cengage Learning.

Hogg, R. V., McKean, J., & Craig, A. T. (2018). Introduction to Mathematical Statistics (8th ed.). Pearson.

Park, C. S., & Park, D. W. (2012). Contemporary Engineering Economics (6th ed.). Pearson.

Sullivan, W. G., Wicks, E. M., & Koelling, C. P. (2014). Engineering Economy (16th ed.). Pearson.

Blank, L. T., & Tarquin, A. J. (2019). Engineering Economy (8th ed.). McGraw-Hill Education.

Riggs, J. L., Bedworth, D. D., & Randhawa, S. U. (2010). Engineering Economics (5th ed.). McGraw-Hill Education.

Park, D. W., & Park, C. S. (2019). Engineering Economic Analysis (14th ed.). Oxford University Press.

Van Vlack, L. H. (1989). Elements of Materials Science and Engineering (6th ed.). Addison-Wesley.

Ashby, M. F., & Jones, D. R. H. (2013). Engineering Materials 1: An Introduction to Properties, Applications, and Design (4th ed.). Butterworth-Heinemann.

Callister Jr., W. D., & Rethwisch, D. G. (2017). Materials Science and Engineering: An Introduction (10th ed.). John Wiley & Sons.

Raja, J. D., & Chawla, K. K. (2016). Engineering Materials: Research, Applications, and Advances. CRC Press.

Garg, R. (2014). Materials Science and Engineering: Concepts, Methodologies, Tools, and Applications (Vol. 3). IGI Global.

Dorf, R. C., & Svoboda, J. A. (2016). Introduction to Electric Circuits (9th ed.). John Wiley & Sons.

Alexander, C. K., & Sadiku, M. N. O. (2017). Fundamentals of Electric Circuits (6th ed.). McGraw-Hill Education.

Nilsson, J. W., & Riedel, S. A. (2014). Electric Circuits (10th ed.). Pearson.

Hayt, W. H., Kemmerly, J. E., & Durbin, S. M. (2014). Engineering Circuit Analysis (8th ed.). McGraw-Hill Education.

Sedra, A. S., & Smith, K. C. (2016). Microelectronic Circuits (7th ed.). Oxford University Press.

Robbins, A. H., & Miller, W. C. (2018). Circuit Analysis: Theory and Practice (5th ed.). Cengage Learning.

Boylestad, R. L., & Nashelsky, L. (2018). Electronic Devices and Circuit Theory (12th ed.). Pearson.

Rashid, M. H. (2013). Microelectronic Circuits: Analysis and Design (2nd ed.). Cengage Learning.

Streetman, B. G., & Banerjee, S. K. (2015). Solid State Electronic Devices (7th ed.). Prentice Hall.

Floyd, T. L., & Buchla, D. M. (2014). Fundamentals of Electric Circuits (5th ed.). Pearson.

Hayt, W. H., Kemmerly, J. E., & Durbin, S. M. (2014). Engineering Circuit Analysis (8th ed.). McGraw-Hill Education.

Nilsson, J. W., & Riedel, S. A. (2014). Electric Circuits (10th ed.). Pearson.

Alexander, C. K., & Sadiku, M. N. O. (2017). Fundamentals of Electric Circuits (6th ed.). McGraw-Hill Education.

Robbins, A. H., & Miller, W. C. (2018). Circuit Analysis: Theory and Practice (5th ed.). Cengage Learning.

Boylestad, R. L., & Nashelsky, L. (2018). Electronic Devices and Circuit Theory (12th ed.). Pearson.

Streetman, B. G., & Banerjee, S. K. (2015). Solid State Electronic Devices (7th ed.). Prentice Hall.

Sedra, A. S., & Smith, K. C. (2016). Microelectronic Circuits (7th ed.). Oxford University Press.

Hsu, S. Y. (2017). Analog and Digital Signal Analysis: From Basics to Applications. Cambridge University Press.

Chen, W. (2017). Linear Networks and Systems: Algorithms and Computational Techniques. Springer.

Van Valkenburg, M. E. (2017). Network Analysis (4th ed.). Pearson.

Proakis, J. G., & Salehi, M. (2013). Communication Systems Engineering (2nd ed.). Pearson.

Haykin, S., & Moher, M. (2010). Modern Wireless Communications (2nd ed.). Pearson.

Lathi, B. P., & Ding, Z. (2016). Modern Digital and Analog Communication Systems (4th ed.). Oxford University Press.

Couch, L. W. (2013). Digital and Analog Communication Systems (8th ed.). Pearson.

Lee, E. A., & Seshia, S. A. (2011). Introduction to Embedded Systems: A Cyber-Physical Systems Approach. MIT Press.

Carlson, A. B., & Crilly, P. B. (2012). Communication Systems (5th ed.). McGraw-Hill Education.

Sklar, B. (2001). Digital Communications: Fundamentals and Applications (2nd ed.). Prentice Hall.

Proakis, J. G., & Proakis, M. (2008). Digital Signal Processing: Principles, Algorithms, and Applications (4th ed.). Prentice Hall.

Couch, L. W. (2016). Digital and Analog Communication Systems (9th ed.). Pearson.

Simon, M. K., Hinedi, S. M., & Lindsey, W. C. (2005). Digital Communication Techniques: Signal Design and Detection. Prentice Hall.

Proakis, J. G., & Manolakis, D. G. (2013). Digital Signal Processing: Principles, Algorithms, and Applications (4th ed.). Pearson.

Oppenheim, A. V., Schafer, R. W., & Buck, J. R. (2019). Discrete-Time Signal Processing (3rd ed.). Pearson.

Mitra, S. K. (2010). Digital Signal Processing: A Computer-Based Approach (4th ed.). McGraw-Hill Education.

Proakis, J. G., & Manolakis, D. G. (2016). Digital Signal Processing: Principles, Algorithms, and Applications (International Edition) (4th ed.). Cengage Learning.

Lyons, R. G. (2011). Understanding Digital Signal Processing (3rd ed.). Pearson.

Smith, S. W. (1997). The Scientist and Engineer's Guide to Digital Signal Processing. California Technical Publishing.

Ingle, V. K., & Proakis, J. G. (2019). Digital Signal Processing Using MATLAB (4th ed.). Cengage Learning.

Hayes, M. H. (1996). Schaum's Outline of Digital Signal Processing. McGraw-Hill Education.

Rabiner, L. R., & Gold, B. (1975). Theory and Application of Digital Signal Processing. Prentice Hall.

Lathi, B. P. (2009). Modern Digital and Analog Communication Systems (4th ed.). Oxford University Press.

Proakis, J. G., & Salehi, M. (2013). Communication Systems Engineering (2nd ed.). Pearson.

Haykin, S., & Moher, M. (2010). Modern Wireless Communications (2nd ed.). Pearson.

Lathi, B. P., & Ding, Z. (2016). Modern Digital and Analog Communication Systems (4th ed.). Oxford University Press.

Couch, L. W. (2013). Digital and Analog Communication Systems (8th ed.). Pearson.

Sklar, B. (2001). Digital Communications: Fundamentals and Applications (2nd ed.). Prentice Hall.

Proakis, J. G., & Proakis, M. (2008). Digital Signal Processing: Principles, Algorithms, and Applications (4th ed.). Prentice Hall.

Carlson, A. B., & Crilly, P. B. (2012). Communication Systems (5th ed.). McGraw-Hill Education.

Simon, M. K., Hinedi, S. M., & Lindsey, W. C. (2005). Digital Communication Techniques: Signal Design and Detection. Prentice Hall.

Couch, L. W. (2016). Digital and Analog Communication Systems (9th ed.). Pearson.

Proakis, J. G., & Salehi, M. (2008). Fundamentals of Communication Systems (2nd ed.). Pearson.

Mohan, N., Undeland, T. M., & Robbins, W. P. (2002). Power Electronics: Converters, Applications, and Design (3rd ed.). John Wiley & Sons.

Rashid, M. H. (2017). Power Electronics: Circuits, Devices, and Applications (4th ed.). Pearson.

Hart, D. W. (2017). Power Electronics (1st ed.). McGraw-Hill Education.

Erickson, R. W., & Maksimovic, D. (2017). Fundamentals of Power Electronics (2nd ed.). Springer.

Rashid, M. H. (2018). Power Electronics Handbook: Devices, Circuits, and Applications (4th ed.). Academic Press.

Ned Mohan, T. M. Undeland, & W. P. Robbins. (2003). Power Electronics: Converters, Applications, and Design (2nd ed.). John Wiley & Sons.

Bimal K. Bose. (2002). Power Electronics and Motor Drives: Advances and Trends (1st ed.). Academic Press.

S. A. Nasar. (1991). Power Electronics: Devices, Circuits, and Industrial Applications. Prentice Hall.

Daniel W. Hart. (1997). Power Electronics. McGraw-Hill.

P. C. Sen. (2009). Modern Power Electronics and AC Drives (1st ed.). Oxford University Press.

Stevenson Jr., W. D., & Grainger, J. J. (2012). Power System Analysis (4th ed.). McGraw-Hill Education.

Glover, J. D., Sarma, M. S., & Overbye, T. J. (2016). Power System Analysis and Design (6th ed.). Cengage Learning.

Bergen, A. R., & Vittal, V. (2018). Power Systems Analysis (3rd ed.). Pearson.

Grainger, J. J., & Stevenson Jr., W. D. (1994). Power System Analysis (2nd ed.). McGraw-Hill Education.

Das, J. C. (2012). Power System Analysis: Short-Circuit Load Flow and Harmonics (2nd ed.). CRC Press.

Elgerd, O. I., & El-Hawary, M. E. (2010). Electric Energy Systems Theory: An Introduction (2nd ed.). McGraw-Hill Education.

Jovcic, D., & Milanovic, J. V. (2016). Power System Stability and Control. Springer.

Nagrath, I. J., & Kothari, D. P. (2017). Power System Engineering (2nd ed.). Tata McGraw-Hill Education.

Stagg, G. W., & El-Abiad, A. H. (1992). Computer Methods in Power System Analysis (2nd ed.). McGraw-Hill Education.

Arrillaga, J., Watson, N. R., & Chen, S. (2018). Power System Harmonic Analysis (2nd ed.). John Wiley & Sons.

Wadhwa, C. L. (2007). Electric Power Systems (5th ed.). Tata McGraw-Hill Education.

Wildi, T. (2017). Electrical Machines, Drives, and Power Systems (6th ed.). Pearson.

Mohan, N., Undeland, T. M., & Robbins, W. P. (2002). Power Electronics: Converters, Applications, and Design (3rd ed.). John Wiley & Sons.

Rashid, M. H. (2017). Power Electronics: Circuits, Devices, and Applications (4th ed.). Pearson.

Wood, A. J., & Wollenberg, B. F. (2012). Power Generation, Operation, and Control (3rd ed.). John Wiley & Sons.

Agrawal, G. K. (2008). Electric Power Systems (2nd ed.). Prentice Hall.

Ilic, M. D., & Galiana, F. D. (2017). Power System Operation and Control. Springer.

Kundur, P. (1994). Power System Stability and Control. McGraw-Hill Education.

Anderson, P. M., & Fouad, A. A. (2003). Power System Control and Stability (2nd ed.). IEEE Press.

Kimbark, E. W. (1992). Power System Stability (Vol. 1). IEEE Press.

Fitzgerald, A. E., Kingsley Jr., C., & Umans, S. D. (2003). Electric Machinery (6th ed.). McGraw-Hill Education.

Wildi, T. (2017). Electrical Machines, Drives, and Power Systems (6th ed.). Pearson.

Chapman, S. J. (2018). Electric Machinery Fundamentals (5th ed.). McGraw-Hill Education.

Sen, P. C. (2006). Principles of Electric Machines and Power Electronics (3rd ed.). Wiley-IEEE Press.

Boldea, I., & Nasar, S. A. (2002). Electric Drives (2nd ed.). CRC Press.

Acha, E., & Agüero, J. R. (2018). Power Electronic Converters for Electric Drives (2nd ed.). John Wiley & Sons.

Rizzoni, G., & Krause, P. C. (2006). Principles and Applications of Electrical Engineering (5th ed.). McGraw-Hill Education.

Hughes, A. (2013). Electric Motors and Drives: Fundamentals, Types and Applications (4th ed.). Newnes.

Fitzgerald, A. E., & Umans, S. D. (2002). Electric Machinery (5th ed.). McGraw-Hill Education.
Huseyin, R. A. (2017). Electrical Machines and Drives: Fundamentals and Advanced Modelling. IET.

Griffiths, D. J. (2013). Introduction to Electrodynamics (4th ed.). Cambridge University Press.

Purcell, E. M., & Morin, D. J. (2013). Electricity and Magnetism (3rd ed.). Cambridge University Press.

Reitz, J. R., Milford, F. J., & Christy, R. W. (2008). Foundations of Electromagnetic Theory (4th ed.). Addison-Wesley.

Cheng, D. K. (1989). Field and Wave Electromagnetics. Addison-Wesley.

Roald K. Wangsness. (1986). Electromagnetic Fields (2nd ed.). Wiley.
Muncaster, R. G. (1998). A-Level Physics (4th ed.). Stanley Thornes Publishers.

Cengel, Y. A., & Boles, M. A. (2014). Thermodynamics: An Engineering Approach (8th ed.). McGraw-Hill Education.

Kulkarni, S. V., & Kulkarni, U. A. (2017). Handbook of Electrical Engineering (5th ed.). McGraw-Hill Education.

Furlani, E. P. (2001). Permanent Magnet and Electromechanical Devices: Materials, Analysis, and Applications. Academic Press.

Duffy, O. B. (2012). Electric Motors and Drives: Principles, Control, Modeling and Simulation. Newnes.

Sedra, A. S., & Smith, K. C. (2014). Microelectronic Circuits (7th ed.). Oxford University Press.

Razavi, B. (2016). Fundamentals of Microelectronics (2nd ed.). John Wiley & Sons.

Boylestad, R. L., & Nashelsky, L. (2018). Electronic Devices and Circuit Theory (11th ed.). Pearson.

Streetman, B. G., & Banerjee, S. K. (2015). Solid State Electronic Devices (7th ed.). Pearson.

Malvino, A. P., & Bates, D. J. (2017). Electronic Principles (8th ed.). McGraw-Hill Education.

Floyd, T. L., & Buchla, D. M. (2018). Electronics Fundamentals: Circuits, Devices, and Applications (9th ed.). Pearson.

Crecraft, D., & Gorham, I. (2012). Electronics: Circuits and Systems (4th ed.). Pearson.

Millman, J., & Halkias, C. C. (2008). Integrated Electronics: Analog and Digital Circuits and Systems. McGraw-Hill Education.

Howe, R., & Sodini, C. (2016). Microelectronics: An Integrated Approach. Pearson.

Rashid, M. H. (2018). Microelectronic Circuits, Devices and Systems. Pearson.

Tocci, R. J., Widmer, N. S., & Moss, G. L. (2014). Digital Systems: Principles and Applications (12th ed.). Pearson.

Mano, M. M., & Ciletti, M. D. (2017). Digital Design (6th ed.). Pearson.

Roth Jr., C. H., & Kinney, L. P. (2014). Fundamentals of Logic Design (7th ed.). Cengage Learning.

Brown, S. V., & Vranesic, Z. G. (2018). Fundamentals of Digital Logic with Verilog Design (3rd ed.). McGraw-Hill Education.

Wakerly, J. F. (2018). Digital Design: Principles and Practices (5th ed.). Pearson.

Floyd, T. L. (2017). Digital Fundamentals (11th ed.). Pearson.

Hayes, J. P. (2018). Introduction to Digital Logic Design. CRC Press.

Katz, R. H., & Borriello, G. (2011). Contemporary Logic Design (2nd ed.). Pearson.

Mano, M. M. (2013). Digital Logic and Computer Design (1st ed.). Pearson.

Lee, S. C., & Seshia, S. A. (2017). Introduction to Embedded Systems: A Cyber-Physical Systems Approach (2nd ed.). MIT Press.

Ogata, K. (2010). Modern Control Engineering (5th ed.). Prentice Hall.

Dorf, R. C., & Bishop, R. H. (2010). Modern Control Systems (12th ed.). Pearson.

Franklin, G. F., Powell, J. D., & Emami-Naeini, A. (2014). Feedback Control of Dynamic Systems (7th ed.). Pearson.

Nise, N. S. (2010). Control Systems Engineering (6th ed.). John Wiley & Sons.

Kuo, B. C. (2010). Automatic Control Systems (9th ed.). Prentice Hall.

Astrom, K. J., & Murray, R. M. (2010). Feedback Systems: An Introduction for Scientists and Engineers. Princeton University Press.

Dorf, R. C., & Bishop, R. H. (2004). Modern Control Systems (9th ed.). Prentice Hall.

Ogata, K. (2018). Discrete-Time Control Systems (3rd ed.). Pearson.

Åström, K. J., & Wittenmark, B. (2013). Adaptive Control (2nd ed.). Dover Publications.

Ljung, L. (2010). System Identification: Theory for the User (2nd ed.). Prentice Hall.

Patterson, D. A., & Hennessy, J. L. (2018). Computer Organization and Design: The Hardware/Software Interface (5th ed.). Morgan Kaufmann.

Tanenbaum, A. S., & Bos, H. (2014). Modern Operating Systems (4th ed.). Pearson.

Stallings, W. (2014). Operating Systems: Internals and Design Principles (8th ed.). Pearson.

Bryant, R. E., & O'Hallaron, D. R. (2016). Computer Systems: A Programmer's Perspective (3rd ed.). Pearson.

Hennessy, J. L., & Patterson, D. A. (2017). Computer Architecture: A Quantitative Approach (6th ed.). Morgan Kaufmann.

Silberschatz, A., Galvin, P. B., & Gagne, G. (2018). Operating System Concepts (10th ed.). John Wiley & Sons.

Tanenbaum, A. S., & Austin, T. (2014). Structured Computer Organization (6th ed.). Pearson.

Hennessey, J. L., & Patterson, D. A. (2017). Computer Organization and Design RISC-V Edition: The Hardware Software Interface (1st ed.). Morgan Kaufmann.

Peterson, L. L., & Davie, B. S. (2011). Computer Networks: A Systems Approach (5th ed.). Morgan Kaufmann.

Tanenbaum, A. S., & Woodhull, A. S. (2015). Operating Systems Design and Implementation (3rd ed.). Prentice Hall.

Kurose, J. F., & Ross, K. W. (2017). Computer Networking: A Top-Down Approach (7th ed.). Pearson.

Peterson, L. L., & Davie, B. S. (2011). Computer Networks: A Systems Approach (5th ed.). Morgan Kaufmann.

Tanenbaum, A. S., & Wetherall, D. J. (2011). Computer Networks (5th ed.). Pearson.

Forouzan, B. A., & Fegan, S. C. (2015). Data Communications and Networking (5th ed.). McGraw-Hill Education.

Stallings, W. (2013). Data and Computer Communications (10th ed.). Pearson.

Comer, D. E. (2014). Computer Networks and Internets (6th ed.). Pearson.

Fall, K. R., & Stevens, W. R. (2012). TCP/IP Illustrated, Volume 1: The Protocols (2nd ed.). Addison-Wesley.

Al-Fuqaha, A., Guizani, M., Mohammadi, M., Aledhari, M., & Ayyash, M. (2015). Internet of Things: A Survey on Enabling Technologies, Protocols, and Applications. IEEE Communications Surveys & Tutorials, 17(4), 2347-2376.

Pahlavan, K., & Krishnamurthy, P. (2018). Principles of Wireless Networks: A Unified Approach (2nd ed.). Prentice Hall.

Safford, D. R., & Andrews, J. G. (2004). Introduction to Broadband Communication Systems (2nd ed.). Prentice Hall.

Sommerville, I. (2016). Software Engineering (10th ed.). Pearson.

Pressman, R. S. (2014). Software Engineering: A Practitioner's Approach (8th ed.). McGraw-Hill Education.

Shaw, M., & Garlan, D. (2009). Software Architecture: Perspectives on an Emerging Discipline (2nd ed.). Prentice Hall.

Boehm, B. W., & Turner, R. (2004). Balancing Agility and Discipline: A Guide for the Perplexed. Addison-Wesley.

Bass, L., Clements, P., & Kazman, R. (2012). Software Architecture in Practice (3rd ed.). Addison-Wesley.

Pfleeger, S. L., & Atlee, J. M. (2010). Software Engineering: Theory and Practice (4th ed.). Prentice Hall.

Beck, K. (1999). Extreme Programming Explained: Embrace Change (1st ed.). Addison-Wesley.

Martin, R. C. (2009). Clean Code: A Handbook of Agile Software Craftsmanship. Prentice Hall.

McConnell, S. (2004). Code Complete: A Practical Handbook of Software Construction (2nd ed.). Microsoft Press.

Sommerville, I. (2010). Software Engineering (9th ed.). Addison-Wesley.

Martin, M. W., & Schinzinger, R. (2012). Ethics in Engineering (4th ed.). McGraw-Hill Education.

Davis, M. (2018). Ethics and the University (2nd ed.). Routledge.

Harris, C. E., Pritchard, M. S., & Rabins, M. J. (2014). Engineering Ethics: Concepts and Cases (5th ed.). Cengage Learning.

Johnson, D. G. (2017). Computer Ethics (5th ed.). Pearson.

Spier, R. E. (2016). Science and Technology Ethics (2nd ed.). Routledge.

Basart, J., & Serra, R. (2016). Engineering Ethics: Challenges and Opportunities. Springer.

Martin, M. W., & Schinzinger, R. (2005). Introduction to Engineering Ethics (2nd ed.). McGraw-Hill Education.

Vesilind, P. A. (2017). Engineering, Ethics, and the Environment (2nd ed.). Cambridge University Press.

May, L. (2012). The Morality of War (Revised ed.). Oxford University Press.

Harris, C. E. (2009). Applied Engineering Ethics. Cengage Learning.

Appendix

Electric and Electronic Symbols

Component group	Component	Recommended symbol	Alternative symbols	Notes
sources	cell	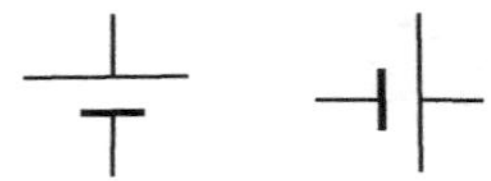	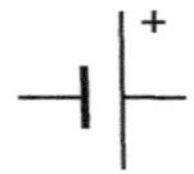	
	battery, DC power supply	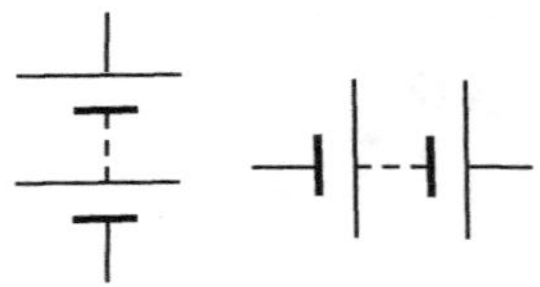	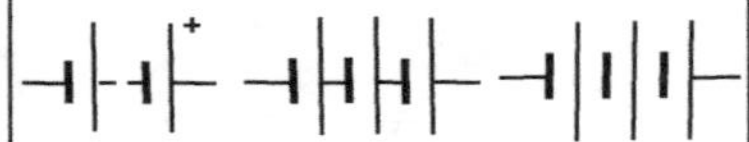	
	variable DC power supply	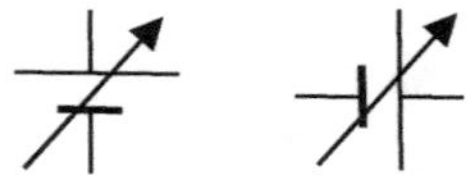		
	AC power supply		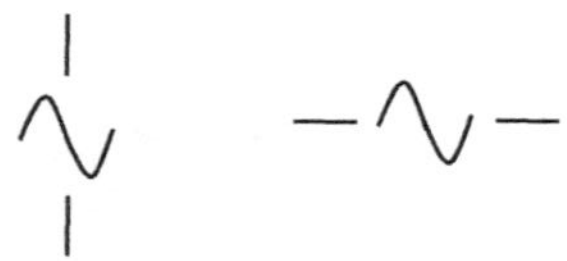	

Component group	Component	Recommended symbol	Alternative symbols	Notes
resistors	resistor	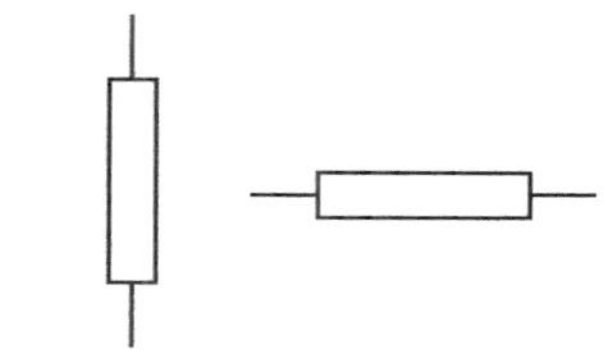		Alternative symbol now regarded as outdated.
	variable resistor, potentiometer, voltage divider	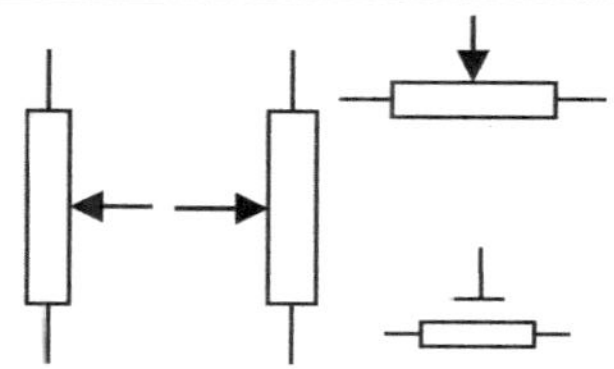	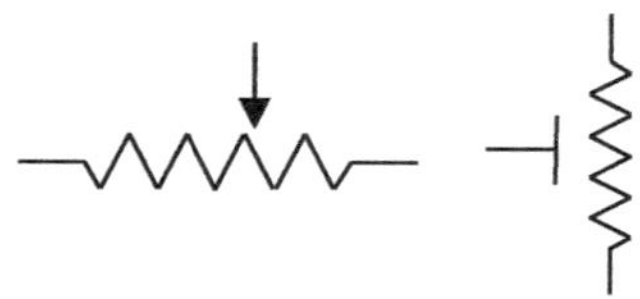	A variable resistor where the variation is achieved by use of a sliding contact.
	variable resistor	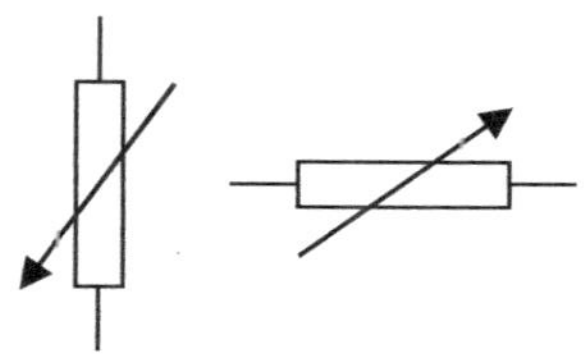		A general symbol for a variable resistor where the means of variation is not specified.

Component group	Component	Recommended symbol	Alternative symbols	Notes
	light-dependent resistor (LDR)	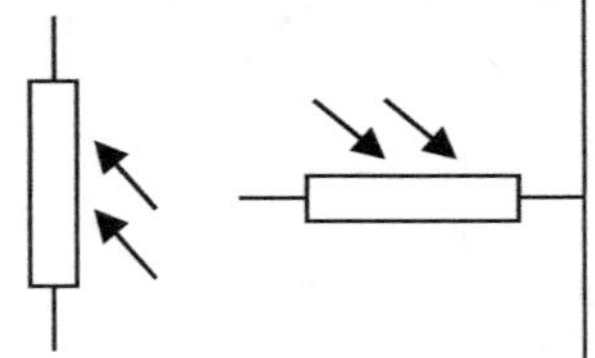	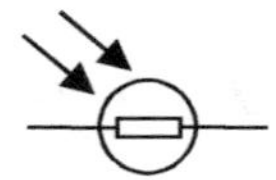	
	filament globe	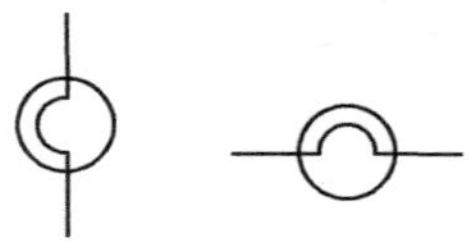	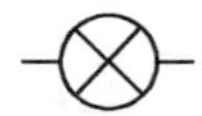	
	thermistor	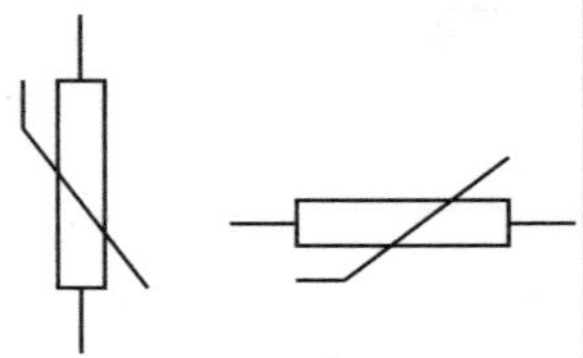		
	fuse	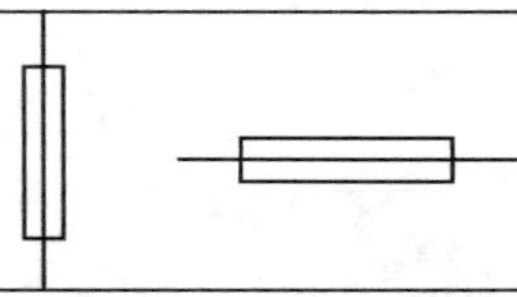		

Component group	Component	Recommended symbol	Alternative symbols	Notes
capacitors	(non-polarised) capacitor			
	variable capacitor	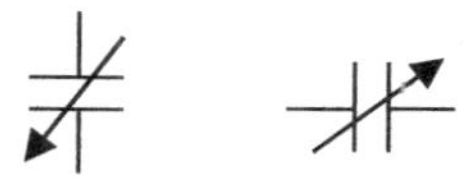		
	polarised capacitor, electrolytic capacitor	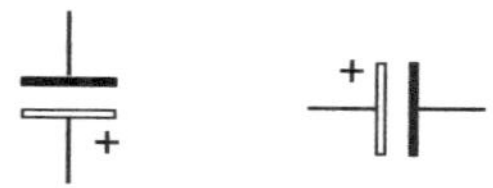	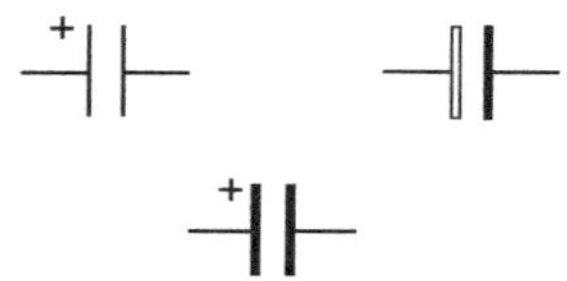	
transformers	iron-cored transformer (one secondary winding)	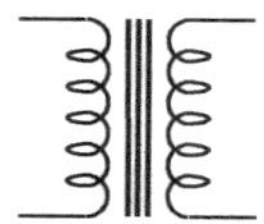		

Component group	Component	Recommended symbol	Alternative symbols	Notes
	iron-cored transformer (one secondary winding – centre-tapped)	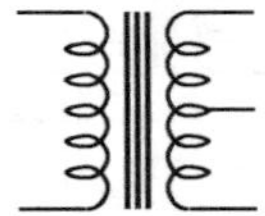		
	iron-cored transformer (two secondary windings)	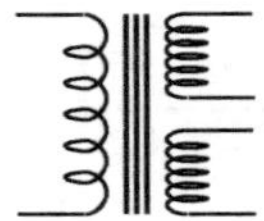		
diodes	junction diode	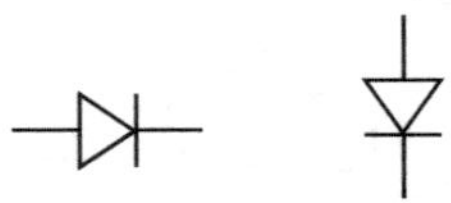	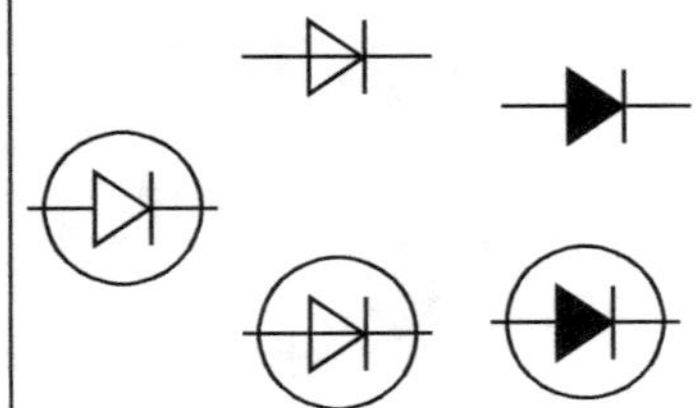	Additional alternative symbols for **all diodes** include those where the single straight-line section of the symbol is shown as a heavier line.

Component group	Component	Recommended symbol	Alternative symbols	Notes
	Zener diode	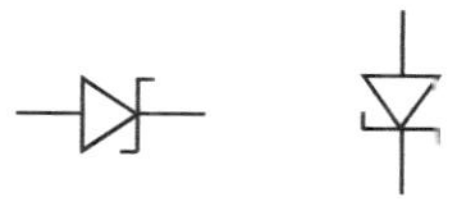	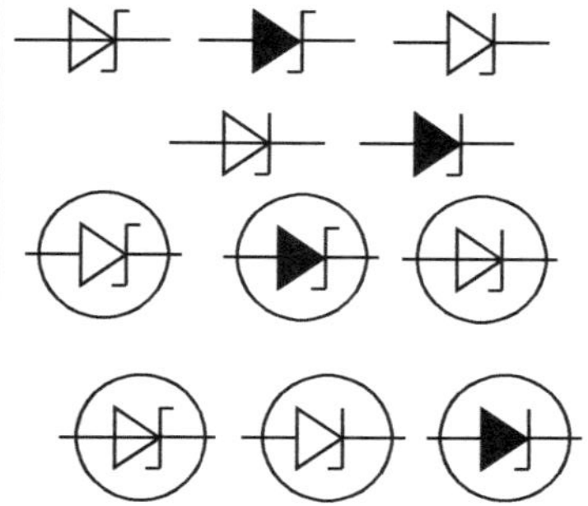	
	photo diode	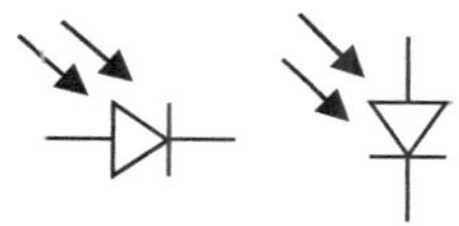	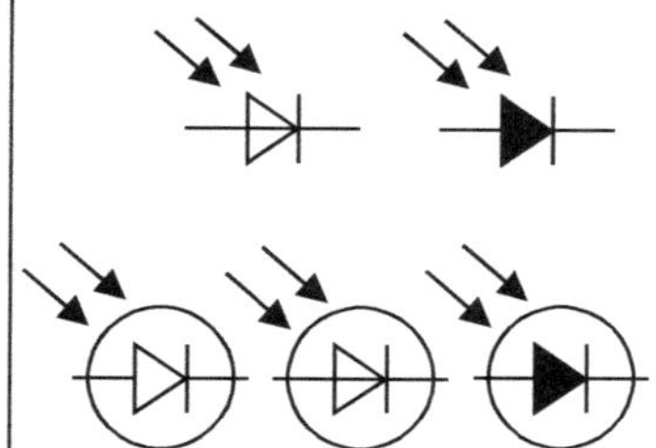	

Component group	Component	Recommended symbol	Alternative symbols	Notes
	light-emitting diode (LED)	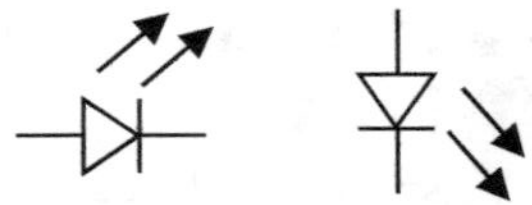	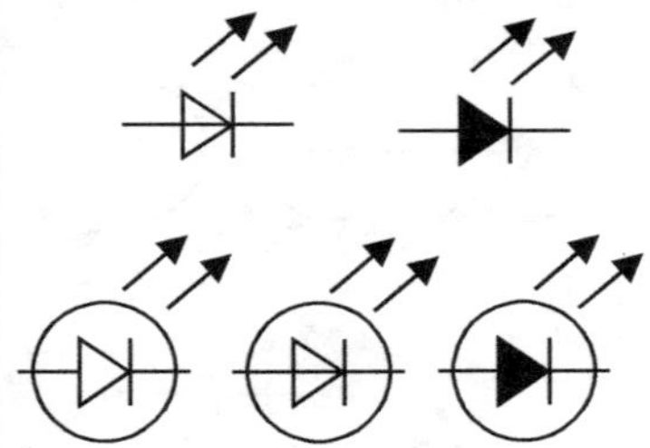	
	diode bridge	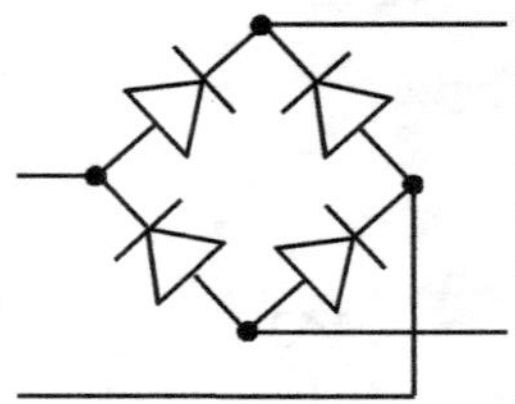 (Note that in an earlier version of this table, this symbol was incorrectly		There are alternative forms using the other alternative symbols for the **junction diode** given above.

Component group	Component	Recommended symbol	Alternative symbols	Notes
		represented.)		
transistors	npn transistor	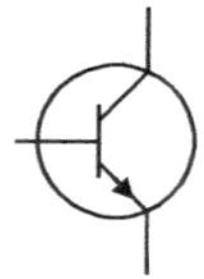	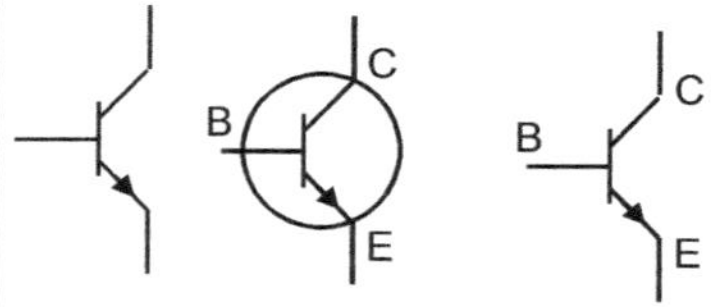	Additional alternative symbols for **all transistors** include those where the vertical straight-line section of the symbol is shown as a heavier line.
	pnp transistor	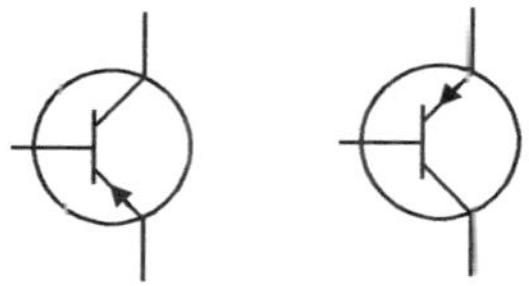	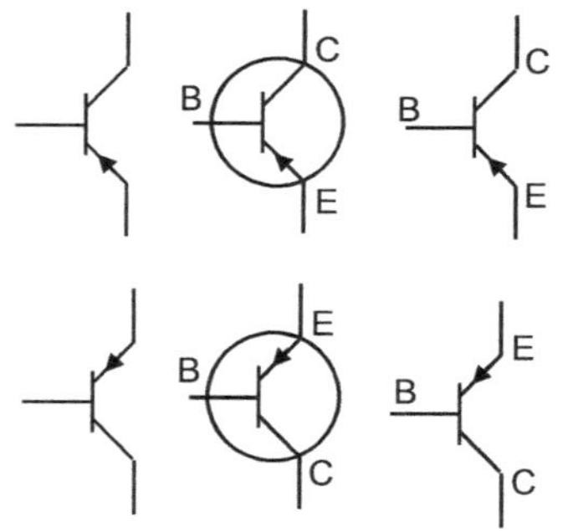	Both orientations of the recommended symbol are used depending on whether the *emitter* (E) is at the bottom or top.

Component group	Component	Recommended symbol	Alternative symbols	Notes
	n-type junction field effect transistor (NJFET)	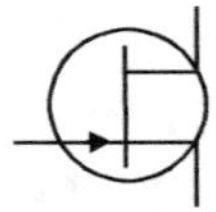	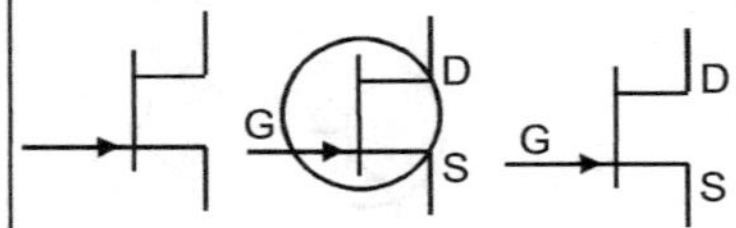	
	p-type junction field effect transistor (PJFET)	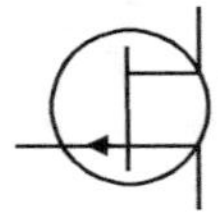	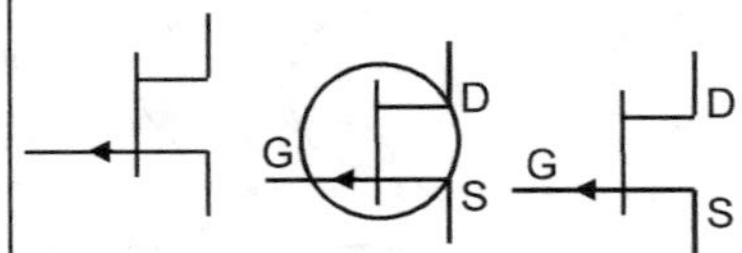	
integrated circuits	integrated circuit (IC)	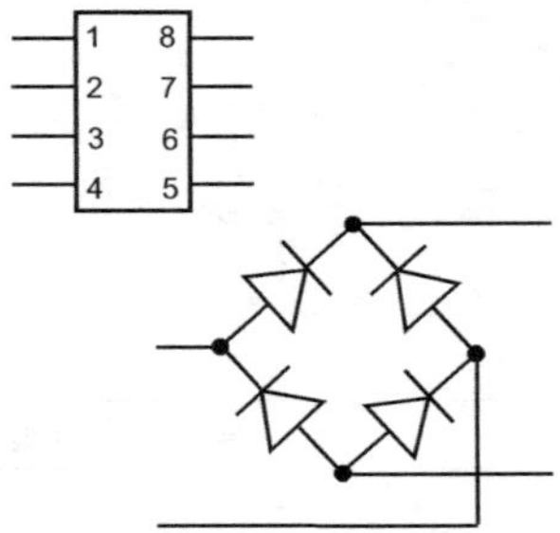	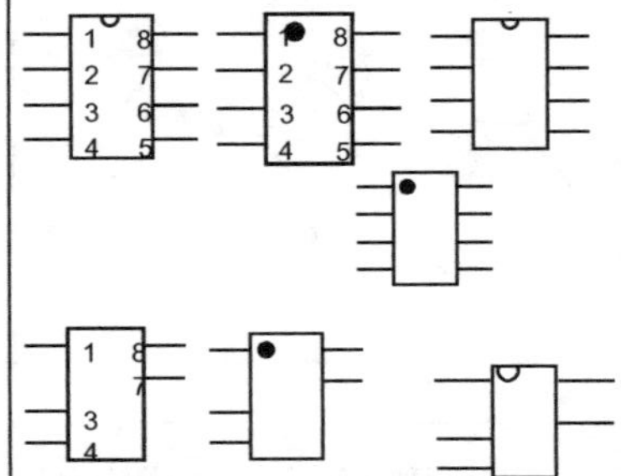	This example shows an IC with 8 connections. Others occur, e.g. 14 and 16. Under alternative symbols identifying marks are shown to indicate the

Component group	Component	Recommended symbol	Alternative symbols	Notes
				numbering of the connections. Other conventions show only the connections used in a given circuit, here 1, 3, 4, 7 and 8.
switches	single pole, single throw (SPST) switch	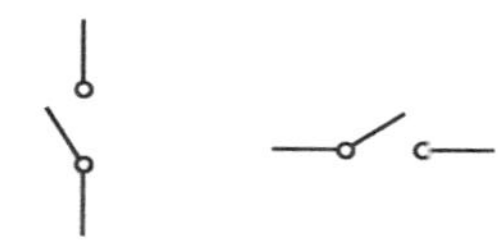		
	single pole, double throw (SPDT) switch	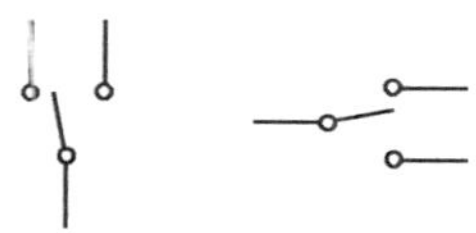		

Component group	Component	Recommended symbol	Alternative symbols	Notes
	double pole, single throw (DPST) switch			
	double pole, double throw (DPDT) switch	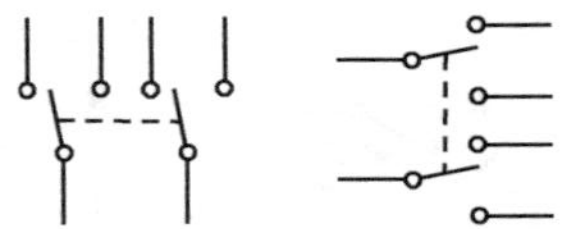	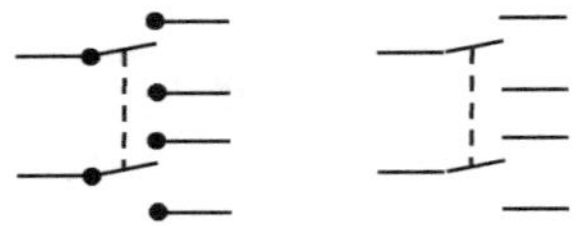	
	normally open (NO) switch	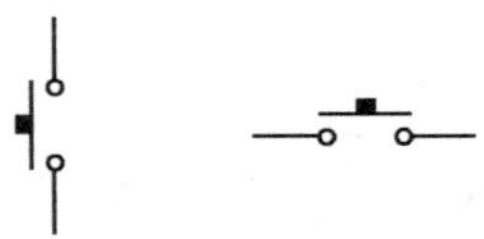		

Component group	Component	Recommended symbol	Alternative symbols	Notes
	normally closed (NC) switch	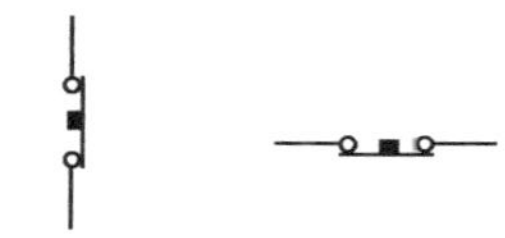		
	relay	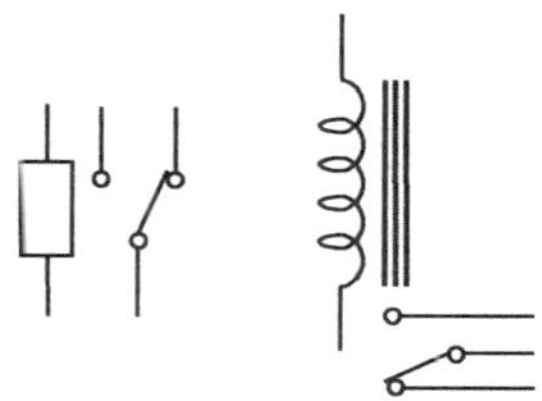		
meters	ammeter	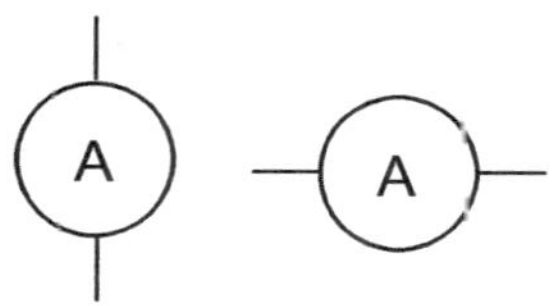	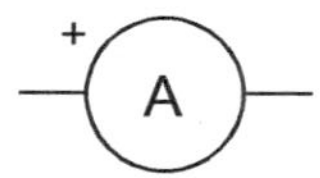 	

Component group	Component	Recommended symbol	Alternative symbols	Notes
	voltmeter	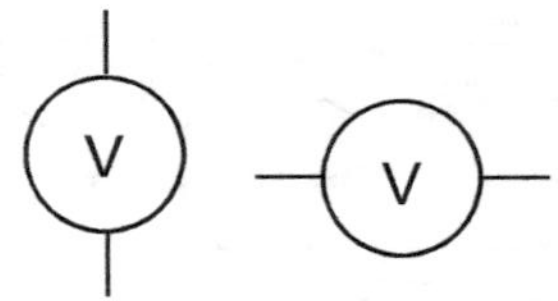	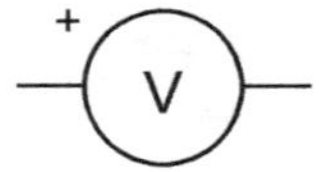	
	galvanometer		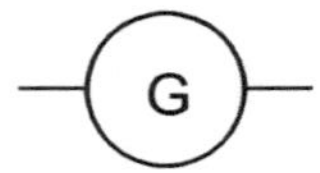	The alternative symbol is not recommended as it is sometimes used to indicate a *generator*.
	cathode ray oscilloscope (CRO)	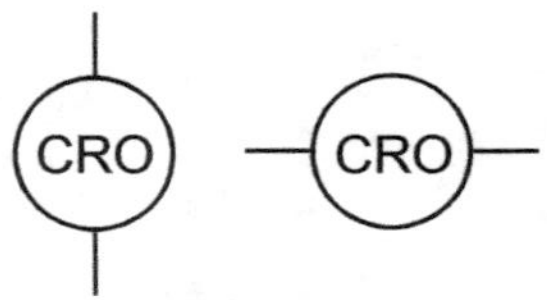	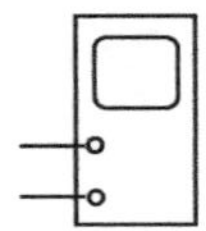	Neither of these encountered symbols is widely used.

Component group	Component	Recommended symbol	Alternative symbols	Notes
amplifiers	voltage amplifier	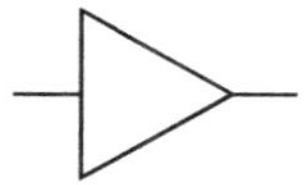		
	operational amplifier (op amp)	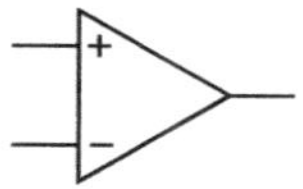	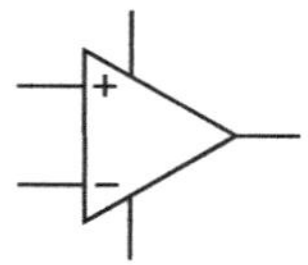	On the alternative symbol the two vertical connections are shown when it is connected to the power supply.
transducers	motor	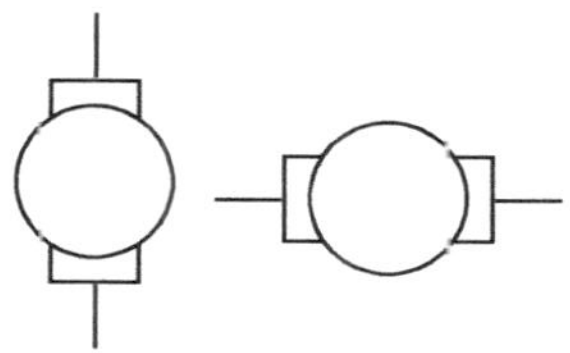	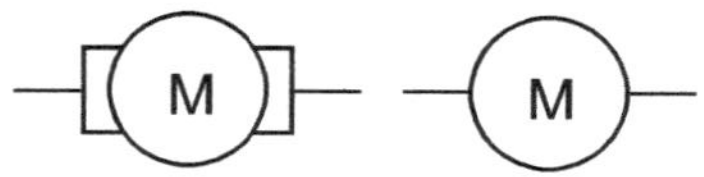	

Component group	Component	Recommended symbol	Alternative symbols	Notes
	microphone	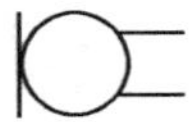		
	loudspeaker	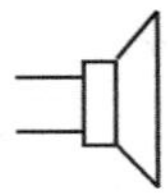		
logic gates	NOT or invert gate	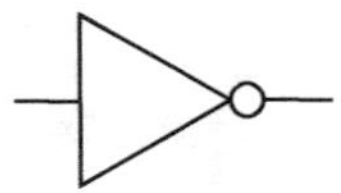	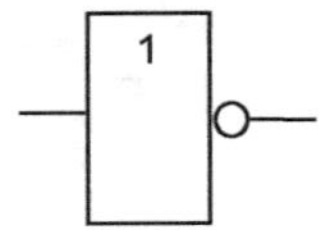	
	OR gate	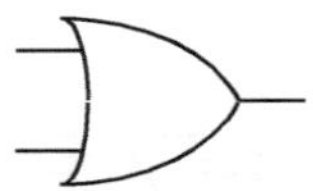	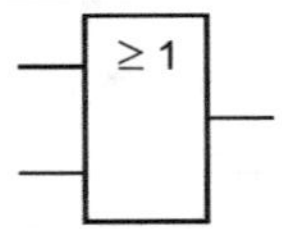	

Component group	Component	Recommended symbol	Alternative symbols	Notes
	XOR (exclusive OR) gate	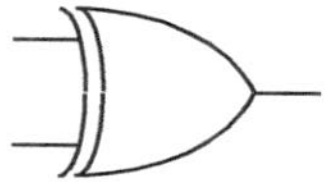	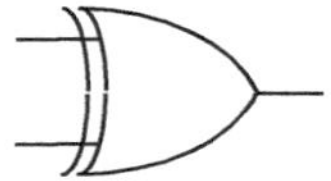 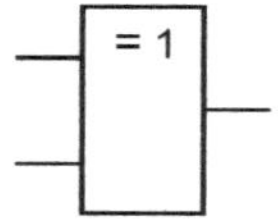	
	NOR gate	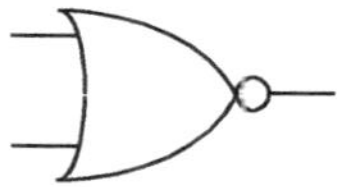	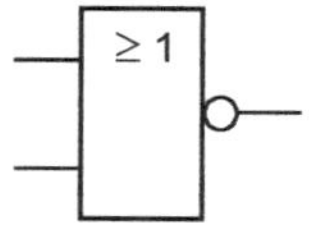	
	AND gate	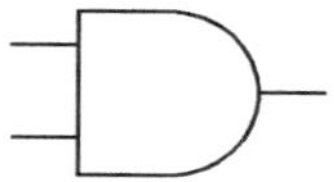	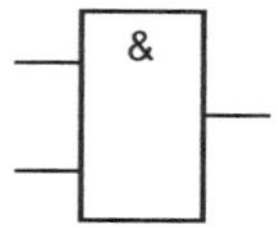	

Component group	Component	Recommended symbol	Alternative symbols	Notes
	NAND gate	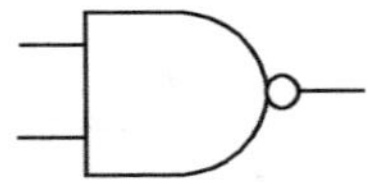	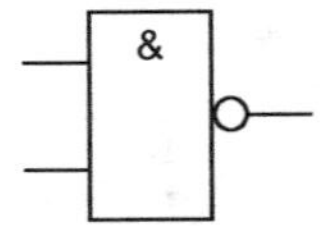 	
flip-flop	T (toggle) flip-flop	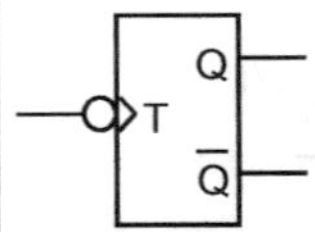 falling edge triggered 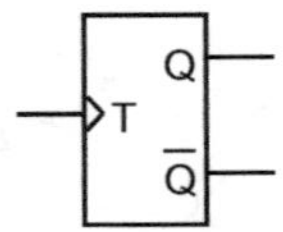 rising edge triggered		
external connections	earth			

Component group	Component	Recommended symbol	Alternative symbols	Notes
	aerial			
circuit connections	non-connected leads		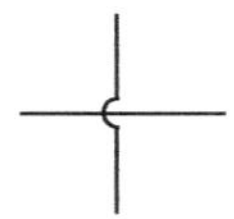	Alternative symbol regarded as outdated.
	connected leads	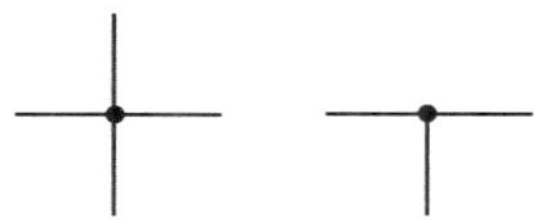		
	dot for junction of connected leads	•		

www.ingramcontent.com/pod-product-compliance
Lightning Source LLC
LaVergne TN
LVHW052208200926
848481LV00001B/58

www.ingramcontent.com/pod-product-compliance
Lightning Source LLC
LaVergne TN
LVHW052208200926
848481LV00001B/58

www.ingramcontent.com/pod-product-compliance
Lightning Source LLC
LaVergne TN
LVHW052208200926
848481LV00001B/58
* 9 7 8 1 9 2 2 6 1 7 7 1 2 *